Chemical Laboratory Practice

Editors

W. Fresenius J. F. K. Huber E. Pungor
G. A. Rechnitz W. Simon Th. S. West

Pretsch Clerc Seibl Simon

Tables of

Spectral Data for Structure Determination of Organic Compounds

Translated from the German by
K. Biemann

Second Edition

Springer-Verlag Berlin Heidelberg NewYork
London Paris Tokyo Hong Kong

P. D. Dr. Ernö Pretsch, Professor Dr. Wilhelm Simon
Eidgenössische Technische Hochschule, Laboratorium für
Organische Chemie, Universitätsstraße 16, CH-8092 Zürich

Professor Dr. Joseph Seibl
Mitterndorferweg 1
A-6380 St. Johann in Tirol

Professor Dr. Thomas Clerc
Pharmazeutisches Institut der Universität, Baltzerstraße 5,
CH-3012 Bern

Translated from the German by
Prof. Dr. Klaus Biemann
Dept. of Chemistry, Massachusetts Inst. of Technology,
Cambridge, MA 02139/USA

ISBN 3-540-51202-0 2. Aufl. Springer-Verlag Berlin Heidelberg NewYork
ISBN 0-387-51202-0 2nd ed. Springer-Verlag NewYork Berlin Heidelberg

ISBN 3-540-12406-3 1. Aufl. Springer-Verlag Berlin Heidelberg NewYork
ISBN 0-387-12406-3 1st ed. Springer-Verlag NewYork Berlin Heidelberg

Printing: Color-Druck Dorfi GmbH, Berlin; Bookbinding: Lüderitz & Bauer, Berlin
52/3020-5432 - Printed on acid -free paper

Preface to the Second English Edition

The Third German Edition of the "Tables of Spectral Data" had been greatly expanded particularly in the area of ^{13}C-NMR and, to a lesser extent, of ^{1}H-NMR. For these two fields, as well as in the chapter on mass spectrometry, some more textual material has been added. These new features, combined with the wide acceptance of the Tables in the English speaking world, made it advisable to also translate the Third German Edition into English. As in the previous translation, this provided an opportunity for a few corrections and further additions. All areas of spectroscopy have now reached a high level of maturity, which assures that the data compiled in this work represent a rather complete condensation of the information available in the literature. It is expected that the book will find good use in the hands of the chemist who is concerned with the correlation of such spectroscopic data with the structure of organic compounds.

Klaus Biemann
Cambridge, MA
June 1989

Preface to the English Edition

Although numerical data are, in principle, universal, the compilations presented in this book are extensively annotated and interleaved with text. This translation of the second German edition has been prepared to facilitate the use of this work, with all its valuable detail, by the large community of English-speaking scientists. Translation has also provided an opportunity to correct and revise the text, and to update the nomenclature. Fortunately, spectroscopic data and their relationship with structure do not change much with time so one can predict that this book will, for a long period of time, continue to be very useful to organic chemists involved in the identification of organic compounds or the elucidation of their structure.

Klaus Biemann
Cambridge, MA, April 1983

Preface to the First German Edition

Making use of the information provided by various spectroscopic techniques has become a matter of routine for the analytically oriented organic chemist. Those who have graduated recently received extensive training in these techniques as part of the curriculum while their older colleagues learned to use these methods by necessity. One can, therefore, assume that chemists are well versed in the proper choice of the methods suitable for the solution of a particular problem and to translate the experimental data into structural information.

Those who are not specialists in any of these techniques and therefore not in continuous contact with the corresponding data may wish to have a compact summary of reference data in a form that can be grasped easily. Even experts appreciate the opportunity to look up in a summary information about compound types with which they are not familiar. The tables compiled in this book are meant to fill this gap. They were compiled for courses and exercises which the authors offered over a ten year period to students at the Federal Institute of Technology (ETH), Zurich, and are thus well suited as a basis for similar courses elsewhere.

Considering such a broad effort there will undoubtedly be some omissions and errors in our presentations. We would be grateful to users of this book for suggestions and criticisms which would help us to keep it up to date. A postcard included in the book may make it easier for the reader to make such comments. We would also be grateful to receive reprints of papers containing information and data which could be incorporated in later editions and thus improve their usefulness.

A book such as this could not be assembled without the help of enthusiastic and knowledgeable collaborators who contributed a good deal to the work. Our special thanks go to Miss I. Port, Dr. D. Wegmann as well as Mr. P. Oggenfuss and Dr. R. Schwarzenbach.

Preface to the Second German Edition

This second edition provided an opportunity to include many additions and correct some errors. Amongst other improvements tables and figures concerning opaque regions and error-signals in the infrared were added. We also appreciate correspondence which led to a number of corrections.

Our special thanks go to Dr. D. Wegmann and Mr. P. Oggenfuss for their very careful cooperation. It is due to their efforts that this new edition was produced in a timely fashion.

Preface to the Third German Edition

In view of the success of the first two editions of our data tables and their translations into various languages, we feel the responsibility to adapt the work to the continuous development of analytical instrumentation and to update the Tables to the present level of information available. The original concept to present a somewhat limited but well balanced set of reference data will be retained.

Therefore, we had to exercise restraint in the selection of additional data to avoid unmanageable growth of the volume. In the interest of the ease of use, we took the risk of a certain redundancy in some of the data if they properly fit into more than one area. To facilitate, if necessary, the access to larger or more specialized collections of data we compiled such a list on page A10-A15.

The major portion of the newly added material concerns the ^{13}C-NMR data, most of which became available only recently. These data were selected and presented in a manner which closely resembles the organization of the ^{1}H-NMR section, so that now all of the listed compound classes are characterized by both of these techniques. The ^{1}H-NMR section was strengthened with the addition of additivity rules and data for compound types (such as naphthalenes and amino acids) which had been underrepresented in the previous editions. To the MS chapter, a brief description of fragmentation processes characteristic of monofunctional compounds was added. In addition to the listing of the typical absorption regions in the IR, the absorption frequencies of selected reference compounds were added.

Finally, in a new section entitled "VARIA" a number of rules of practical importance are outlined.

1) Calculation of the number of double bond equivalents from a molecular formula.

2) Conversion of chemical shifts from external to internal reference.

3) Prediction of the type of spin-spin interactions based on the symmetry properties and rapid changes in the conformation of molecules.

We would like to thank our readers for comments which led to corrections and changes in the new edition and, at the same time, ask for continuous criticism and suggestions, which we greatly appreciate.

We are particularly indebted to Dr. D. Wegmann for the expertise with which she very carefully prepared the manuscript and to Mrs. M. Schlatter for the exceptional layout and composition of the Tables. Both have contributed substantially to the success of the work.

Zürich, December 1985

Table of Contents

Introduction A5

Abbreviations and Symbols A20

Summary Tables B5

- ^{13}C-NMR B5
- ^{1}H-NMR B15
- IR B35
- UV/VIS B65

Combination Tables B75

- Alkanes, Cycloalkanes B75
- Alkenes, Cycloalkenes B85
- Alkynes B95
- Aromatics B105
- Heteroaromatics B115
- Halogen Compounds B125
- Alcohols, Phenols B135
- Ethers B145
- Amines B155
- Nitro Compounds B165
- Thiols, Thioethers B175
- Aldehydes B185
- Ketones B195
- Carboxylic Acids B205
- Carboxylic Esters, Lactones B215
- Amides, Lactams B235

^{13}C-Nuclear Magnetic Resonance Spectroscopy C5

- Alkanes C5
- Additivity Rules for Substituted Alkanes C10
- Methyl Groups C30
- Monosubstituted Alkanes C40
- n-Octanes C46
- Alicycles C47
- Methylcyclohexanes C50
- Monosubstituted Cyclohexanes C70
- Condensed Alicyles C72
- Alkenes C80
- Vinyl Compounds C90
- 1,2-Disubstituted Alkenes C96
- Enols C97
- Cycloalkenes C100
- Allenes C105
- Alkynes C110

Aromatic Hydrocarbons C115
Substituent Effects in Benzene C120
Substituent Effects in Naphthalene C126
5-Membered Heteroaromatic Rings C135
Substituent Effects in Pyridine C140
6-Membered Heteroaromatic Rings C155
Condensed Heteroaromatics C160
Halogen Compounds C167
Alcohols C170
Ethers C172
Amines C174
Nitro- and Nitroso Compounds C178
Nitramines C179
Thiols C179
Thioethers, Disulfides C180
Sulfonium Salts, Sulfoxides, Sulfones, Sulfonic Acids C181
Acetyl Derivatives C183
Additivity Rules for Aldehydes, Ketones C184
Additivity Rules for Carboxylic Acids and Esters C184
Additivity Rules for Amides C185
Carbonyl Groups C186
Aldehydes C187
Ketones C188
Quinones C190
Carboxylic Acids C191
Carboxylate Anions C192
Carboxylic Acid Esters C193
Lactones C195
Amides C196
Lactams C197
Acid Anhydrides C198
Acid Halides C198
Carbonic Acid Derivatives C199
Nitriles, Isonitriles C200
Imines, Oximes, Isocyanates C201
Hydrazones, Carbodiimides C202
Sulfur Containing Carbonyl Derivatives C203
Thiocyanates, Isothiocyanates C204
Amino Acids C205
Monosaccharides C210
Pyrimidine Bases and Nucleosides C211
Purine Bases and Nucleosides C212
Phosphorous Compounds C215
$^{13}C-^{1}H$ Coupling Constants C220
$^{13}C-^{13}C$ Coupling Constants C240
$^{13}C-^{19}F$ Coupling Constants C245
Solvent Spectra C250

Proton Resonance Spectroscopy H5

Monosubstituted Alkanes H5
Polysubstituted Alkanes, Additivity Rule H15
Coupling in Aliphatic Compounds H20
Aromatic Substituted Alkanes H30
Aliphatic Halogen Compounds H45
Alcohols H50
Ethers H60
Amines H75
Nitro Compounds H90
N-Nitroso-, Azo- and Azoxy Compounds H90
Thiols H95

Thioethers H95
Other Sulfur Compounds H110
Aldehydes H120
Ketones H125
Carboxylic Acids H135
Esters H140
Lactones H145
Amides H150
Lactams H160
Imides H165
Acid Halides and Anhydrides H170
Carbonic Acid Derivatives H170
Oximes, Imines, Hydrazones and Azines H175
Nitriles, Isonitriles, Cyanates and Isocyanates H180
Saturated Alicyclics H185
Alkenes H205
Alkynes H225
Unsaturated Alicyclics H230
Aromatics H245
Heteroaromatics H265
Amino Acids H351
$^{19}F-^{1}H$ Coupling Constants H357
$^{31}P-^{1}H$ Coupling Constants H360
Solvent Spectra H365

Infrared Spectroscopy I5

Alkyl Groups I5
Alkenyl Groups I20
Alkynyl Groups I40
Aromatic Compounds I45
Compounds of Type X≡Y I65
Compounds of Type X=Y=Z I70
Alcohols and Phenols I85
Ethers I90
Peroxides and Hydroperoxides I95
Amines I100
Halogen Compounds I110
Aldehydes I120
Ketones I125
Esters and Lactones I135
Amides, Lactams, Imides and Hydrazides I145
Carbonic Acid Derivatives I155
Carboxylic Acids I165
Amino Acids I175
Acid Halides I180
Anhydrides I185
Compounds with C=N Groups I190
Oximes I195
Compounds with N=N Groups I200
Nitrites and Nitroso Compounds I205
Nitrates, Nitro Compounds and Nitramines I210
Mercaptans, Thioethers and Disulfides I215
Compounds with C=S Groups I220
Compounds with C=O Groups I225
Phosphorous Compounds I235
Silicon Compounds I250
Boron Compounds I260
Interferences, Opaque Regions, Suspension Media I265

Mass Spectrometry M5

Mass Correlation Tables M5
Isotope Patterns of All Naturally Occurring Elements in the Periodic Table M55
Mass and Abundance of the Isotopes of All Naturally Occurring Elements M60
Calculation of Isotope Distributions M90
Isotope Patterns of Various Combinations of Chlorine and Bromine M100
Isotopic Abundances of Various Combinations of Chlorine and Bromine M105
Indications of Structural Type M115
Indications of the Presence of Heteroatoms M125
Rules for the Determination of Relative Molecular Weight . . M135
Metastable Peaks M145
Solvent Spectra M155
Frequent Contaminants M170
Matrices for FAB M170
Typical Fragmentations of Monofunctional Compounds M205

UV/VIS (Spectroscopy in the Ultraviolet or Visible Region of the Spectrum) U5

Correlation Between Wavelength of Absorbed Radiation and Observed Color U5
Simple Chromophores U10
α,β-Unsaturated Carbonyl Compounds (extended Woodward rules) U20
Dienes and Polyenes (Woodward-Fieser rules) U30
Aromatic Carbonyl Compounds (Scott rules) U40
Aromatics U50
Reference Spectra U60

Varia V1

Calculation of the Number of Double Bond Equivalents V1
Corrections for Volume Susceptibility V2
Effect of Symmetry and Rapid Conformational Changes of Molecules on their NMR Spectra V4
Properties of Spin Systems V10

Subject Index Z1

Introduction

The following collection of data is intended to serve as an aid in the interpretation of ^{13}C- and ^{1}H-nuclear magnetic resonance, infrared, mass and electron excitation spectra. It is to be viewed as an addition to texts and reference works dealing with these spectroscopic techniques. It is designed for those who are routinely faced with the task to interprete this type of spectral information. The use of this book for the interpretation of spectra requires only the knowledge of basic principles of these techniques, but its content is structured in a way that it will also serve as a reference work for the specialist.

To aid rapid access to relevant data the following codes are listed at the top of the appropriate pages:

COMB: Summary of characteristic spectroscopic data arranged by structural elements.
(pages B5 through B245)

^{13}C-NMR: ^{13}C-nuclear magnetic resonance spectral data ordered by compound types.
(pages C5 through C265)

^{1}H-NMR: Proton magnetic resonance spectral data.
(pages H5 through H370)

IR: Infrared absorption frequencies ordered by functional groups or compound types.
(pages I5 through I280)

MS: Tables and suggestions for the interpretation of mass spectra.
(pages M5 through M255)

UV/VIS: Tables, suggestions and reference spectra for rationalization of electron excitation spectra.
(pages U5 through U155)

VARIA: Instructions and tables for calculating the number of double bond equivalents, volume susceptibility corrections and predicting the types of spin-spin interactions.
(pages V1 to V10)

The tables are arranged as much as possible in an analogous manner for all methods, but the details of the presentation are dictated by the characteristics of the individual technique.

The summaries presented on pages B5 through B70 facilitate the recognition of first-cut conclusions concerning the structure by those users less familiar with such interpretations, particularly in those cases where no ancillary information is available. The tables presented on pages B75 through B245 make it possible to check in a second step the suspected structural elements for the most important compound types. The remaining tables make it possible to predict the spectroscopic characteristics of a proposed compound. They also serve as a reference compilation for the correlation of the structure of organic compounds with the corresponding spectral data.

Because a large part of the tabulated data is from our own measurements and the rest is based on a large body of literature data a comprehensive reference to published sources is not generally included. Whenever possible the data refer to conventional modes and conditions of measurement. For example, the chemical shifts for NMR spectra were determined generally in deuterochloroform or carbon tetrachloride. The wave numbers (IR) refer to solvents of low polarity, such as chloroform or carbon disulfide. Mass spectral data were recorded at an electron energy of 70 eV. Most of the data were taken from the following sources:

Asahi Research Center, Handbook of Proton-NMR Spectra and Data, Vol. 1-5. Tokyo: Academic Press, 1985.

Batterham, T.J.: NMR spectra of simple heterocycles. New York - London - Sydney - Toronto: Wiley-Interscience, 1973.

Bhacca, N.S., Hollis, D.P., Johnson, L.F., Pier, E.A., Shoolery, J.N.: NMR spectra catalog. Varian Associates, 1962 and 1963.

Bremser, W., Ernst, L. Franke, B.: Carbon-13 NMR spectral data. Weinheim: Verlag Chemie, 3rd. ed., 1981.

Bremser, W., Franke, B., Wagner, H.: Chemical Shift Ranges in Carbon-13 NMR Spectroscopy. Weinheim: Verlag Chemie, 1982.

Bruegel, W.: Handbook of NMR spectral parameters, Vol. 1-3. London - Philadelphia - Rheine: Heyden, 1979.

Clerc, J.T., Pretsch, E.: Kernresonanzspektroskopie, Teil I: Protonenresonanz. Frankfurt am Main: Akademische Verlagsgesellschaft, 2nd. ed., 1973.

Clerc, J.T., Pretsch, E., Sternhell, S.: ^{13}C-Kernresonanzspektroskopie. Frankfurt am Main: Akademische Verlagsgesellschaft, 1973.

Colthup, N.B., Daly, L.H., Wiberley, S.E.: Introduction to infrared and raman spectroscopy. New York - London: Academic Press, 1964.

Cross, A.D.: Introduction to practical infrared spectroscopy. London: Butterworths, 1960.

Dokumentation der Molekülspektroskopie. Institut für Spektrochemie und angewandte Spektroskopie, Dortmund. Weinheim/Bergstrasse: Verlag Chemie.

Dolphin, D., Wick, A.E.: Tabulation of infrared spectral data. New York York - London - Sydney - Toronto: John Wiley & Sons, 1977.

Fachinformationszentrum Energie, Physik, Mathematik GmbH: ^{13}C-NMR Datenbank (accessible from computer terminals), D-7514 Eggenstein-Leopoldshafen 2.

Graselli, J.G., Ritchey, W.M. (Eds.): Atlas of spectral data and physical constants for organic compounds. Cleveland: CRC Press, 1975.

Hediger, H.J.: Infrarotspektroskopie. Frankfurt am Main: Akademische Verlagsgesellschaft, 1971.

Jackman, L.M., Sternhell, S.: Applications of NMR spectroscopy in organic chemistry. Oxford - London - Edinburgh- New York - Toronto - Sydney - Paris - Braunschweig: Pergamon Press, 2nd ed., 1969.

Johnson, L.F., Jankowski, W.C.: Carbon-13 NMR spectra. Huntington - New York: R.E. Krieger Publ. Co., 1978.

Kirba-Kartei. Eglisau /Schweiz: Verlag Dr. H.J. Hediger.

NMR Spectra Catalog. Philadelphia /Pennsylvania, Sadtler Research Laboratories.

Seibl, J.: Massenspektrometrie. Frankfurt am Main: Akademische Verlagsgesellschaft, 1970.

Socrates, G.: Infrared characteristic group frequenties. Chichester - New York - Brisbane - Toronto: John Wiley & Sons, 1980.

Yukawa, Y.: Handbook of organic structural analysis. New York - Amsterdam: W.A. Benjamin, Inc., 1965.

Abbreviations and Symbols

al	aliphatic
ar	aromatic
as	asymmetric
ax	axial
comb	combination frequency
δ	IR: deformation frequency NMR: chemical shift
eq	equatorial
γ	skeletal vibration
gem	geminal
hal	halogen
ip	"in plane" vibration
oop	"out of plane" vibration
st	stretching vibration
sy	symmetric

COMB

Summary of the Regions of the Chemical Shifts for Carbon in Various Bonding Environments

(δ in ppm relative to TMS)

The multiplicity of "off-resonance" uncoupled first order spectra is indicated by the following abbreviations: S = singlet, D = doublet, T = triplet, Q = quadruplet

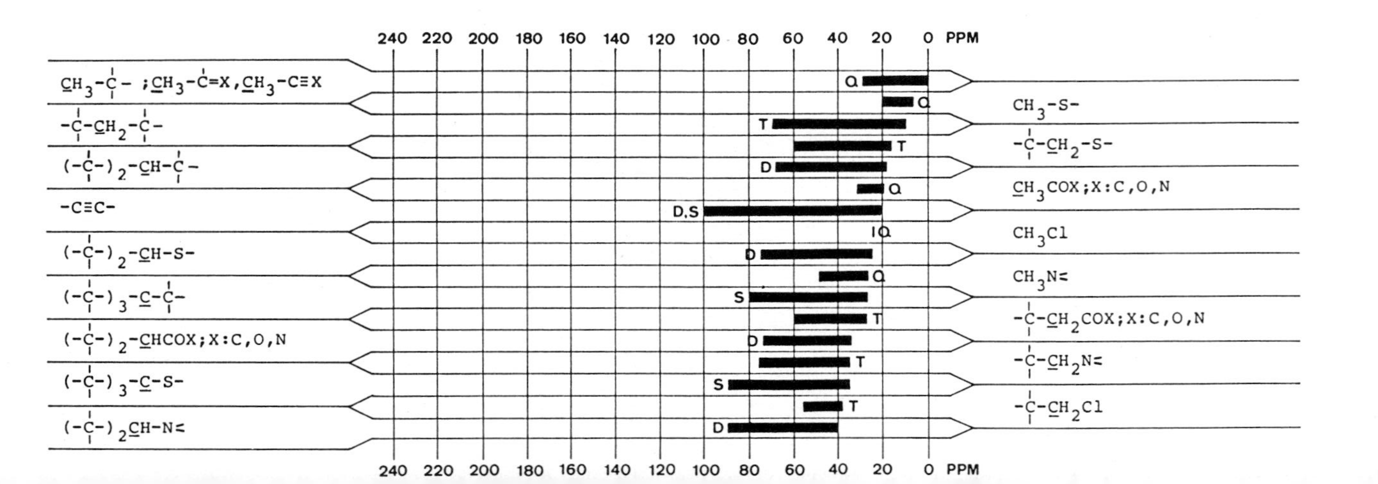

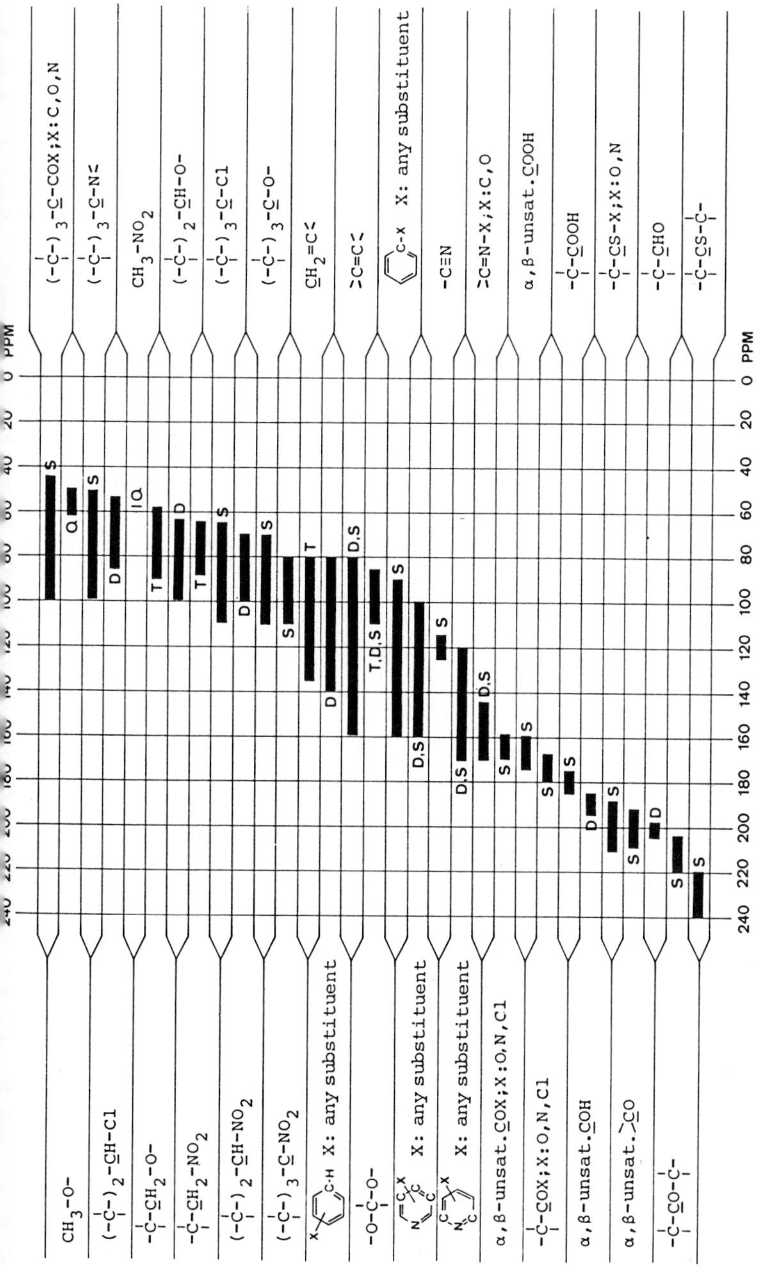

PPM: 240 220 200 180 160 140 120 100 80 60 40 20 0

- $(-\overset{|}{\underset{|}{C}}-)_3-\underline{C}-COX$; X: C, O, N — S
- CH_3-O- — Q
- $(-\overset{|}{\underset{|}{C}}-)_3-\underline{C}-N<$ — S
- $(-\overset{|}{\underset{|}{C}}-)_2-\underline{C}H-Cl$ — D
- CH_3-NO_2 — Q
- $-\overset{|}{\underset{|}{C}}-\underline{C}H_2-O-$ — T
- $(-\overset{|}{\underset{|}{C}}-)_2-\underline{C}H-O-$ — D
- $-\overset{|}{\underset{|}{C}}-\underline{C}H_2-NO_2$ — T
- $(-\overset{|}{\underset{|}{C}}-)_3-\underline{C}-Cl$ — S
- $(-\overset{|}{\underset{|}{C}}-)_2-\underline{C}H-NO_2$ — D
- $(-\overset{|}{\underset{|}{C}}-)_3-\underline{C}-O-$ — S
- $(-\overset{|}{\underset{|}{C}}-)_3-\underline{C}-NO_2$ — S
- $\underline{C}H_2=C<$ — T
- X-C₆H₄-C·H X: any substituent — D
- $>C=C<$ — D,S
- $-O-\overset{|}{\underset{|}{C}}-O-$ — T,D,S
- C₆H₅-C-X X: any substituent — S
- pyridine C-X X: any substituent — D,S
- $-C\equiv N$ — S
- pyridine C-X X: any substituent — D,S
- $>C=N-X$; X: C, O — D,S
- α,β-unsat.$\underline{C}$OX; X: O, N, Cl — S
- α,β-unsat.$\underline{C}$OOH — S
- $-\overset{|}{\underset{|}{C}}-\underline{C}OX$; X: O, N, Cl — S
- $-\overset{|}{\underset{|}{C}}-\underline{C}OOH$ — S
- α,β-unsat.$\underline{C}$OH — D
- $-\overset{|}{\underset{|}{C}}-\underline{C}S-X$; X: O, N — S
- α,β-unsat.$>\underline{C}O$ — S
- $-\overset{|}{\underset{|}{C}}-\underline{C}HO$ — D
- $-\overset{|}{\underset{|}{C}}-\underline{C}O-\overset{|}{\underset{|}{C}}-$ — S
- $-\overset{|}{\underset{|}{C}}-\underline{C}S-\overset{|}{\underset{|}{C}}-$ — S

PPM: 240 220 200 180 160 140 120 100 80 60 40 20 0

Summary of the Regions of the ^{1}H Chemical Shifts of Protons in Various Bonding Environments (δ in ppm relative to TMS)

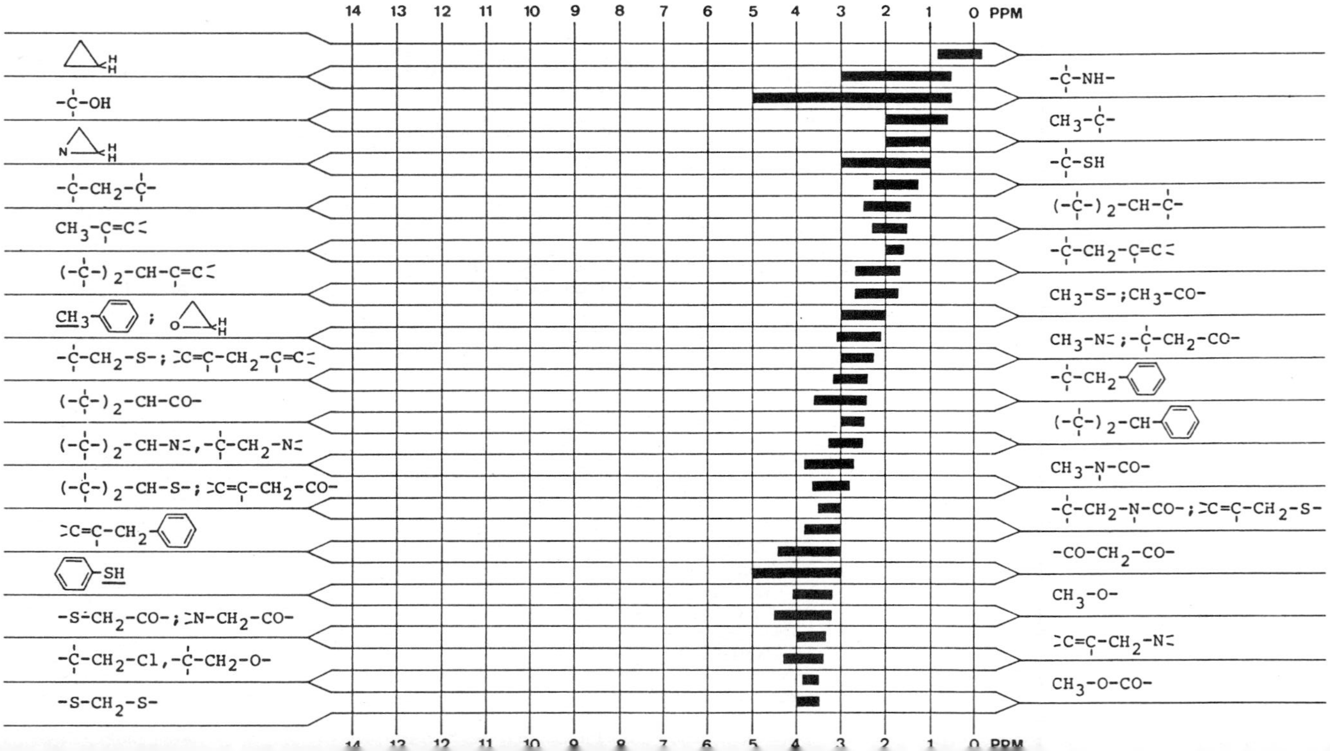

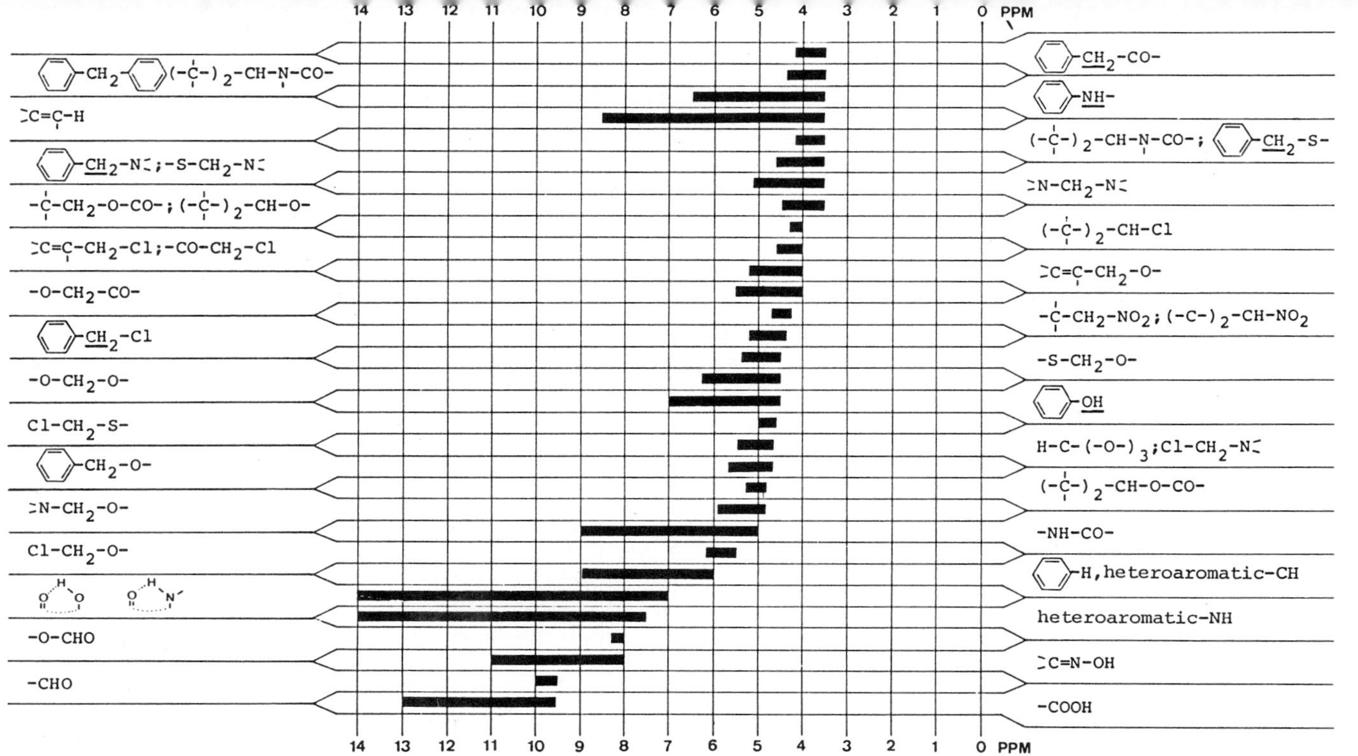
PPM
14 13 12 11 10 9 8 7 6 5 4 3 2 1 0
-CH2-CO-
-NH-
(-C-)2-CH-N-CO-; -CH2-S-
>N-CH2-N<
(-C-)2-CH-Cl
>C=C-CH2-O-
-C-CH2-NO2; (-C-)2-CH-NO2
-S-CH2-O-
-OH
H-C-(-O-)3; Cl-CH2-N<
(-C-)2-CH-O-CO-
-NH-CO-
-H, heteroaromatic-CH
heteroaromatic-NH
>C=N-OH
-COOH
-CH2- (-C-)2-CH-N-CO-
>C=C-H
-CH2-N<; -S-CH2-N<
-C-CH2-O-CO-; (-C-)2-CH-O-
>C=C-CH2-Cl; -CO-CH2-Cl
-O-CH2-CO-
-CH2-Cl
-O-CH2-O-
Cl-CH2-S-
-CH2-O-
>N-CH2-O-
Cl-CH2-O-
-O-CHO
-CHO

Summary of Ranges of Coupling Constants Between Differently Bound Protons (|J| in Hz)

X and Y represent different substituents.

Geminal coupling:

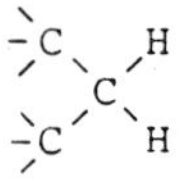

$(-C)_2CH_2$	10 - 14
$XYCH_2$	0 - 20
cyclopropane CH_2	2 - 5
$C=CH_2$	0 - 4

Vicinal coupling:

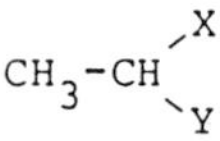

CH_3-CHXY	6 - 8
H-C-C-H	0 - 18
cyclohexane H_{ax}, H_{eq}	ax,ax: 6 - 13; ax,eq: 2 - 4; eq,eq: 2 - 4
H-C=C-H (cis)	5 - 14
H-C=C-H (trans)	12 - 18
benzene ortho H,H	6 - 10

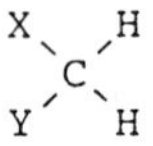

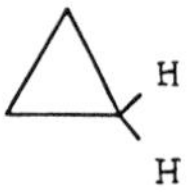

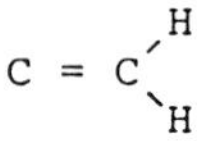

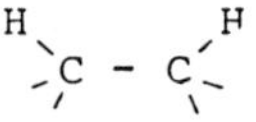

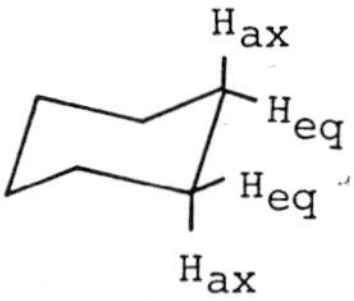

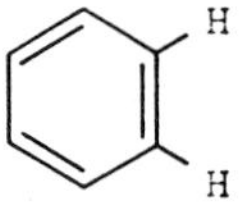

"Long-range" coupling:

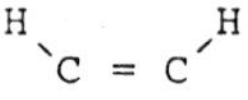

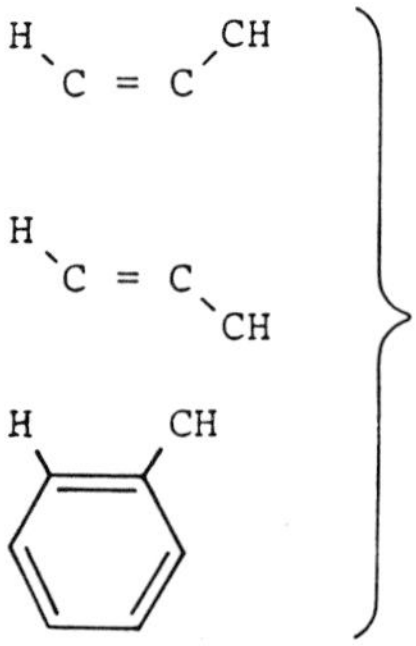

H-C=C-CH (cis)	0 - 2.5
H-C=C-CH (trans)	0 - 2.5
benzene H / ortho CH	0 - 2.5
CH - C ≡ CH	2 - 3
CH - C ≡ C - CH	2 - 3

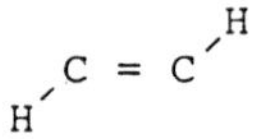

Summary of the Most Important Strong IR Absorption Bands (in cm^{-1})

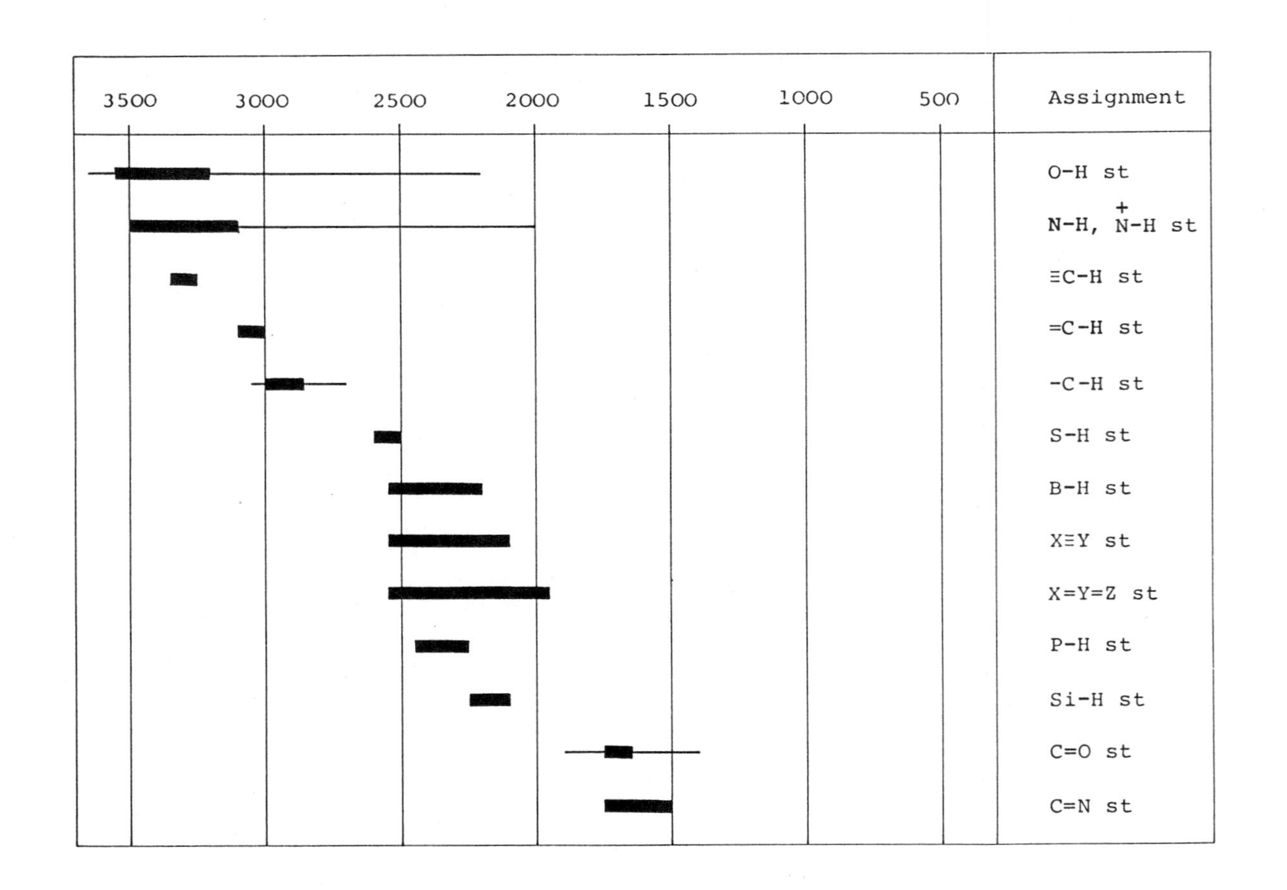

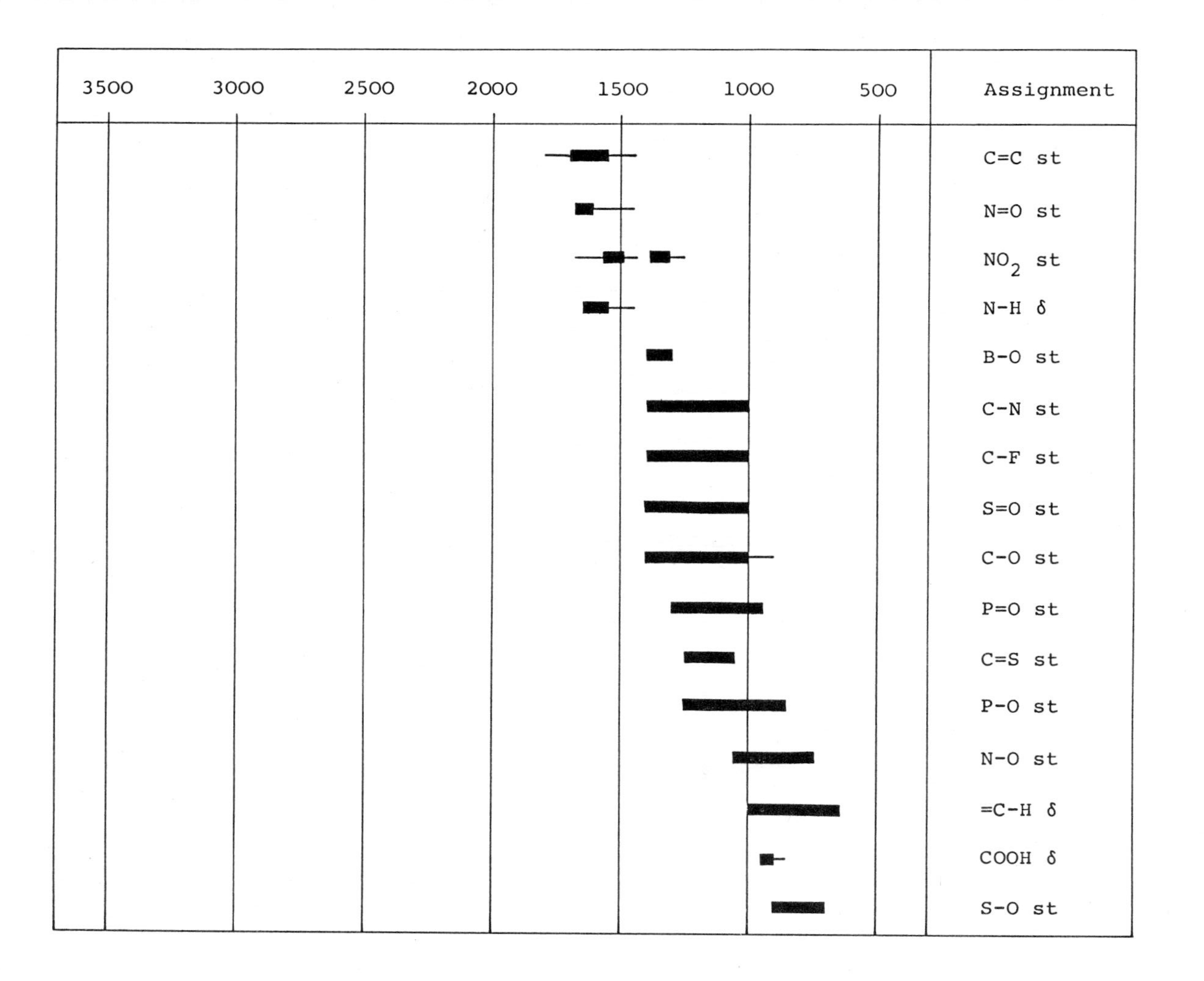
3500
3000
2500
2000
1500
1000
500
Assignment
C=C st
N=O st
NO_2 st
N-H δ
B-O st
C-N st
C-F st
S=O st
C-O st
P=O st
C=S st
P-O st
N-O st
=C-H δ
COOH δ
S-O st

Summary of IR Absorption Bands of >C=O Groups (in cm^{-1})

X: any substituent

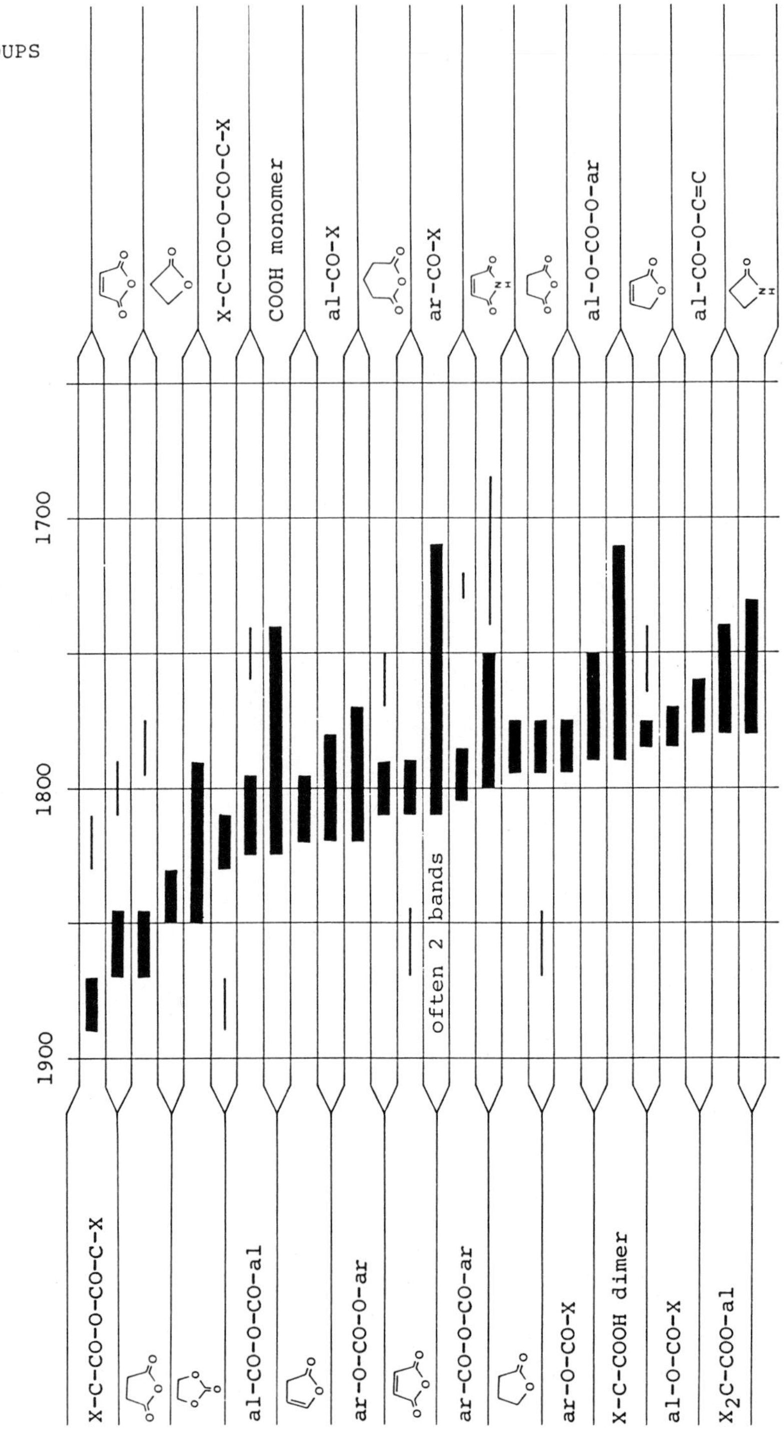

1900 1800 1700

X-C-CO-O-CO-C-X

al-CO-O-CO-al

ar-O-CO-O-ar

often 2 bands

ar-CO-O-CO-ar

ar-O-CO-X

X-C-COOH dimer

al-O-CO-X

X_2C-COO-al

X-C-CO-O-CO-C-X

COOH monomer

al-CO-X

ar-CO-X

al-O-CO-O-ar

al-CO-O-C=C

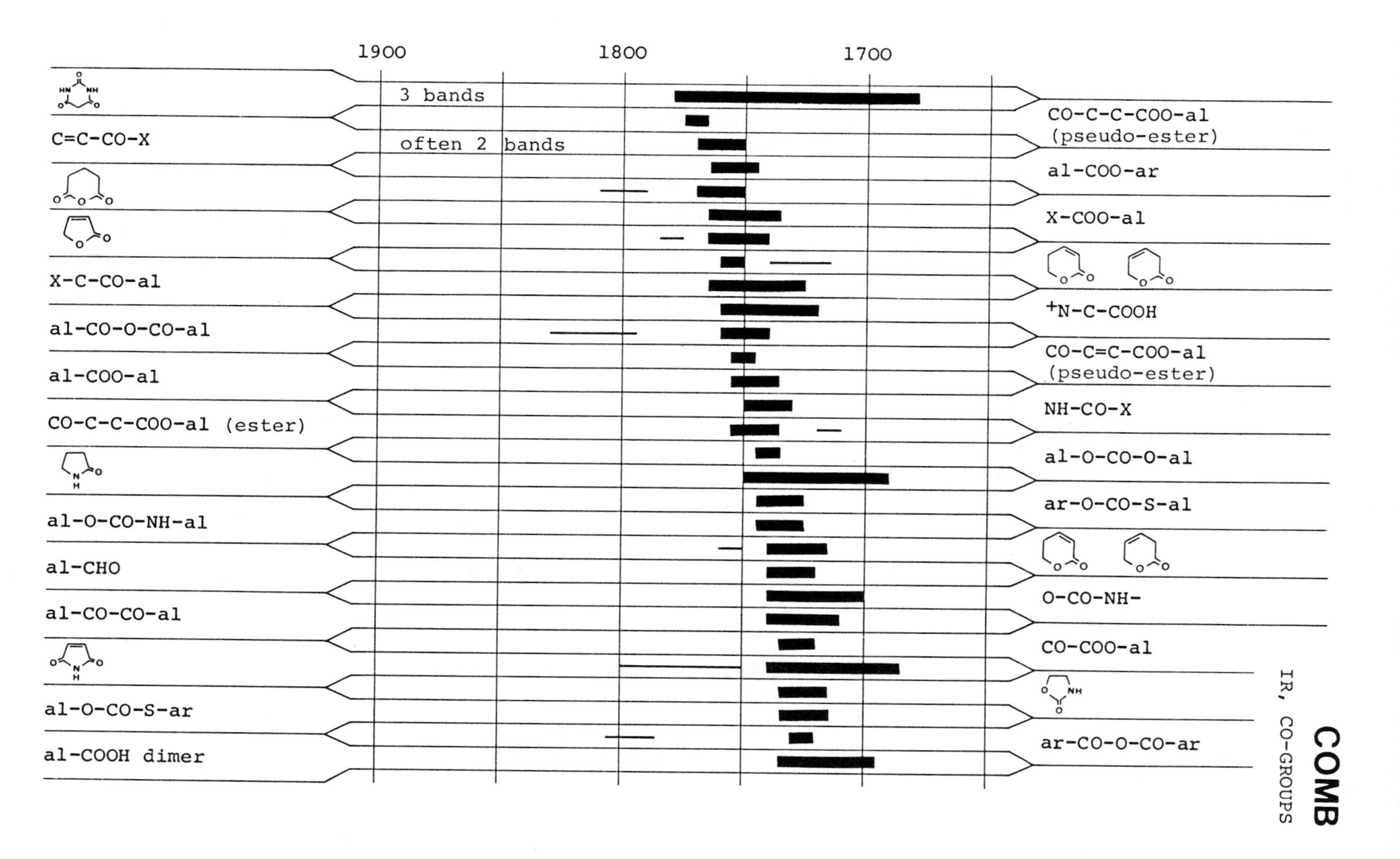

1900
1800
1700
3 bands
often 2 bands
C=C-CO-X
X-C-CO-al
al-CO-O-CO-al
al-COO-al
CO-C-C-COO-al (ester)
al-O-CO-NH-al
al-CHO
al-CO-CO-al
al-O-CO-S-ar
al-COOH dimer
CO-C-C-COO-al
(pseudo-ester)
al-COO-ar
X-COO-al
+N-C-COOH
CO-C=C-COO-al
(pseudo-ester)
NH-CO-X
al-O-CO-O-al
ar-O-CO-S-al
O-CO-NH-
CO-COO-al
ar-CO-O-CO-ar

1800 1700 1600

C=C-COO-al

ar-COO-al

al-CO-C-CO-al (diketone)

CO-C=C-COO-al (ester)

al-O-CO-NH-al

al-CO-C-C-CO-al

al-CO-al

COOH···H

CO-C-C-COO-al (ketone)

CO-C-COO-al (ester)

C=C-COOH, ar-COOH dimer

al-O-CO-S-al

ar-CHO

C=C-CHO

CO-C-COO-al (ketone)

ar-CO-al

CO-NH_2 (in solution)

CO-NH (in solution)

al-S-CO-NH-al

NH / C=O 6-ring and larger

C=C-CO-al

O= =O (quinones)

CO-NH (solid)

X-C-COO^-

CO-C=C-COO-al (ketone)

ar-CO-ar

C=C-CO-C=C

HN NH / O (cyclic urea) ; NH-CO-NH

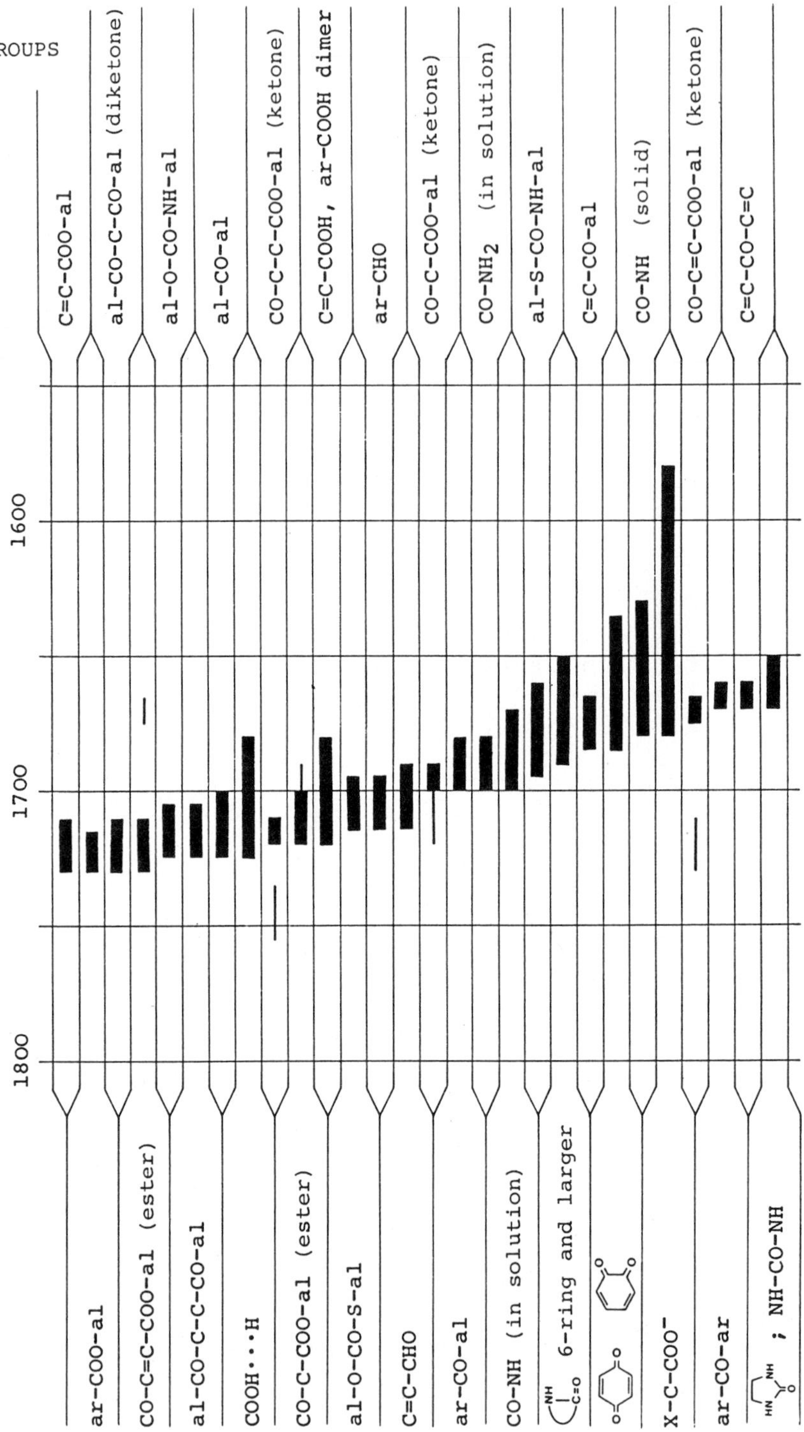

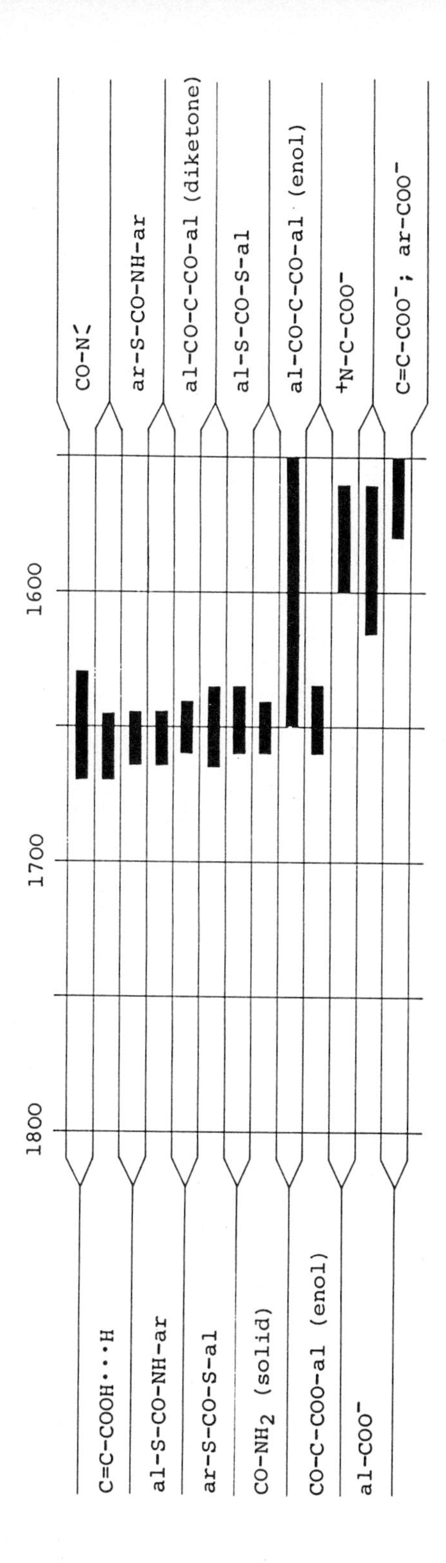

CO-N<
ar-S-CO-NH-ar
al-CO-C-CO-al (diketone)
al-S-CO-S-al
al-CO-C-CO-al (enol)
$^+$N-C-COO$^-$
C=C-COO$^-$; ar-COO$^-$
1600
1700
1800
C=C-COOH···H
al-S-CO-NH-ar
ar-S-CO-S-al
CO-NH_2 (solid)
CO-C-COO-al (enol)
al-COO$^-$

UV/VIS-Absortion Bands of Various Compound Types

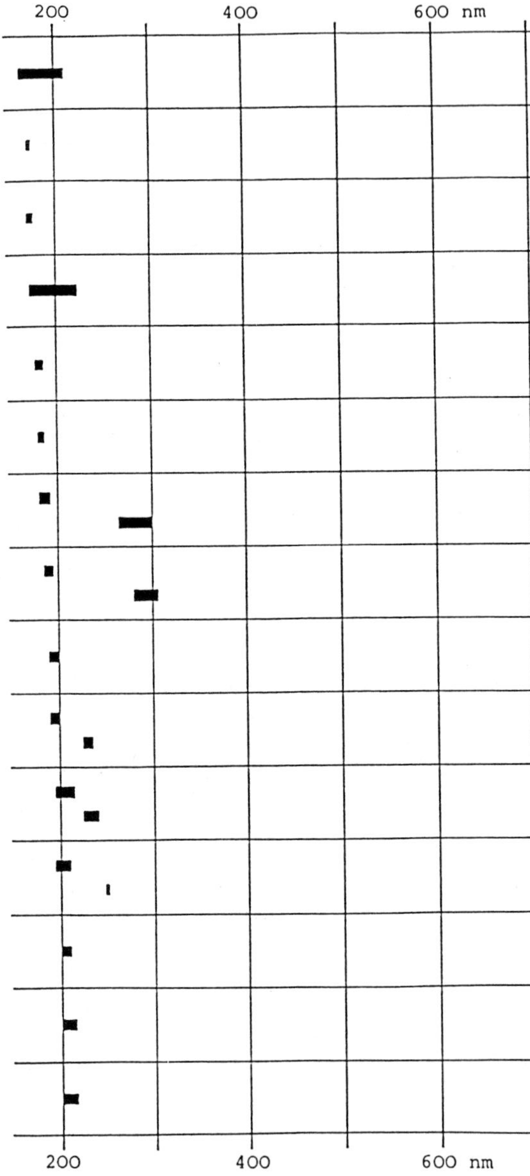

200 400 600 nm

Compound type	Transition (log ε)
$A_2C=CA_2$	π→π* (3-4)
oxirane (epoxide)	n→σ* (3.6)
R-Cl	n→σ* (2.4)
A-≡-A	π→π* (3.7-4)
R-OH	n→σ* (2.5)
R-O-R	n→σ* (3.5)
$R_2C=O$	π→π* (3-4) n→π* (1-2)
$R(H)C=O$	π→π* (2) n→π* (0.9-1.4)
$R-NH_2$	n→σ* (3.5)
R-SH	n→σ* (3.2) n→σ* (2.2 Sch)
R-S-R	n→σ* (3-3.6) n→σ* (2-3 Sch)
R-S-S-R	n→σ* (3-4) n→σ* (2.6)
R-Br	n→σ* (2.5)
$R(HO)C=O$	n→π* (1.7)
$R(RO)C=O$	n→π* (1.7)

200 400 600 nm

A: alkyl or H R: alkyl sh: shoulder

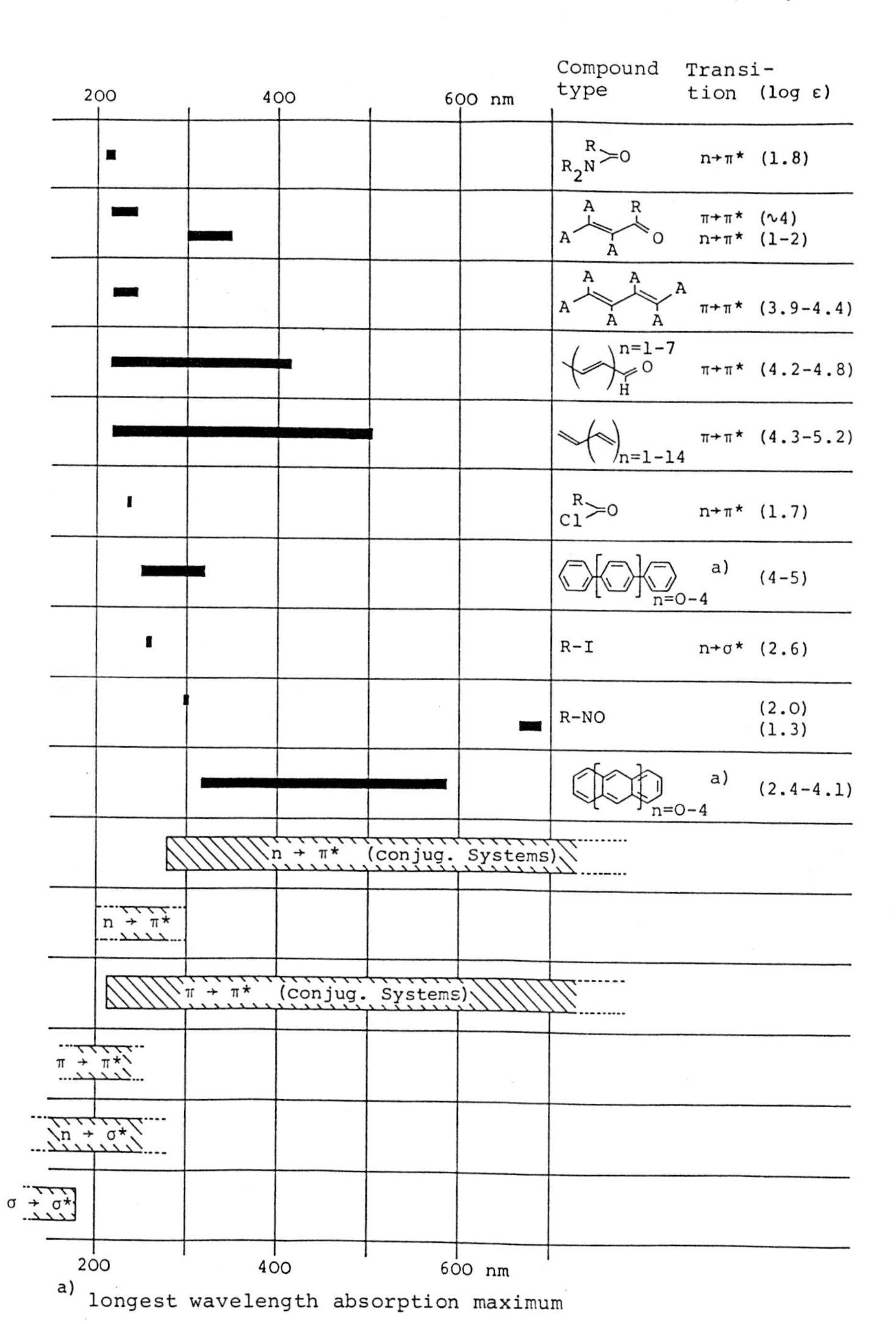

a) longest wavelength absorption maximum

Characteristic Spectroscopic Data for Alkanes and Saturated Alicyclics

	Assignment	Range	Comments	Details see page:
^{13}C-NMR			reference data: alkanes	C5
			cycloalkanes	C50
	$-CH_3$, $-CH_2-$, $>CH-$, $>C<$	5-60 ppm	$-CH_3$, $-CH_2-$, $>CH-$ and $>C<$ can be differentiated by partial ("off resonance") uncoupled spectra or based on relaxation times	
			found outside of the quoted range if present in three-membered rings	
^{1}H-NMR			reference data, alkanes: chemical shifts	H5
			coupling constants	H20
			cycloalkanes: chemical shifts and coupling constants	H185
	$-CH_3$	0.8-1.2 ppm		
	$-CH_2-$, $>CH-$	1.1-1.8 ppm	found outside of the quoted range if present in three-membered rings	

IR			reference data	I5
	$\geq$CH st	3000-2840 cm^{-1}	found outside the quoted range if present in three-membered rings	
	$-CH_3$ δ as } $-CH_2-$ δ	∿1460 cm^{-1}		
	$-CH_3$ δ sy	∿1380 cm^{-1}	doublet for geminal methyl groups	
	$-CH_2-$ γ	770-720 cm^{-1}	in $C-(CH_2)_n-C$ if $n \geq 4$ at ∿720 cm^{-1}	
MS	Molecular ion		n-alkanes: weak; isoalkanes: very weak } $m/z = 14n + 2$ monocycloalkanes: medium $m/z = 14n$	
	Fragments		n-alkanes: local maxima at $14n + 1$, intensity variations smooth, minimum at $M^{+\cdot} - 15$	M115 M205
			isoalkanes: local maxima at $14n + 1$, intensity distribution irregular, relative maxima due to fragmentation at branching points with charge retention at most substituted C.	
			monocycloalkanes: local maxima at $14n - 1$, intensity distribution irregular, relative maxima due to cleavage at ring.	
	Rearrangements		n-alkanes: non-specific isoalkanes, monocycloalkanes } elimination of alkanes { $m/z = 14n$, $m/z = 14n - 2$	
UV			no absorption above 200 nm	

Characteristic Spectroscopic Data for Alkenes and Saturated Alicyclics

	Assignment	Range	Comments	Details see page:
^{13}C-NMR	C=C	100-150 ppm	reference data	C80,C100
	C-(C=C)	10-40 ppm	considerable differences between $C{-}C{=}C{<}^{X}_{H}$ and $C{-}C{=}C{<}^{H}_{X}$	C10
^{1}H-NMR	H-(C=C)	4.5-6 ppm	reference data, additivity rules	H205
			coupling constants |J|: geminal 0 - 3 Hz; cis 5-12 Hz; trans 12-18 Hz	
	CH_3-(C=C)	~1.7 ppm	reference data	H5,H225
	-CH_2-(C=C)	~2.0 ppm	alkenes	H5
			cycloalkenes	H230
			considerable differences between -CH_2-C-C=C- (~1.5 ppm) and CH_2-C-C=C (ring closed via C_n) n≥1 (~1.8 ppm, see H230)	
			coupling constants |J|: CH_2-CH=C ~7 Hz; CH_2-CH=C (ring closed via C_n) n=2 ~0.5 Hz, n=3 ~1.5 Hz, n=4 ~4 Hz	
			"long-range" coupling constants: see H205	

IR			reference data	I20
	H-C(=C) st	3100-3000 cm^{-1}		
	C=C st	1690-1635 cm^{-1}		
	H-C(=C) δ oop	1000-675 cm^{-1}		
			$-CH_2-(C=C)$ δ at 1440 cm^{-1} (see I10)	
MS	Molecular ion		medium abundant, m/z = 14n, in monocyclic alkenes at 14n - 2	M205
	Fragments		local maxima at 14n - 1, in monocyclic alkenes at 14n - 3	M115
			double bonds can be localized only in special cases	
	Rearrangements		unspecific; in cyclohexenes Retro-Diels-Alder-reaction: 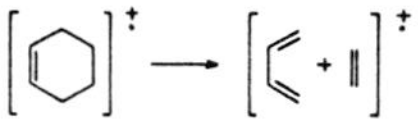	
			specific for: 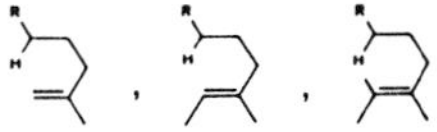 , , following scheme	
			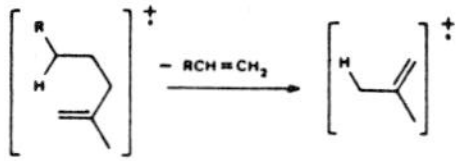 - RCH=CH₂	
UV	C=C π→π*	<210 nm (log ε = 3-4)	for isolated double bonds; for highly substituted double bonds often endabsorption	
	$(C=C)_2$ π→π*	215-280 nm (log ε=3.5-4.5)	reference data	U30

Characteristic Spectroscopic Data of Alkynes

	Assignment	Range	Comments	Details see page:
^{13}C-NMR			reference data	C110
	C≡C	65-85 ppm	coupling constant H-C≡^{13}C: ∿50 Hz, generally recognizable in partially ("off resonance") uncoupled spectra	
	C-(C≡C)	0-30 ppm		C10
^{1}H-NMR	H-C≡C	1.5-3 ppm	reference data	H225
			coupling constants |J|: CH-C≡CH ∿3 Hz; CH-C≡C-CH ∿3 Hz	
	CH_3-C≡C	∿1.8 ppm		
	-CH_2-C≡C	∿2.2 ppm		H225
	>CH-C≡C	∿2.6 ppm		
IR			reference data	I40
	H-(C≡C) st	3340-3250 cm^{-1}	sharp	
	C≡C st	2260-2100 cm^{-1}	sometimes very weak	

MS	Molecular ion		weak, for 1-alkynes up to C_7 often absent	M210
	Fragments / Rearrangements		vary in extent between alkanes and aromatics	M115
UV	C≡C $\pi\to\pi^*$	<210 nm ($\log \varepsilon = 3.7-4.0$)	endabsorption, often a few weak bands <240 nm	

Characteristic Spectroscopic Data of Aromatic Hydrocarbons

	Assignment	Range	Comments	Details see page:
^{13}C-NMR			reference data	C115
	arC	120-150 ppm	also in polycyclic aromatic hydrocarbons	C115-C130
	arCH	110-130 ppm		
	C-(arC)	10-60 ppm		C10
^{1}H-NMR			reference data	H245
	H-(arC)	6.8-7.5 ppm	in polycyclic aromatic hydrocarbons up to ∿9 ppm	
			coupling constants: J_{ortho} ∿7 Hz, J_{meta} ∿2 Hz, J_{para} <1 Hz (in routine-spectra generally not detectable)	H245
	CH_3-(arC)	∿2.3 ppm	often band broadening due to "long-range" coupling with aromatic protons (see H245)	H5,H30
	$-CH_2$-(arC)	∿2.6 ppm		H5,H30
	>CH-(arC)	∿2.9 ppm		H5
			geminal coupling in $-CH_2$-(arC): see H20	
IR			reference data	I45
	arC-H st	3080-3030 cm^{-1}	often multiple bands, weak	
	comb	2000-1650 cm^{-1}	very weak	
	arC-C st	∿1600 cm^{-1}, ∿1500 cm^{-1}, ∿1450 cm^{-1}	often split, sometimes not all three bands observable	
	arC-H δ oop	900-650 cm^{-1}	strong, frequently multiplicity of bands	

MS	Molecular ion		abundant, often base peak	M210
	Fragments		m/z = 39, 50-53, 63-65, 75-78; dehydrations; $M^{+\cdot} - 26$, $M^{+\cdot} - 39$; frequently doubly charged fragment ions; benzylic cleavage	M120
			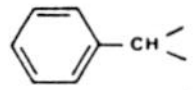 (benzyl structure) m/z = 90-92	
			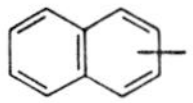 (naphthyl structure) m/z = 127	
			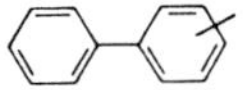 (biphenyl structure) m/z = 152, 153	
			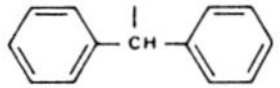 (diphenylmethyl structure) m/z = 152, 165	
	Rearrangements		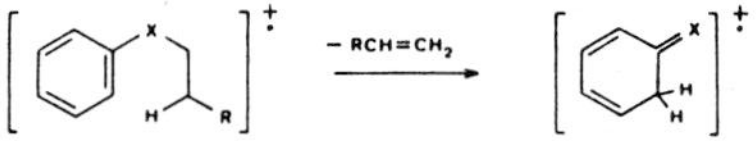 (structure) $\xrightarrow{-RCH=CH_2}$ (structure)	
UV			reference data	U50,U40, U110-U125
		~200-210 nm (log ε = ~4); ~260 nm (log ε = ~2.4)	in benzene and alkyl benzene	

Characteristic Spectroscopic Data of Heteroaromatics

	Assignment	Range	Comments	Details see page:
^{13}C-NMR			reference data	C135-C166
	arC-(arX), arC-(arC)	100-160 ppm		
^{1}H-NMR			reference data	H265-H350
	H-(arC)	6-9 ppm	coupling in 6-membered rings similar to aromatic hydrocarbons (see H275); for 5-membered ring heteroaromatics see H265	
			H-(arN): 7-14 ppm, strongly solvent dependent, generally broad	
IR			reference data	I45
	arC-H st	3100-3000 cm^{-1}	frequently multiple bands, weak	
	arC-C st	∿1600 cm^{-1}, ∿1500 cm^{-1}, ∿1450 cm^{-1}	often split, sometimes not all three bands observable	
	arC-H δ oop	1000-650 cm^{-1}	often strong, frequently multiplicity of bands arN-H st: 3500-2800 cm^{-1}	

MS			hints for heteroatoms	M125
	Molecular ion		abundant, often base peak	
	Fragments		m/z 39, 50-53  ... , (pyridinium ion) : m/z 78 benzyl-analogous cleavage for heteroaromatics γ-cleavage  [2-(CH₂CH₂R)pyridine]⁺˙ —(- R˙)→ [2-methylenepyridinium]⁺	
	Rearrangements		for N-heteroaromatics loss of HCN 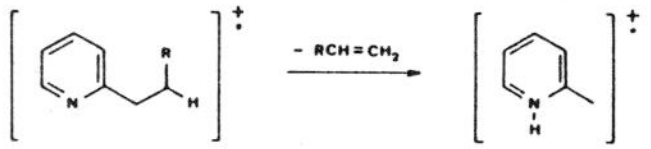 [2-(CH₂CHRH)pyridine]⁺˙ —(- RCH=CH₂)→ [2-methylpyridine, N-H]⁺˙ for O-heteroaromatics loss of CO for S-heteroaromatics loss of CS; m/z 45 (CHS^+)	
UV			reference data	U130-U155

^{13}C-NMR

Characteristic Spectroscopic Data of Halogen Compounds

Assignment	Range	Comments	Details see page:
alC-F	70-100 ppm	reference data	C40,C167
		additivity rules (poor for polyhalogenated compounds)	C10
		CF_3: ∿115 ppm	
⪀C-F	125-175 ppm	additivity rules	C90
C(=C-F)	65-115 ppm		
arC-F	135-165 ppm	additivity rules	C120,C130
arC-(C-F)	105-135 ppm		
		$^{19}F-^{13}C$ coupling constants	C169,C245
alC-Cl	30-60 ppm	reference data	C40,C167
		additivity rules (poor for polyhalogenated compounds)	C10
⪀C-Cl	100-150 ppm	additivity rules	C90
arC-Cl	120-150 ppm	additivity rules	C120,C130
alC-Br	10-45 ppm	reference data	C40,C167
		additivity rules (poor for polyhalogenated compounds)	C10
⪀C-Br	90-140 ppm	additivity rules	C90
arC-Br	110-140 ppm	additivity rules	C120,C130
alC-I	-20 bis +30 ppm	reference data	C40,C167
		additivity rules (poor for polyhalogenated compounds)	C10
⪀C-I	60-110 ppm	additivity rules	C90
arC-I	85-115 ppm	additivity rules	C120,C130

^{1}H-NMR	$-CH_2-F$	~4.3 ppm	reference data	H5
			$^{19}F-^1H$ coupling constants	H357
			additivity rules: olefins aromatics	H215 H255
	$-CH_2-Cl$ $-CH_2-Br$	~3.5 ppm	reference data	H5
	$-CH_2-I$	~3.1 ppm	additivity rules: aliphatics olefins aromatics	H15 H215 H255
IR			reference data	I110
	C-F st	1400-1000 cm^{-1}	strong	
	C-Cl st	<830 cm^{-1}		
	C-Br st	<700 cm^{-1}		
	C-I st	<600 cm^{-1}		
MS	Molecular ion		for saturated aliphatic halogen compounds frequently weak, for polyhalogenated compounds often absent	M245
			isotope pattern of chlorine and bromine	M100,M105
	Fragments		R-C┼hal > R┼C-hal	M120
			CF_3: m/z 69; $M^{+\cdot}$-50 or fragment$^+$-50 (CF_2)	
			indications of halogen atoms	M130
	Rearrange-ments		for fluorine and chlorine compounds -HF or -HCl, respectively	
UV			reference data	U10,U70, U75
	$n \rightarrow \sigma^*$	≤280 nm (log ε = ~2.5)	for C-Br and C-Cl generally only endabsorption, for C-F no absorption	

Characteristic Spectroscopic Data of Alcohols and Phenols

	Assignment	Range	Comments	Details see page:
^{13}C-NMR			reference data, additivity rules: aliphatic, alicyclic, aromatic	C10,C40, C170; C70; C120-C130; C140
	C-(OH)	50-100 ppm	shift with respect to C-(H) about +40 to +50 ppm	
	C-(C-OH)	10-60 ppm	hardly any shift with respect to C-(C-CH_3)	
	C-(C-C-OH)	10-60 ppm	shift with respect to C-(C-C-CH_3) about -5 ppm	
	arC-(OH)	135-155 ppm	shift with respect to arC-(H) about +25 ppm, for carbon atoms in ortho- and para-position about -10 ppm	
^{1}H-NMR	H-(O)		reference data	H50
		0.5-5 ppm	aliphatic, alicyclic } position and peak shape strongly dependent on experimental conditions	
		5-8 ppm	aromatic } position and peak shape strongly dependent on experimental conditions	
	-CH_2-(OH), >CH-(OH)	3.5-4 ppm	reference data, additivity rules: aliphatic, alicyclic	H15,H55; H195
	H-(ar-OH)	6.5-7.5 ppm	for C-aromatics shift with respect to H-(ar-H): ortho ∿ -0.6 ppm; meta ∿ -0.1 ppm; para ∿ -0.5 ppm	H255

IR	-OH st C-O(H) st	3650-3200 cm^{-1} 1260-970 cm^{-1}	reference data band position and shape depends on degree of association strong	I85
MS	Molecular ion		aliphatic: low abundance, often missing for primary or highly branched alcohols; peaks at highest mass are then often due to $M^{+\cdot}-18$ or $M^{+\cdot}-15$ aromatic: abundant	M220
	Fragments		hints for oxygen aliphatic: m/z 31, 45, 59..; $M^{+\cdot}-18$, $M^{+\cdot}-33$, $M^{+\cdot}-46$ primary alcohols: m/z 31 > m/z 45 ∿ m/z 59 secondary and tertiary alcohols: local maxima due to α-cleavage: $R\text{-}CH(\dot{O}H^{+})\text{-}R \xrightarrow{-R\cdot} R\text{-}CH{=}OH^{+}$ aromatic: $[ar\text{-}O]^{+\cdot}$; $M^{+\cdot}-28$ (CO), $M^{+\cdot}-29$ (CHO), generally accompanied by rearrangement peaks	M115, M120 M125
	Rearrangements		aliphatic: elimination of H_2O from $M^{+\cdot}$ and products of α-cleavage; elimination of H_2O followed by elimination of olefins olefinic: vinylcarbinols: spectra similar to ketones allyl alcohols: specific, aldehyde elimination: $[R_1\text{-}CH{=}CH\text{-}CH_2\text{-}CH(OH)\text{-}R_2]^{+\cdot} \xrightarrow{-R_2CHO} [R_1\text{-}CH{=}CH\text{-}CH_3]^{+\cdot}$ aromatic: if ortho-substitution suitable for: [ortho-substituted phenol]$^{+\cdot}$ → [ketonic cyclohexadiene with =Y]$^{+\cdot}$ + H–Z Y–Z : –CO-OR, C-Hal, O-R u.ä.	
UV			aliphatic: no absorption above 200 nm aromatic: in alkaline solution shift to longer wavelength and more intense	U80, U85

Characteristic Spectroscopic Data of Ethers

	Assignment	Range	Comments	Details see page:
^{13}C-NMR			reference data, additivity rules: aliphatic	C10,C40,C172
			olefinic	C90,C172
			aromatic	C120-C130
				C172
	C-(O)	50-100 ppm	for oxiranes outside the normal range	
	C-(C-O) C-(C-C-O)	10-60 ppm	O-C-O: 85-110 ppm	
	(C)=C-(O)	115-165 ppm	shift with respect to (C)=C-(C) about +15 ppm	
	C=(C-O)	70-120 ppm	shift with respect to C=(C-C) about -30 ppm	
	arC-(O)	135-155 ppm	shift with respect to arC-(H) about +25 ppm, for ortho- or para-carbon atoms about -10 ppm	
^{1}H-NMR			reference data, additivity rules: aliphatic	H15,H60
			cyclic	H65
			olefinic	H215
			aromatic	H255
	CH_3-(O)	3.3-4 ppm	singlet	
	-CH_2-(O)	3.4-4.2 ppm	O-CH_2-O: ∿4.5-6 ppm	
	>CH-(O)	3.5-4.3 ppm	CH-$(O)_3$: ∿5-6 ppm	
	H-(C(=C)-O)	5.7-7.5 ppm	shift with respect to H-(C(=C)-H) about +1.2 ppm	
	H-(C=C-O)	3.5-5 ppm	shift with respect to H-(C=CH) about -1 ppm	
	H-(ar-O)	6.6-7.6 ppm	for C-aromatics shifted with respect to H-(ar-H): ortho ∿ -0.5 ppm meta ∿ -0.1 ppm para ∿ -0.4 ppm	

IR			reference data	I90
	CH_3-(O) st, CH_2-(O) st	2880-2815 cm^{-1}	O-CH_2-O: 2880-2750 cm^{-1}, two bands	
	C-O-C st as	1310-1000 cm^{-1}	strong, sometimes two bands	
MS	Molecular ion		aliphatic: low abundance; tendency to protonate aromatic: abundant	M220
	Fragments		indications for oxygen aliphatic: m/z = 31,45,59,73..; base peak of aliphatic ethers generally due to fragmentation of the bond α to the ether bond: $[R_1 \overset{\alpha}{-} C-O-R_2]^{+\cdot} \xrightarrow{-R_1^\cdot} C=\overset{+}{O}-R_2$ or due to heterolytic cleavage of the C-O bond, especially for polyethers: $[R_1-O-R_2]^{+\cdot} \xrightarrow{-R_1-O\cdot} R_2^+$ arylalkyl ethers: preferential loss of the alkyl chain diaryl ethers: preferential loss of CO(28) from $M^{+\cdot}$ and/or $[M-H]^+$ as well as $ar_1-O \mid ar_2$	M115,M120 M125
	Rearrangements		for aliphatic ethers frequently elimination of alcohol for aromatic ethyl and higher ethers: $[C_6H_5-O-CH_2-CH(H)-R]^{+\cdot} \xrightarrow{-R-C=C} [C_6H_5-OH]^{+\cdot}$	
UV			aliphatic: no absorption above 200 nm aromatic	U80,U105

Characteristic Spectroscopic Data of Amines

	Assignment	Range	Comments	Details see page:
^{13}C-NMR			reference data, additivity rules: aliphatic, alicyclic, aromatic	C10,C45,C174 C70,C176 C120-C130 C140,C176
	C-(N)	25-80 ppm	shift with respect to C-(H) about 20 to 30 ppm	
	C-(C-N)	10-60 ppm	shift with respect to C-(C-C) about +2 ppm	
	C-(C-C-N)	10-60 ppm	shift with respect to C-(C-C-C) about -2 ppm	
	arC-(N)	130-150 ppm	shift with respect to arC-(H) about +20 ppm, for ortho- or para-carbon atoms about -10 to -15 ppm	
^{1}H-NMR	H-(N)		reference data	H75
		0.5-4.0 ppm	aliphatic, alicyclic } couplings: H80	
		2.5-5.0 ppm	aromatic } couplings: H80	
	$H-(N^+)$	6.0-7.0 ppm		
	CH_3-(N)	2.3-3.1 ppm	reference data, additivity rules: aliphatic, cyclic, olefinic, aromatic	H15,H75 H85,H195 H215 H255
	$-CH_2$-(N)	2.5-3.5 ppm		
	>CH-(N)	3.0-3.7 ppm		
	H-(ar-N)	6.0-7.5 ppm	for C-aromatics shifts with respect to H-(ar-H): ortho ∿ -0.8 ppm, meta ∿ -0.2 ppm, para ∿ -0.7 ppm	H255,H315
	$H-(C-N^+)$	3.2-4.0 ppm	couplings: H80	
	$H-(ar-N^+)$	7.5-8.0 ppm	for C-aromatics shifts with respect to H-(ar-H): ortho ∿ +0.7 ppm, meta ∿ +0.4 ppm, para ∿ +0.3 ppm	H255

IR			reference data	I100
	$-NH$ st	3500-3200 cm^{-1}	position of band depends on extent of association	
	$-N^{+}H$ st	3000-2000 cm^{-1}	broad	
	$-NH$ δ	1650-1550 cm^{-1}	medium	
	$-N^{+}H$ δ	1600-1460 cm^{-1}	frequently weak	
MS	Molecular ion		odd number if odd number of nitrogens; low abundance; tendency to protonate	M225
	Fragments			M115
			indications for nitrogen	M125
			base peak due to amine cleavage $\left[(R_1)(R_2)N-CH_2-R_3 \right]^{+\cdot} \xrightarrow{-R_3^{\cdot}} (R_1)(R_2)\overset{+}{N}=CH_2$	
			fragments of even mass relatively frequent typical: m/z 30	
	Rearrangements		elimination of olefins following amine cleavage: $(R_1)(R_2)\overset{+}{N}=CH_2 \longrightarrow R_1-\overset{+}{N}H=CH_2$	
			elimination of amines from $M^{+\cdot}$: $\left[(R_1)(R_2)N-CH_2-R_3 \right]^{+\cdot} \xrightarrow{-R_1R_2NH} \left[R_3-CH \right]^{+\cdot}$	
UV			aliphatic: endabsorption	U75
			aromatic: in acidic solution shifted to lower wavelength and less intense	U80,U85, U100

Characteristic Spectroscopic Data of Nitro Compounds

	Assignment	Range	Comments	Details see page:
^{13}C-NMR			reference data, additivity rules: aliphatic alicyclic olefinic aromatic	C10,C45 C178 C70 C90 C120-C130 C140,C178
	C-(NO_2)	55-110 ppm	C-(C-NO_2): shift with respect to C-(C-C) about -6 ppm	
	arC-(NO_2)	130-150 ppm	frequently broad due to fast spin-spin relaxation; shift with respect to arC-(H) about +20 ppm, for ortho-carbon atoms about -5 ppm, for para-carbon atoms about +6 ppm	
^{1}H-NMR			reference data, additivity rules: aliphatic alicyclic olefinic aromatic	H15,H90 H195 H215 H255,H315
	H-(C-NO_2)	4.2-4.6 ppm		
	H-(ar-NO_2)	7.5-8.5 ppm	for C-aromatics shifted with respect to H-(ar-H): ortho ∿ +1.0 ppm meta ∿ +0.3 ppm para ∿ +0.4 ppm	

IR			reference data	I210
	NO_2 st as	1660-1490 cm^{-1}	strong to very strong	
	NO_2 st sy	1390-1260 cm^{-1}		
MS	Molecular ion		aliphatic: low abundance or absent	M250
			aromatic: abundant; odd numbered if odd number of nitrogens	
	Fragments		indications for nitrogen	M125
			$M^{+\cdot}$ -16, -46	
	Rearrangements		$M^{+\cdot}$ -30, m/z = 30, $M^{+\cdot}$ -17, -47	
			oxygen transfer to substituents in ortho position	
UV			for aliphatic nitro compounds weak absorption (log ε < 2) around 275 nm	U15,U60
			aromatic	U50,U75

Characteristic Spectroscopic Data of Thiols and Thioethers

	Assignment	Range	Comments	Details see page:
^{13}C-NMR			reference data, additivity rules: aliphatic	C10,C45 C179,C180
			alicyclic	C70
			olefinic	C90
			aromatic	C120,C130 C179,C180
	C-(S)	5-60 ppm	no appreciable shift relative to C-(C)	
	arC-(S)	120-140 ppm		
^{1}H-NMR	H-(S-alC)	1.0-2.0 ppm	couplings: H95	
	H-(S-arC)	2.0-4.0 ppm		
	H-(C-S)	2.0-3.2 ppm	reference data, additivity rules: aliphatic	H15,H95
			cyclic	H100,H195
			olefinic	H215
			aromatic	H255
IR			reference data	I215
	-SH st	2600-2540 cm^{-1}	frequently weak	

MS	Molecular ion		stronger than for alcohols or ethers	M245
			^{34}S-isotope peak; at $M^{+\cdot} + 2 \geq 4{,}5\%$	M60
	Fragments			
			indications for sulfur	M130
			sulfide cleavage: $[R_1\text{-S-CH}_2\text{-R}_2]^{+\cdot} \xrightarrow{-R_2^{\cdot}} [R_1\text{-S=CH}_2]^{+}$	
			sulfur peaks: m/z = 47, 61 ... $C_nH_{2n+1}S$	M120
	Rearrangements		$M^{+\cdot}$ -33, -34; m/z = 34, 35, 48	
			elimination of olefin after sulfide cleavage	
UV	n → σ*	<225 nm (log ε = 3-4)	in aliphatic compounds	U70,U75
	n → σ*	220-250 nm (log ε =2-3)	aromatic	U95,U105

Characteristic Spectroscopic Data of Aldehydes

	Assignment	Range	Comments	Details see page:
^{13}C-NMR	-CHO ($-C{\lesssim}^{O}_{H}$)	190-205 ppm	doublet in partially ("off resonance") decoupled spectra coupling constant $^{13}C-CHO$: 20-50 Hz, in partially ("off resonance") decoupled spectra usually recognizable	C184,C186 C187
	C-(CHO)	30-70 ppm		C10,C45 C187
	C-(C-CHO)	5-50 ppm	shift with respect to C-(C-CH_3) about -10 ppm	
	>C=C-(CHO), C=(C-CHO)	110-160 ppm		C90,C187
	ar-(CHO)	120-150 ppm		C120-C130 C140,C187
^{1}H-NMR	H-(C=O)	9.0-10.5 ppm	coupling: H-C-CHO 0-3 Hz H-C=C-CHO ∿ 8 Hz	H120
	H-(C-CHO)	2.0-2.5 ppm		H15,H120
	CH=CH-(CHO)	5.5-7.0 ppm		H215
	H-(ar-CHO)	7.2-8.0 ppm	in C-aromatics shifted with respect to H-(ar-H): ortho ∿ +0.6 ppm meta ∿ +0.2 ppm para ∿ +0.3 ppm	

IR	comb C=O st	2900-2700 cm^{-1} 1765-1645 cm^{-1}	reference data two weak bands aliphatic: ∿1730 cm^{-1} conjugated: ∿1690 cm^{-1}	I120
MS	Molecular ion		aliphatic: moderate aromatic: abundant	M215
	Fragments		hints for oxygen $M^{+\cdot}$ -1 (for aliphatic aldehydes only to C_7), $M^{+\cdot}$ -29	M125
	Rearrange-ments		aliphatic: m/z = 44, $M^{+\cdot}$ -44 $[R-CH(-CH_2-CH_2-)C(H)=O]^{+\cdot} \longrightarrow R-CH=CH_2 + [CH_2CHOH]^{+\cdot}$	
UV	$n \rightarrow \pi^*$	270-310 nm (log ε: ∿1)	for saturated aldehydes α,β-unsaturated aldehydes benzaldehydes	B65 U20,U65 U40,U95

Characteristic Spectroscopic Data of Ketones

	Assignment	Range	Comments	Details see page:
^{13}C-NMR	>C=O	195-220 ppm		C183,C184 C186 C188-C190 C10,C45
	C-(C=O)	25-70 ppm		
	C-(C-C=O)	5-50 ppm	shift with respect to C-(C-C) about -6 ppm	
	>=C-(C=O)	105-160 ppm		C90,C188
	C=(C-C=O)			
	ar-(C=O)	120-150 ppm		C120-C130 C140,C189
^{1}H-NMR	H-(C-C=O)	2.0-3.6 ppm	H-(C-CO-al): 2.0-2.6 ppm H-(C-CO-ar): 2.5-3.6 ppm	H15,H125
	CH=CH-(C=O)	5.5-7.0 ppm		H215
	H-(ar-C=O)	7.2-8.0 ppm	for C-aromatics shifted with respect to H-(ar-H): ortho ∿ +0.6 ppm meta ∿ +0.1 ppm para ∿ +0.2 ppm	H255

IR	C=O st	1775-1650 cm^{-1}	reference data aliphatic: ∿1715 cm^{-1} cyclic: 6-membered and larger rings ∿1715 cm^{-1} smaller rings > 1750 cm^{-1} conjugated: 1690-1665 cm^{-1}	I125
MS	Molecular ion		aliphatic: moderate aromatic: abundant	M210
	Fragments		hints for oxygen ketone cleavages: $[R_1-C(=O)-R_2]^{+\cdot} \longrightarrow R_1CO^+, R_1^+, R_2CO^+, R_2^+$	M115,M120 M125
	Rearrangements		$[R_1CH(H)CH_2CH_2C(=O)R_2]^{+\cdot} \xrightarrow{-R_1CH=CH_2} \dot{C}H_2-C(=\overset{+}{O}H)-R_2$ may occur more than once	
UV	$\pi \rightarrow \pi^*$	<200 nm (log ε = 3-4)	for saturated ketones	U10,U65
	$n \rightarrow \pi$	250-300 nm (log ε = 1-2)		
			α,β-unsaturated ketones	U20,U60
			phenyl ketones	U40,U90, U105

Characteristic Spectroscopic Data of Carboxylic Acids

	Assignment	Range	Comments	Details see page:
^{13}C-NMR	$-C{\lesssim}^{O}_{OH}$	170-185 ppm	in $-COO^-$ shift with respect to -COOH by 0 to +8 ppm	C184, C186 C191
	C-(COOH)	20-70 ppm		C10, C45
	C-(C-COOH)	5-50 ppm	shift with respect to C-(C-CH_3) about -6 ppm	
	$\gtrless$C-(COOH) }	105-160 ppm		C90, C191
	C=(C-COOH) }			
	ar-(COOH)	120-150 ppm		C120-C130 C191
^{1}H-NMR	H-(O-CO)	10.0-13.0 ppm	position and peak shape strongly dependent on experimental conditions	H135
	H-(C-COOH)	2.0-2.6 ppm		H15, H135
	CH=CH-(COOH)	5.2-7.5 ppm	shift with respect to CH=CH-(H): geminal ∿ +0.9 ppm cis ∿ +1.0 bis +1.4 ppm trans ∿ +0.3 bis +0.7 ppm	H215
	H-(ar-COOH)	7.5-8.5 ppm	for C-aromatics shifted with respect to H-(ar-H): ortho ∿ +0.8 ppm meta ∿ +0.2 ppm para ∿ +0.3 ppm	H255

IR			reference data	I165
	COO-H st	3550-2500 cm^{-1}	broad	
	C=O st	1800-1650 cm^{-1}	aliphatic: ∿1715 cm^{-1} conjugated: ∿1695 cm^{-1} in COO^- two bands at 1580 cm^{-1} and 1420 cm^{-1}	
	CO-OH δ oop	∿920 cm^{-1}		
MS	Molecular ion		aliphatic: moderate, for long chains more abundant; tendency to protonate aromatic: abundant	M230
	Fragments			M115
			hints for oxygen $M^{+\cdot}$ -17 (for aromatic carboxylic acids abundant), $M^{+\cdot}$ -45	M125
	Rearrangements		aliphatic: $M^{+\cdot}$ -18, m/z = 60,61 aromatic: ortho-effect: 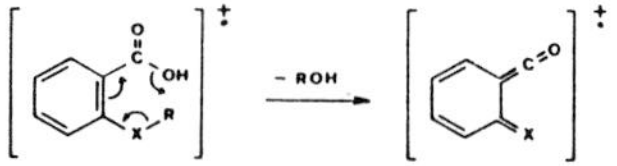 X : CH_2, S, N, O etc.	
UV	n → π*	<220 nm (log ε = 1-2)	for saturated acids	U10
			α,β-unsaturated acids	U20,U60, U65
			benzoic acids	U40,U90, U95

Characteristic Spectroscopic Data of Carboxylic Esters and Lactones

^{13}C-NMR

Assignment	Range	Comments	Details see page:
$-C(=O)O-$	165-180 ppm	shift with respect to -COOH by -5 to -10 ppm	C184,C186 C193-C195
C-(COO-)	20-70 ppm		C10,C45
C-(C-COO-)	5-50 ppm	shift with respect to C-(C-C) about -6 ppm	
C-(OOC-)	50-100 ppm	shift with respect to C-(OH) about +2 to +10 ppm	
C-(C-OOC-)	10-60 ppm	shift with respect to C-(C-OH) about -4 ppm	
$>$C=C-(COO-) C=(C-COO-)	105-160 ppm		C90,C193
$>$C=C-(OOC-)	100-150 ppm		
C=(C-OOC-)	80-130 ppm		
ar-(COO-)	120-150 ppm		C120-C130 C140,C193
ar-(OOC-)	100-160 ppm	shift with respect to ar-(H): α ∿ +20 ppm ortho ∿ - 6 ppm meta ∿ + 1 ppm para ∿ - 2 ppm	C120-C130 C194

^{1}H-NMR

		reference data, additivity rules: aliphatic olefinic aromatic	H15, H140 H215 H255
H-(C-COO-)	2.0-2.5 ppm	CH_3COO-: $\sim$2.0 ppm; -CH_2COO-: $\sim$2.3 ppm; >CHCOO-: $\sim$2.5 ppm	H140
H-(C-OOC-)	3.5-5.3 ppm	CH_3OOC-: 3.5-3.9 ppm; -CH_2OOC-: 4.0-4.5 ppm; >CHOOC-: 4.8-5.3 ppm	
CH=CH-(COO-)	5.2-7.5 ppm	shift with respect to CH=CH-(H): geminal $\sim$ +0.8 ppm cis $\sim$ +1.1 ppm trans $\sim$ +0.5 ppm	H215
=C(H)(OOC-) HC=(C-OOC-)	6.0-8.0 ppm 4.5-6.0 ppm	shift with respect to CH=CH-(H): geminal $\sim$ +2.1 ppm cis $\sim$ -0.4 ppm trans $\sim$ -0.6 ppm	
H-(ar-COO-)	7.5-8.5 ppm	in C-aromatics shift with respect to H-(ar-H): ortho $\sim$ +0.7 ppm meta $\sim$ +0.1 ppm para $\sim$ +0.2 ppm	H255
H-(ar-OOC-)	6.8-7.5 ppm	in C-aromatics shift with respect to H-(ar-H): ortho $\sim$ -0.2 ppm meta $\sim$ 0 ppm para $\sim$ -0.1 ppm	

Characteristic Spectroscopic Data of Carboxylic Esters and Lactones

(continued)

	Assignment	Range	Comments	Details see page:
IR			reference data	I135
	C=O st	1745-1730 cm^{-1}	strong; quoted range holds for aliphatic esters	
			shifted to higher wavenumbers for hal-C-COO-, -COO-C=C, -COO-ar and for small-ring lactones	
			shifted to lower wavenumbers for C=C-COO- and ar-COO-	
	C-O st	1330-1050 cm^{-1}	two bands, at least one of them strong	
MS	Molecular ion		esters: aliphatic: low abundance, tendency to protonate	M235
			aromatic: abundant	
			lactones: aliphatic: medium to low abundance, tendency to protonate	M240
			aromatic: strong	
	Fragments		indications for oxygen	M115
			esters: $M^{+\cdot}$ - RO, $M^{+\cdot}$ - ROCO	M125
			lactones: loss of α-substituents (attached to ether-carbon), decarbonylation, for aromatic lactones double decarbonylation	

Rearrange- ments		<u>esters:</u> elimination of olefin from the alcohol moiety: $[R_1\text{-C(=O)-O-CH}_2\text{-CH(H)-}R_2]^{+\cdot} \longrightarrow [R_1\text{-COOH} + R_2\text{-CH=CH}_2]^{+\cdot}$ elimination of the alcohol side chain with trans- fer of two hydrogens: $[R_1COOR_2]^{+\cdot} \longrightarrow R_1C(=\overset{+}{O}H)OH$ aliphatic esters: elimination of the alkyl chain of the acid moiety as an olefin: $[R_1\text{-CH(H)-CH}_2\text{-CH}_2\text{-C(=O)-}OR_2]^{+\cdot} \xrightarrow{-\,R_1\text{-CH=CH}_2} \dot{C}H_2\text{-C(=}\overset{+}{O}H\text{)-}OR_2$ aromatic esters: ortho elimination: $[\text{o-(HX)C}_6H_4\text{-C(=O)-OR}]^{+\cdot} \xrightarrow{-\,ROH} [\text{C}_6H_4(\text{=C=O})(\text{=X})]^{+\cdot}$ <u>lactones:</u> $M^{+\cdot} - 18$	
n → π*	<220 nm (log ε = 1-2)	for aliphatic esters	U10
		C=C-COO-	U20
		COO-C=C-C=C	U30
		C_6H_5-COO-	U40

UV

Characteristic Spectroscopic Data of Amides and Lactams

	Assignment	Range	Comments	Details see page:
^{13}C-NMR	C(=O)N	165-180 ppm		C185,C186 C196
	C-(CON)	20-70 ppm		C10,C45
	C-(C-CON)	5-50 ppm	shift with respect to C-(C-C) about -6 ppm	
	C-(NCO)	25-80 ppm	shift with respect to C-(NH) about -1 to -2 ppm	C10
	C-(C-NCO)	10-60 ppm		
	ar-(CON)	120-150 ppm		C120-C126 C130,C140
	ar-(NCO)	110-150 ppm	shift with respect to ar-(H): α ∿ +11 ppm ortho ∿ -10 ppm meta ∿ 0 ppm para ∿ - 6 ppm	C120,C130 C196
^{1}H-NMR	H-(NCO)	5-10 ppm	frequently broad to very broad; splitting due to H-N-C-H coupling often only recognizable in the CH-signal, see H150	H150
	H-(C-CON)	2.0-2.5 ppm		H5
	H-(C-NCO)	2.7-4.8 ppm	CH_3-NCO 2.7-3.0 ppm -CH_2-NCO 3.1-3.5 ppm CH-NCO 3.8-4.8 ppm	H155

	CH=CH-(CON)	5.2-7.5 ppm	shift with respect to CH=CH-(H): geminal ∿ +1.4 ppm cis ∿ +1.0 ppm trans ∿ +0.5 ppm	H215
	$=C\langle^{H}_{NCO}$	6.0-8.0 ppm	shift with respect to CH=CH-(H): geminal ∿ +2.1 ppm cis ∿ -0.6 ppm trans ∿ -0.7 ppm	
	HC=(C-NCO)	4.5-6.0 ppm		
	H-(ar-CON)	7.5-8.5 ppm	for C-aromatics shift with respect to H-(ar-H): ortho ∿ +0.6 ppm meta ∿ +0.1 ppm para ∿ +0.2 ppm	H255
	H-(ar-NCO)	6.8-7.5 ppm	for C-aromatics shift with respect to H-(ar-H): ortho ∿ 0 ppm meta ∿ 0 ppm para 0 to -0.3 ppm	
IR			reference data	I145
	N-H st	3500-3100 cm^{-1}	position and shape of band depends on degree of association; frequently multiplicity of bands	
	C=O st (amide I)	1700-1650 cm^{-1}	strong; range listed holds for amides as well as δ- and larger lactams for β- and γ-lactams higher	
	N-H δ N-C=O st sy (amide II)	1630-1510 cm^{-1}	frequently strong; missing in tertiary amides and in lactams	

Characteristic Spectroscopic Data of Amides and Lactams

(continued)

	Assignment	Range	Comments	Details see page:
MS	Molecular ion		aliphatic: moderate abundance, tendency to protonate	M240
			aromatic: abundant	
	Fragments		aliphatic amides	M120
			indications for nitrogen	M125
			<u>amides:</u> amide cleavage resulting in acyl ion, followed by loss of CO; unusually large number of fragments of even mass	
			<u>lactams:</u> loss of α-substituents, loss of CO	
	Rearrangements		<u>amides:</u> elimination of the amine moiety elimination of an olefin from the amine or acid moiety in analogy to esters (see B230)	
			<u>lactams:</u> $M^{+\cdot} - 18$	
UV	$n \rightarrow \pi^*$	<220 nm (log ε = 1-2)	for aliphatic amides and lactams	U10
			aromatic	U85

^{13}C-Chemical Shifts in Alkanes

(δ in ppm relative to TMS; see also p. C10)

	δ
CH_4	- 2.3

	δ
CH_3	7.3
CH_3	

	δ
CH_3	15.4
CH_2	15.9
CH_3	

	δ
CH_3	13.0
CH_2	24.8
CH_2	
CH_3	

	δ
CH_3	24.1
CH	25.0
$(CH_3)_2$	

	δ
CH_3	14.2
CH_2	22.8
CH_2	34.8
CH_2	
CH_3	

	δ
CH_3	11.8
CH_2	32.0
CH	30.1
$(CH_3)_2$	22.3

	δ
CH_3	31.3
C	27.7
$(CH_3)_3$	

	δ
CH_3	14.2
CH_2	23.1
CH_2	32.2
CH_2	
CH_2	
CH_3	

	δ
CH_3	14.1
CH_2	23.1
CH_2	32.4
CH_2	29.5
CH_2	
CH_2	
CH_3	

	δ
CH_3	14.1
CH_2	22.8
CH_2	32.1
CH_2	29.5
CH_2	
CH_2	
CH_2	
CH_3	

The ^{13}C-chemical shifts in alkanes can be estimated using the additivity rule stated on p. C10.

For cycloalkanes see p. C50.

Estimation of the ^{13}C-Chemical Shifts in Aliphatic Compounds

(δ in ppm relative to TMS)

$$\delta = -2.3 + \sum_i Z_i + \sum_j S_j + \sum_k K_k$$

	Substituent		Increment Z_i for substituents in position			
			α	β	γ	δ
	-H		0.0	0.0	0.0	0.0
C	-C≡	(*)	9.1	9.4	-2.5	0.3
	-▽(O)	(*)	21.4	2.8	-2.5	0.3
	-C=C-	(*)	19.5	6.9	-2.1	0.4
	-C≡C-		4.4	5.6	-3.4	-0.6
	-Phenyl		22.1	9.3	-2.6	0.3
HAL	-F		70.1	7.8	-6.8	0.0
	-Cl		31.0	10.0	-5.1	-0.5
	-Br		18.9	11.0	-3.8	-0.7
	-I		-7.2	10.9	-1.5	-0.9
O	-O-	(*)	49.0	10.1	-6.2	0.3
	-O-CO-		56.5	6.5	-6.0	0.0
	-O-NO		54.3	6.1	-6.5	-0.5
N	-N<	(*)	28.3	11.3	-5.1	0.0
	$-N^{+}\equiv$	(*)	30.7	5.4	-7.2	-1.4
	$-NH_3^+$		26.0	7.5	-4.6	0.0
	$-NO_2$		61.6	3.1	-4.6	-1.0
	-NC		31.5	7.6	-3.0	0.0
S	-S-	(*)	10.6	11.4	-3.6	-0.4
	-S-CO-		17.0	6.5	-3.1	0.0
	-SO-	(*)	31.1	7.0	-3.5	0.5
	$-SO_2-$	(*)	30.3	7.0	-3.7	0.3
	$-SO_2Cl$		54.5	3.4	-3.0	0.0
	-SCN		23.0	9.7	-3.0	0.0
O=C<	-CHO		29.9	-0.6	-2.7	0.0
	-CO-		22.5	3.0	-3.0	0.0
	-COOH		20.1	2.0	-2.8	0.0
	$-COO^-$		24.5	3.5	-2.5	0.0
	-COO-		22.6	2.0	-2.8	0.0
	-CON<		22.0	2.6	-3.2	-0.4
	-COCl		33.1	2.3	-3.6	0.0
	-CS-N<		33.1	7.7	-2.5	0.6
	-C=NOH syn		11.7	0.6	-1.8	0.0
	-C=NOH anti		16.1	4.3	-1.5	0.0
	-CN		3.1	2.4	-3.3	-0.5
	-Sn≤		-5.2	4.0	-0.3	0.0

Steric Corrections S

Observed ^{13}C-center	Number of substituents other than H at the α-atom (for each of the α-substituents which are marked (*) on p. C10)			
	1	2	3	4
primary (CH_3)	0.0	0.0	-1.1	-3.4
secondary (CH_2)	0.0	0.0	-2.5	-6.0
tertiary (CH)	0.0	-3.7	-8.5	-10.0
quaternary (C)	-1.5	-8.0	-10.0	-12.5

Conformation Corrections K for γ-Substituents

conformation		K
synperiplanar	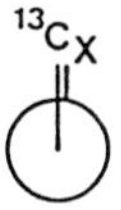 ^{13}C X	-4.0
synclinal	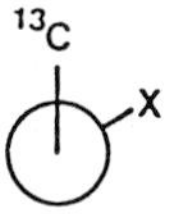 ^{13}C X	-1.0
anticlinal	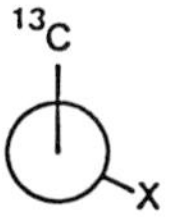 ^{13}C X	0.0
antiperiplanar	 ^{13}C X	2.0
not fixed	—	0.0

Example: Estimation of the ^{13}C-chemical shifts for N-t-butoxycarbonylalanine

$$\underset{d}{(CH_3)_3}-\underset{c}{C}-O-CO-NH-\underset{a}{\overset{\overset{b}{CH_3}}{\overset{|}{CH}}}-COOH$$

(a)		(b)	
base value:	-2.3	base value:	-2.3
1αC	9.1	1αC	9.1
1αCOOH	20.1	1βCOOH	2.0
1αNH	28.3	1βNH	11.3
1βCOO	2.0	1γCOO	-2.8
1δC	0.3	S(p,3)	-1.1
S(t,2)	-3.7		
estimated:	53.8	estimated:	16.2
determined:	49.0	determined:	17.3

(c)		(d)	
base value:	-2.3	base value:	-2.3
3αC	27.3	1αC	9.1
1αOCO	56.5	2βC	18.8
1γNH	-5.1	1βOCO	6.5
1δC	0.3	1δNH	0.0
3S(q,1)	-4.5	S(p,4)	-3.4
estimated:	72.2	estimated:	28.7
determined:	78.1	determined:	28.1

The ^{13}C-chemical shifts estimated for aliphatic hydrocarbons on the basis of the additivity rule differ in general by less than about 5 ppm from the experimental values. Larger discrepancies may be expected for highly branched systems (particularly for quaternary carbon atoms). For chemical shifts exceeding about 90 ppm the deviations can be so large as to render the rule useless.

One can also use the chemical shift of a reference compound as the base value if its structure is closely related to that assumed for the unknown. The increments corresponding to the structural elements missing in the reference compound are then added to the base value, while those of structural elements present in the reference but absent in the unknown are subtracted.

Example: Estimation of the ^{13}C-chemical shifts for (a) and (b) in N-t-butoxycarbonylalanine (see p. C20) using the chemical shifts of valine as base values (a', b').

$(CH_3)_2$-CH-CH(NH_2)-COOH (b' = CH, a' = CH)

(a')	61.6
(b')	30.2

(a)

base value:	61.6
+1βCOO	2.0
+1δC	0.3
+S(t,2)	-3.7
-2βC	-18.8
-S(t,3)	9.5
estimated:	50.9
determined:	49.0

(b)

base value:	30.2
+1γCOO	-2.8
+S(p,3)	-1.1
-2αC	-18.2
-S(t,3)	9.5
estimated:	17.6
determined:	17.3

^{13}C-Chemical Shifts for Methyl Groups
(δ in ppm relative to TMS)

Substituent X	δ_{CH_3-X}	Substituent X	δ_{CH_3-X}
-H	-2.3	-cyclopentyl	20.5
$-CH_3$	7.3	-cyclohexyl	23.1
$-CH_2CH_3$	15.4	$-CH=CH_2$	18.7
$-CH(CH_3)_2$	24.1	$-C\equiv CH$	3.7
$-C(CH_3)_3$	31.3	-phenyl	21.4
$-(CH_2)_6CH_3$	14.1	-α-naphthyl	19.1
$-CH_2$phenyl	15.7	-β-naphthyl	21.5
$-CH_2F$	15.8	C -2-pyridyl	24.2
$-CH_2Cl$	18.7	-3-pyridyl	18.0
$-CH_2Br$	19.1	-4-pyridyl	20.6
$-CH_2I$	20.4	-2-furyl	13.7
$-CHCl_2$	31.6	-2-thienyl	14.7
$-CHBr_2$	31.8	-2-pyrrolyl	11.8
$-CCl_3$	46.3	-2-indolyl	13.4
C $-CBr_3$	49.4	-3-indolyl	9.8
$-CH_2OH$	18.2	-4-indolyl	21.6
$-CH_2OCH_3$	14.7	-5-indolyl	21.5
$-CH_2OCH_2CH_3$	15.4	-6-indolyl	21.7
$-CH_2OCH=CH_2$	14.6	-7-indolyl	16.6
$-CH_2O$phenyl	14.9	H -F	71.6
$-CH_2OCOCH_3$	14.4	A -Cl	25.6
$-CH_2NH_2$	19.0	L -Br	9.6
$-CH_2NHCH_3$	14.3	-I	-24.0
$-CH_2N(CH_3)_2$	12.8	-OH	50.2
$-CH_2NO_2$	12.3	$-OCH_3$	60.9
$-CH_2SH$	19.7	$-OCH_2CH_3$	57.6
$-CH_2SO_2CH_3$	6.7	$-OCH(CH_3)_2$	54.9
$-CH_2SO_3H$	8.0	O $-OC(CH_3)_3$	49.4
$-CH_2CHO$	5.2	$-OCH_2CH=CH_2$	57.4
$-CH_2COCH_3$	7.0	-Ocyclohexyl	55.1
$-CH_2COOH$	9.6	$-OCH=CH_2$	52.5

	Substituent X	δ_{CH_3-X}		Substituent X	δ_{CH_3-X}
	-Ophenyl	54.8		$-SO_2CH_2CH_3$	39.3
	$-OCOOCH_3$	54.9		$-SO_2Cl$	52.6
	$-OCOCH_3$	51.5	S	$-SO_3H$	39.6
O	-OCOcyclohexyl	51.2		$-SO_3Na$	41.1
	$-OCOCH=CH_2$	51.5		-CHO	31.2
	-OCOphenyl	51.8		$-COCH_3$	30.7
	$-OSO_2OCH_3$	59.1		$-COCH_2CH_3$	27.5
	$-OSO_2$ptoluyl	56.3		$-COCCl_3$	21.1
	$-NH_2$	28.3		$-COCH=CH_2$	25.7
	$-NH_3^+$	26.5		-COcyclohexyl	27.6
	$-NHCH_3$	38.2	O	-COphenyl	25.7
	-NHcyclohexyl	33.5	‖	-COOH	21.7
	-NHphenyl	30.2	C	$-COO^-$	24.4
	$-N(CH_3)_2$	47.5	/\	$-COOCH_3$	20.6
	-N-pyrrolidinyl	42.7		$-COOCOCH_3$	21.8
N	-N-piperidinyl	47.7		$-CONH_2$	22.3
	$-N(CH_3)$phenyl	39.9		$-CON(CH_3)_2$	21.5
	-1-pyrrolyl	35.9		-COSH	32.6
	-1-imidazolyl	32.2		$-COSCH_3$	30.2
	-1-pyrazolyl	38.4		$-COCOCH_3$	23.2
	-1-indolyl	32.1		-COCl	33.6
	$-NHCOCH_3$	26.1		-COBr	39.1
	$-N(CH_3)CHO$	36.5;31.5		$-COSi(CH_3)_3$	35.7
	-NC	26.8		-CN	1.7
	-NCS	29.1			
	$-NO_2$	61.2			
	-SH	6.5			
	$-SCH_3$	19.3			
S	$-SC_8H_{17}$	15.5			
	-Sphenyl	15.6			
	$-SSCH_3$	22.0			
	$-SOCH_3$	40.1			
	$-SO_2CH_3$	42.6			

^{13}C-Chemical Shifts in Monosubstituted Alkanes

(δ in ppm relative to TMS)

	Substituent	Methyl	Ethyl		n-Propyl			Isopropyl		t-Butyl	
		$-CH_3$	$-CH_2$	$-CH_3$	$-CH_2$	$-CH_2$	$-CH_3$	$-CH$	$-CH_3$	$-C$	$-CH_3$
	-H	- 2.3	7.3	7.3	15.4	15.9	15.4	15.9	15.4	25.0	24.1
C	$-CH=CH_2$	18.7	27.4	13.4	36.2	22.4	13.6	32.3	22.1	33.8	29.4
	$-C\equiv CH$	3.7	12.3	13.8	20.6	22.2	13.4	20.3	22.8	27.4	31.1
	-phenyl	21.4	29.1	15.8	38.3	24.8	13.8	34.3	24.0	34.6	31.4
H	-F	71.6	80.1	15.8	85.2	23.6	9.2	87.3	22.6	93.5	28.3
	-Cl	25.6	39.9	18.9	46.8	26.3	11.6	53.7	27.3	66.7	34.6
A	-Br	9.6	27.6	19.4	35.6	26.4	13.0	44.8	28.5	62.1	36.4
L	-I	-24.0	- 1.6	20.6	9.1	27.0	15.3	20.9	31.2	43.0	40.4
	-OH	50.2	57.8	18.2	64.2	25.9	10.3	64.0	25.3	68.9	31.2
O	$-OCH_3$	60.9	67.7	14.7	74.5	23.2	10.5	72.6	21.4	72.7	27.0
	$-OCH_2CH_3$	57.6	66.0	15.4	72.5	23.2	10.7				
	$-OCH(CH_3)_2$	54.9						68.5	23.0	73.0	28.5
	$-OC(CH_3)_3$	49.4						63.5	25.2	76.3	33.8
	-O-phenyl	54.8	63.2	14.9	69.4	22.8	10.6	69.3	22.0		
	$-OCOCH_3$	51.5	60.4	14.4	66.2	22.4	10.5	67.5	21.9	79.9	28.1
	-OCO-phenyl	51.8	60.8	14.4	66.4	22.2	10.5	68.2	21.9	80.7	28.2
	$-OSO_2$-p-toluyl	56.3	66.9	14.7							

	Substituent	Methyl	Ethyl		n-Propyl			Isopropyl		t-Butyl	
		$-CH_3$	$-CH_2$	$-CH_3$	$-CH_2$	$-CH_2$	$-CH_3$	$-CH$	$-CH_3$	$-C$	$-CH_3$
N	$-NH_2$	28.3	36.9	19.0	44.6	27.4	11.5	43.0	26.5	47.2	32.9
	$-NHCH_3$	38.2	45.9	14.3	54.0	23.2	12.5	50.5	22.5	50.4	28.2
	$-N(CH_3)_2$	47.6	53.6	12.8	61.8	20.6	11.9	55.5	18.7	53.6	25.4
	$-NHCOCH_3$	26.1	34.4	14.6	40.7	22.5	11.1	40.5	22.3	49.9	28.6
	$-NC$	26.8	36.4	15.3	43.4	22.9	11.0	45.5	23.4	54.0	30.7
	$-NO_2$	61.2	70.8	12.3	77.4	21.2	10.8	78.8	20.8	85.2	26.9
S	$-SH$	6.5	19.1	19.7	26.4	27.6	12.6	29.9	27.4	41.1	35.0
	$-SCH_3$	19.3									
	$-SSCH_3$	22.0	31.8	14.7							
	$-SOCH_3$	40.1									
	$-SO_2CH_3$	42.6	48.2	6.7	56.3	16.3	13.0	53.5	15.2	57.6	22.7
	$-SO_2Cl$	52.6	60.2	9.1	67.1	18.4	12.1	67.6	17.1	74.2	24.5
	$-SO_3H$	39.6	46.7	8.0	53.7	18.8	13.7	52.9	16.8	55.9	25.0
O=C<	$-CHO$	31.3	36.7	5.2	45.7	15.7	13.3	41.1	15.5	42.4	23.4
	$-COCH_3$	30.7	35.2	7.0	45.2	17.5	13.5	41.6	18.2	44.3	26.5
	$-CO$-phenyl	25.7	31.7	8.3	40.4	17.7	13.8	35.2	19.1	43.5	27.9
	$-COOH$	21.7	28.5	9.6	36.2	18.7	13.7	34.1	18.8	38.7	27.1
	$-COOCH_3$	20.6	27.2	9.2	35.6	18.9	13.8	34.1	19.1	38.7	27.3
	$-CONH_2$	22.3	29.0	9.7				34.9	19.5		
	$-COCl$	33.6	41.0	9.3	48.9	18.8	13.0	46.5	19.0	49.4	27.1
	$-CN$	1.7	10.8	10.6	19.3	19.0	13.3	19.8	19.9	28.1	28.5

^{13}C-Chemical Shifts in Substituted n-Octanes

(δ in ppm relative to TMS)

Substituent X	$X-CH_2$	$-CH_2$	$-CH_2$	$-CH_2$	$-CH_2$	$-CH_2$	$-CH_2$	$-CH_3$
-H	14.1	22.8	32.1	29.5	29.5	32.1	22.8	14.1
C $-CH=CH_2$	34.5	~29.6	~29.6	~29.6	~29.6	32.2	23.0	13.9
-phenyl	36.2	31.7	~29.6	~29.6	~29.6	32.1	22.8	14.1
-F 1)	84.2	30.6	25.3	29.3	29.3	31.9	22.7	14.1
HAL -Cl	45.1	32.8	27.0	29.0	29.2	31.9	22.8	14.1
-Br	33.8	33.0	28.3	28.8	29.2	31.8	22.7	14.1
-I	6.9	33.7	30.6	28.6	29.1	31.8	22.6	14.1
-OH	63.1	32.9	25.9	29.5	29.4	31.9	22.8	14.1
O $-OC_8H_{17}$	71.1	30.0	26.3	29.6	29.4	32.0	22.8	14.1
-ONO	68.3	29.2	26.0	29.3	29.3	31.9	22.7	14.0
$-NH_2$	42.4	34.1	27.0	29.6	29.4	31.9	22.7	14.1
N $-N(CH_3)_2$	60.1	29.5*	~27.9*	~27.7*	29.7*	32.0	22.8	14.4
$-NO_2$	75.8	26.2	27.9	~29.6	~29.6	31.4	22.6	14.0
-SH	24.7	34.2	28.5	29.2	29.1	31.9	22.7	14.1
S $-SCH_3$	34.5	29.0	29.4	29.4	29.4	31.9	22.8	14.1
$-SOC_8H_{17}$	52.6	~29.1	~29.1	~29.1	~29.1	31.8	22.7	14.1
-CHO	44.0	22.2	~29.3	~29.3	~29.3	31.9	22.7	14.1
O=C $-COCH_3$	43.7	24.1	~29.5	~29.5	~29.5	32.0	22.8	14.1
-CO-phenyl	38.6	24.4	29.5	29.5	29.5	31.9	22.7	14.0
-COOH	34.2	24.8	~29.3	~29.3	~29.3	31.9	22.7	14.1
$-COOCH_3$	34.2	25.1	29.3	29.3	29.3	31.9	22.8	14.1
$-CONH_2$	35.5	25.4	29.1	29.1	29.1	31.6	22.3	14.0
-COCl	47.2	25.1	28.5	29.1	29.1	31.8	22.7	14.1
-CN	17.2	25.5	~29.9	~29.9	~29.9	31.8	22.7	14.0

1) $|J|_{CF}$ = 164.8 Hz, $|J|_{C-CF}$ = 18.3 Hz, $|J|_{C-C-CF}$ = 6.2 Hz,

$|J|_{C-C-C-C-F}$ = 0 Hz

* assignment uncertain

^{13}C-Chemical Shifts in Saturated Alicycles
(δ in ppm relative to TMS)

$(CH_2)_n$

n	δ	n	δ
3	- 2.8	12	23.8
4	22.9	13	26.2
5	25.6	14	25.2
6	27.1	15	27.0
7	28.8	16	26.9
8	26.8	18	27.5
9	26.0	20	28.0
10	25.1	30	29.3
11	26.3	40	29.4
		72	29.7

Unsaturated alicycles: see p. C100.
Condensed alicycles: see p. C72.

Additivity Rule for Estimating ^{13}C-Chemical Shifts for Methyl Groups in Methylcyclohexanes

(δ in ppm relative to TMS) see: D.K. Dalling, D.M. Grant, J. Am. Chem. Soc. 89, 6612 (1967), 94, 5318 (1972)

base value: 18.8 23.1

increments for methyl substituents (example: see p. C60)

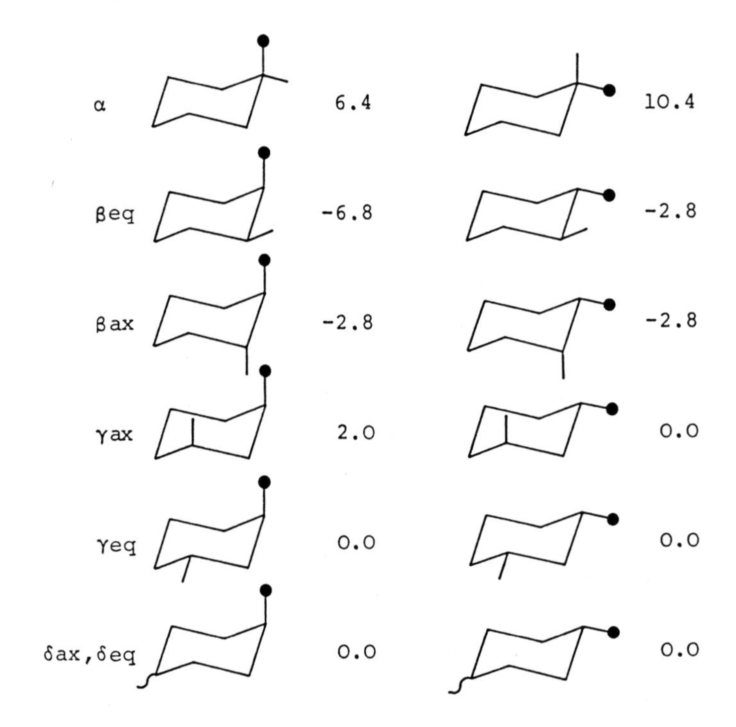

Additivity Rule for Estimating ^{13}C-Chemical Shifts for Ring Carbons in Methyl-Substituted Cyclohexanes

(δ in ppm relative to TMS)

base value: 27.1

increments:

- for equatorial C-substituents

αeq	6.0
βeq	9.0
γeq	0.0
δeq	-0.2

- for axial C-substituents

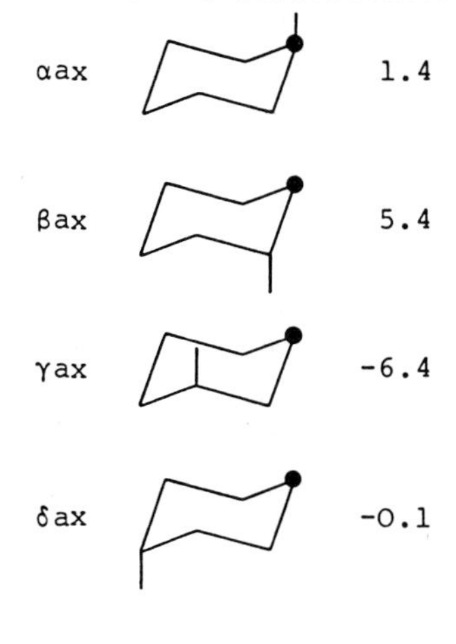

αax	1.4
βax	5.4
γax	-6.4
δax	-0.1

correction terms:

- for geminal disubstitution

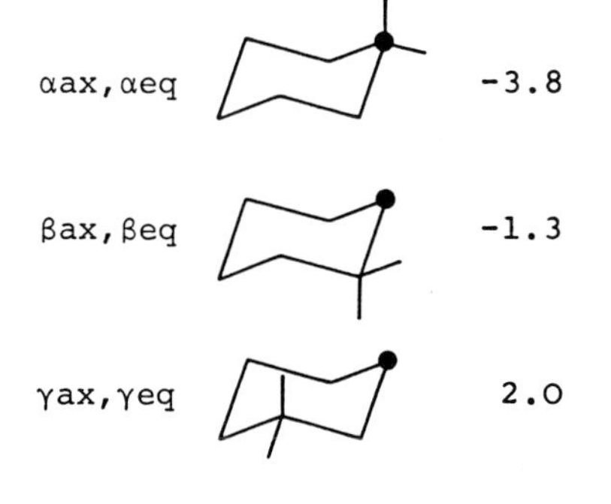

αax,αeq	-3.8
βax,βeq	-1.3
γax,γeq	2.0

- for vicinal disubstitution

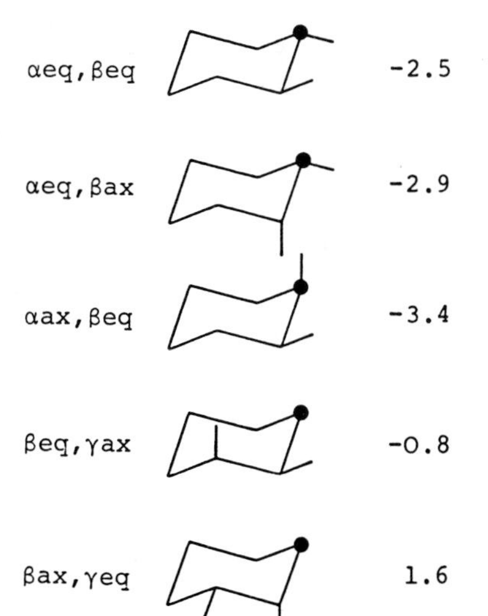

αeq,βeq	-2.5
αeq,βax	-2.9
αax,βeq	-3.4
βeq,γax	-0.8
βax,γeq	1.6

Example: Estimation of the ^{13}C-chemical shifts in 1-cis-2-cis-4-trimethylcyclohexane:

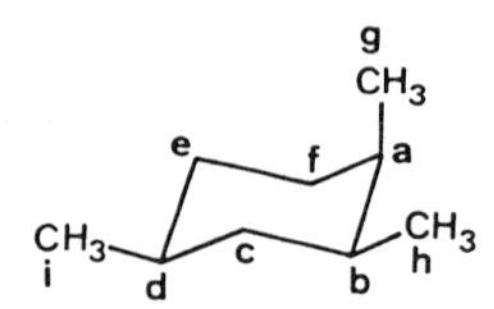

(a)		
base value:	27.1	
1αax	1.4	
1βeq	9.0	
1δeq	-0.2	
1αax,βeq	-3.4	
estimated:	33.9	
determined:	33.7	(34.1)

(b)	
base value:	27.1
1αeq	6.0
1βax	5.4
1γeq	0.0
1αeq,βax	-2.9
estimated:	35.6
determined:	35.5

(c)	
base value:	27.1
2βeq	18.0
1γax	-6.4
1βeq,γax	-0.8
estimated:	37.9
determined:	38.0

(d)	
base value:	27.1
1αeq	6.0
1γeq	0.0
1δax	-0.1
estimated:	33.0
determined:	32.9

(e)	
base value:	27.1
1βeq	9.0
1γax	-6.4
1δeq	-0.2
estimated:	29.5
determined:	29.3

(f)		
base value:	27.1	
1βax	5.4	
2γeq	0.0	
1βax,γeq	1.6	
estimated:	34.1	
determined:	33.7	(34.1)

(g)

base value:	18.8
$1CH_3\beta eq$	-6.8
$1CH_3\delta eq$	0.0
estimated:	12.0
determined:	11.7

(h)

base value:	23.1
$1CH_3\beta ax$	-2.8
$1CH_3\gamma eq$	0.0
estimated:	20.3
determined:	20.3

(i)

base value:	23.1
$1CH_3\gamma eq$	0.0
$1CH_3\delta ax$	0.0
estimated:	23.1
determined:	23.0

The ^{13}C-chemical shifts in substituted alicyclics can be estimated using the additivity rule for aliphatic hydrocarbons (see p. C10). For this it is advisable to take as base value the chemical shifts for the unsubstituted (or similarly substituted) alicyclic compound. The discrepancies between experimental and calculated values are generally somewhat larger than those found for aliphatic compounds (see p. C20).

SUBSTITUTED CYCLOHEXANES

^{13}C-Chemical Shifts in Monosubstituted Cyclohexanes

(δ in ppm relative to TMS)

	Substituent X	^{13}C-chemical shift for a	b	c	d
C	-H	27.1	27.1	27.1	27.1
	$-CH_3$	33.4	36.0	27.1	27.0
	$-CH_2CH_3$	40.2	33.7	27.1	27.4
	$-CH_2CH_2CH_2CH_3$	38.4	34.1	27.1	27.3
	$-C(CH_3)_3$	48.8	28.1	27.7	27.1
	-cyclohexyl	44.3	30.8	27.4	27.4
	-phenyl	45.1	34.9	27.4	26.7
HAL	-F[1]	90.5	33.1	23.5	26.0
	-Cl	59.8	37.2	25.2	25.6
	-Br	52.6	37.9	26.1	25.6
	-I	31.8	39.8	27.4	25.5
O	-OH	70.0	36.0	25.0	26.4
	$-OCH_3$	78.6	32.3	24.3	26.7
	$-OCOCH_3$	72.3	32.2	24.4	26.1
N	$-NH_2$	51.1	37.7	25.8	26.5
	$-NHCH_3$	58.7	32.7	25.7	26.8
	$-N(CH_3)_2$	64.3	29.2	26.5	26.9
	$-NH_3^+Cl^-$	51.8	32.2	24.6	25.2
	-N=C=N-cyclohexyl	55.7	35.0	24.8	25.5
	$-NO_2$	84.6	31.4	24.7	25.5
S	-SH	38.5	38.5	26.8	25.9
O=C<	-CHO	50.1	26.1	25.2	25.2
	$-COCH_3$	51.5	29.0	26.6	26.3
	-COOH	43.7	29.6	26.2	26.6
	$-COO^-$	47.2	30.9	26.9	26.9
	$-COOCH_3$	43.4	29.6	26.0	26.4
	-COCl	55.4	29.7	25.5	25.9
	-CN	28.3	30.1	24.6	25.8

1) $|J|_{CF}$ = 171 Hz, $|J|_{C-CF}$ = 19 Hz, $|J|_{C-C-CF}$ = 5 Hz

Effect of Equatorial and Axial Substitutents on the Chemical Shift in Cyclohexane

(δ in ppm, relative to the shift of 27.0 ppm for cyclohexane)

	Substituent	Position 1		Position 2		Position 3		Position 4	
		eq.	ax.	eq.	ax.	eq.	ax.	eq.	ax.
C	$-CH_3$	5.9	1.4	9.0	5.4	0.0	-6.4	0.2	-0.1
	$-CH_2CH_3$	13.0	8.5	6.0	3.0	-0.4	-5.6	0.1	0.1
	$-CH(CH_3)_2$	17.6	14.1	3.0	3.2	-0.2	-5.4	0.3	0.1
	$-C(CH_3)_3$	21.6		0.7		-0.1		0.1	
	$-CH=CH_2$	15.1	10.0	5.3	3.0	-1.0	-5.8	0.0	0.0
	-C≡CH	1.7	1.0	5.1	3.0	-1.8	-5.8	-2.6	-1.3
HAL	-F	64.5	61.1	5.6	3.1	-3.4	-7.2	-2.5	-2.0
	-Cl	32.8	33.1	10.6	6.9	-0.1	-6.6	-1.8	-1.0
	-Br	25.1	28.4	11.8	7.9	1.4	-5.5	-1.4	-0.6
	-I	3.6	11.3	13.4	9.0	2.3	-4.2	-1.7	-0.9
O	-OH	43.9	38.5	8.6	6.2	-1.8	-6.5	-0.9	0.0
	$-OCH_3$	52.9	47.9	5.1	3.0	-2.3	-5.9	-0.9	-0.4
	-OCO-phenyl	45.8	42.0	4.5	2.3	-2.9	-6.7	-2.3	-2.3
	$-OSi(CH_3)_3$	43.5	39.1	9.0	6.1	-2.3	-7.2	-2.0	-2.0
N	$-NH_2$	24.2	20.4	10.5	6.8	-1.2	-7.0	-0.9	0.0
	$-N_3$	32.5	29.8	4.5	2.0	-2.5	-6.9	-2.5	-1.8
	-NC	24.9	23.3	6.7	3.5	-2.6	-6.9	-1.8	-1.8
	-NCS	28.3	25.8	6.9	4.3	-2.5	-6.4	-2.2	-2.2
	-SH	11.1	8.9	10.7	6.1	-0.6	-7.6	-2.4	-1.3
	-CHO	23.2	19.6	-1.3	-2.3	-1.7	-4.3	0.0	0.0
	$-COOCH_3$	16.3	12.1	2.5	0.7	-1.1	-3.9	-0.7	-0.3
	-CN	0.7	-0.6	2.2	0.4	-2.5	-5.1	-2.6	-2.0

^{13}C-Chemical Shifts in Some Saturated Condensed Alicyclics

(δ in ppm relative to TMS)

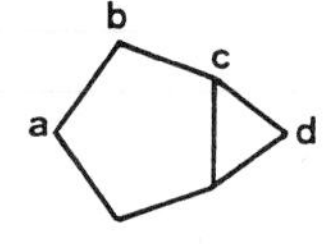

(a) 20.2
(b) 27.6
(c) 16.7
(d) 5.8

(a) 21.5
(b) 23.9
(c) 9.4
(d) 10.3

(a) 22.9
(b) 28.1
(c) 33.3
(d) 24.6

(a) 26.5
(b) 31.8
(c) 45.4
(d) 29.4

(a) 23.8
(b) 28.0
(c) 39.9
(d) 29.9
(e) 22.6

(a) 27.1
(b) 32.4
(c) 47.3
(d) 31.7
(e) 22.1

(a) 36.8
(b) 29.7
(c) 24.5

(a) 44.0
(b) 34.6
(c) 27.1

(a) 26.8
(b) 24.5
(c) 38.7
(d) 42.6
(e) 38.7
(f) 24.1

(a) 27.5
(b) 22.0
(c) 37.6
(d) 32.7

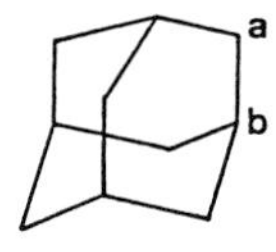

(a) 37.9
(b) 28.5

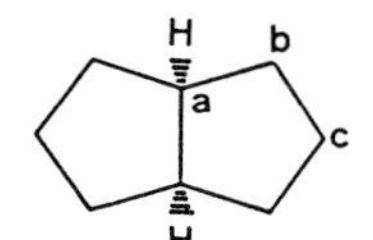

(a) 43.3
(b) 34.3
(c) 26.4

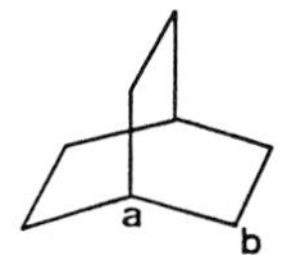

(a) 24.6
(b) 26.7

(a) 36.5
(b) 29.8
(c) 38.5

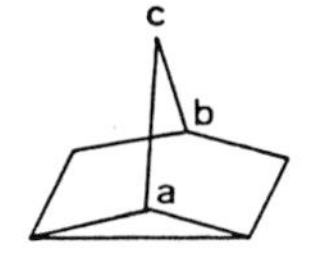

(a) 9.9
(b) 29.7
(c) 33.2

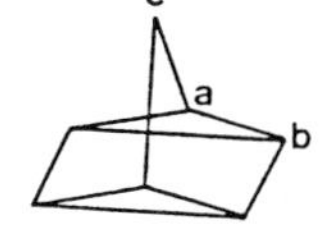

(a) 23.2
(b) 15.0
(c) 32.2

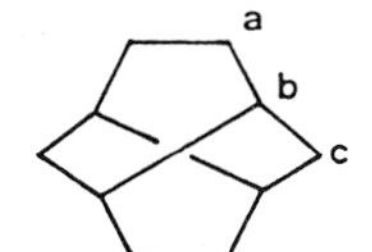

(a) 24.4
(b) 28.4
(c) 28.8

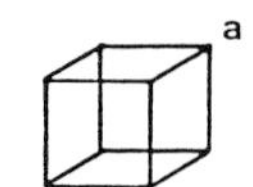

(a) 47.3

Unsaturated condensed alicyclics: see p. C100.

Substituted bicyclic systems: cf. J.K. Whitesell, M.S. Minton, Stereochemical Analysis of Alicyclic Compounds by C-13 NMR Spectroscopy, Chapman and Hall, London, 1987.

Alkenes

The ^{13}C-chemical shifts of the carbons of C=C double bonds range from ca. 80 - 160 ppm. In unsubstituted alkenes they can be estimated quite accurately (see p. C85).

To estimate the ^{13}C-chemical shifts in substituted alkenes one can use the substituent effects listed for the ^{13}C-chemical shifts in vinyl groups (see the example on p. C95).

The ^{13}C-chemical shifts of sp^3-hybridized carbons in the vicinity of double bonds can be estimated using the additivity rule outlined on p. C10. The conformational correction factor K differs widely for γ-substituents of cis *vs.* trans-disubstituted olefins because the conformation is fixed by the double bond. It is thus quite easy to assign the correct isomeric structure.

Estimation of the ^{13}C-Chemical Shifts of sp^2-Hybridized Carbons in Unsubstituted Alkenes (δ in ppm relative to TMS)

C-C-C-C'=C-C-C-C

γ' β' α' ↑ α β γ

base value: 123.3

increments for C-substituents:

- of the C-atom under consideration (C)		- at the neighboring C-Atom (C')	
α	10.6	α'	-7.9
β	4.9	β'	-1.8
γ	-1.5	γ'	1.5

steric corrections:

- for each pair of cis α,α'-substituents:	-1.1
- for a pair of geminal α,α-substituents:	-4.8
- for a pair of geminal α',α'-substituents:	2.5
- if one or more β-substituents are present:	2.3

Example: Estimation of the chemical shifts in cis-4-methyl-2-pentene:

CH_3-CH=CH-CH(CH_3)$_2$ (a = CH, C-2; b = CH, C-3)

(a)		(b)	
base value:	123.3	base value:	123.3
1αC	10.6	1αC	10.6
1α'C	-7.9	2βC	9.8
2β'C	-3.6	1α'C	-7.9
correction:		correction:	
cis α,α'	-1.1	cis α,α'	-1.1
		1β-substituent	2.3
estimated:	121.3	estimated:	137.0
determined:	121.8	determined:	138.8

SUBSTITUTED ALKENES

Effect of a Substituent on the ^{13}C-Chemical Shifts in Vinyl Compounds

(δ in ppm relative to TMS)

$$X-\underset{1}{C}H=\underset{2}{C}H_2 \qquad \delta_{C_i} = 123.3 + Z_i$$

	Substituent X	Z_1	Z_2
	-H	0.0	0.0
C	$-CH_3$	12.9	- 7.4
	$-CH_2CH_3$	17.2	- 9.8
	$-CH_2CH_2CH_3$	15.7	- 8.8
	$-CH(CH_3)_2$	22.7	-12.0
	$-CH_2CH_2CH_2CH_3$	14.6	- 8.9
	$-C(CH_3)_3$	26.0	-14.8
	$-CH_2Cl$	10.2	- 6.0
	$-CH_2Br$	10.9	- 4.5
	$-CH_2I$	14.2	- 4.0
	$-CH_2OH$	14.2	- 8.4
	$-CH_2OCH_2CH_3$	12.3	- 8.8
	$-CH=CH_2$	13.6	- 7.0
	-C≡CH	- 6.0	5.9
	-phenyl	12.5	-11.0
HAL	-F	24.9	-34.3
	-Cl	2.8	- 6.1
	-Br	- 8.6	- 0.9
	-I	-38.1	7.0
O	$-OCH_3$	29.4	-38.9
	$-OCH_2CH_3$	28.8	-37.1
	$-OCH_2CH_2CH_2CH_3$	28.1	-40.4
	$-OCOCH_3$	18.4	-26.7
N	$-N(CH_3)_2$	28.0*	-32.0*
	$-N^+(CH_3)_3$	19.8	-10.6
	-N-pyrrolidonyl	6.5	-29.2
	$-NO_2$	22.3	- 0.9
	-NC	- 3.9	- 2.7
S	$-SCH_2$-phenyl	18.5	-16.4
	$-SO_2CH=CH_2$	14.3	7.9
O=C<	-CHO	15.3	14.5
	$-COCH_3$	13.8	4.7
	-COOH	5.0	9.8
	$-COOCH_2CH_3$	6.3	7.0
	-COCl	8.1	14.0
	-CN	-15.1	14.2
	$-Si(CH_3)_3$	16.9	6.7
	$-SiCl_3$	8.7	16.1

* estimated values

The values listed on p. C90 can also be used to estimate the ^{13}C-chemical shifts of C=C double bonds with more than one substituent.

Example: Estimation of the ^{13}C-chemical shifts in 1-bromo-1-propene:

$$\underset{\mathbf{a}}{Br-CH}=\underset{\mathbf{b}}{CH}-CH_3$$

(a)			(b)		
base value:	123.3		base value:	123.3	
Z_1 (Br)	-8.6		Z_2 (Br)	-0.9	
Z_2 (CH_3)	-7.4		Z_1 (CH_3)	12.9	
estimated:	107.3		estimated:	135.3	
determined:	108.9	(cis)	determined:	129.4	(cis)
	104.7	(trans)		132.7	(trans)

Measured and Estimated (in Parentheses) Chemical Shifts in Polysubstituted Alkenes

(δ in ppm relative to TMS)

Compound		Measured (Estimated)
$(NC)_2C^a=C^b(N(CH_3)_2)_2$	(a)	39.1 (29.1)
	(b)	171.0 (207.7)
$C^aH_2=C^b(N(CH_3)_2)_2$	(a)	69.2 (59.3)
	(b)	163.0 (179.3)
$(CH_3)_2N(H)C^a=C^b(H)NO_2$	(a)	151.0 (150.4)
	(b)	111.4 (113.6)
$C^aH_2=C^b(OCH_3)_2$	(a)	54.7 (45.5)
	(b)	167.9 (182.1)

^{13}C-Chemical Shifts in cis- and trans- 1,2-Disubstituted Alkenes

(δ in ppm relative to TMS)

Substituent	δ_{cis}	δ_{trans}
$-CH_3$	123.3	124.5
$-CH_2CH_3$	131.2	131.3
-Cl	118.1	119.9
-Br	116.4	109.4
-I	96.5	79.4
$-OCH_3$	130.3	135.2
-COOH	130.4	134.2
$-COOCH_3$	130.1	133.5
-CN	120.8	120.2

^{13}C-Chemical Shifts in a few Enols

(δ in ppm relative to TMS)

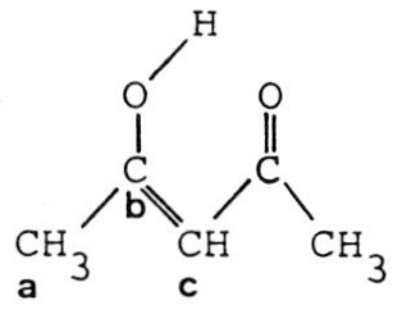

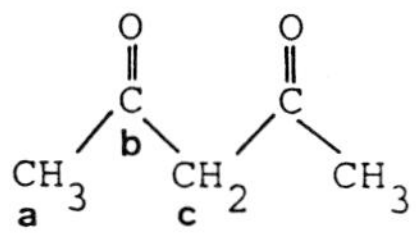

	Enol	Ketone
(a)	22.5	28.5
(b)	190.5	201.1
(c)	99.0	56.6

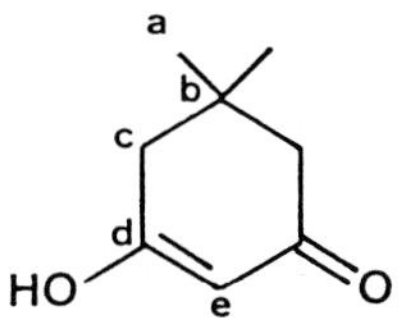

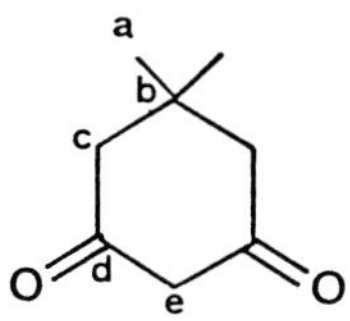

	Enol	Ketone
(a)	28.3	28.3
(b)	32.8	31.0
(c)	46.2	54.2
(d)	191.1	203.6
(e)	103.3	57.3

^{13}C-Chemical Shifts in Unsaturated Alicyclics

(δ in ppm relative to TMS)

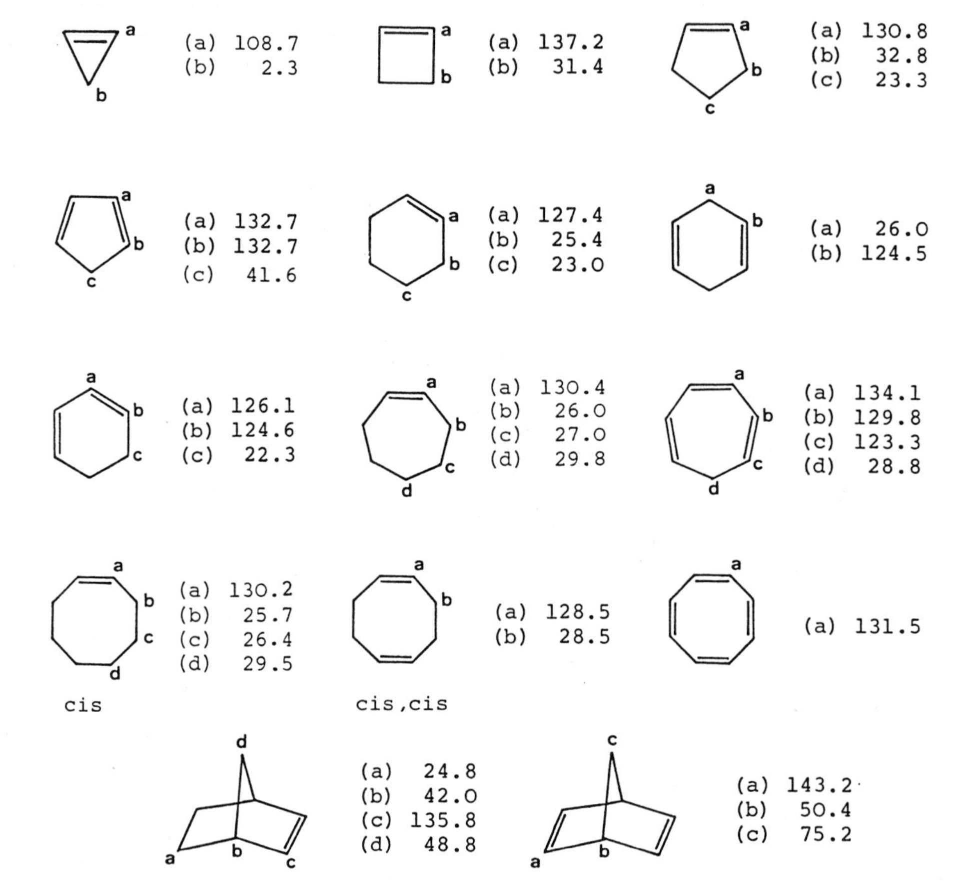

Non-condensed saturated alicyclics: see p. C47.

Condensed saturated alicyclics: see p. C72.

(a) 25.3
(b) 32.8
(c) 143.9
(d) 125.9
(e) 124.2

(a) 23.6
(b) 29.5
(c) 136.8
(d) 125.5
(e) 129.0

<u>^{13}C-Chemical Shifts in Aliphatic Dienes</u> (δ in ppm relative to TMS)

(a) allenes

$\overset{b}{\ }\overset{a}{\ }$
$CH_2=C=CH_2$

(a) 74.8
(b) 213.5

Monosubstituted allenes: see R.H.A.M. Janssen, R.J.J.Ch. Lousberg, M.J.A. De Bie, Rec. Trav. <u>100</u>, 85 (1981).

(b) conjugated dienes

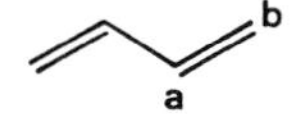

(a) 136.9
(b) 116.3

^{13}C-Chemical Shifts in Some Aromatic Hydrocarbons

(δ in ppm relative to TMS)

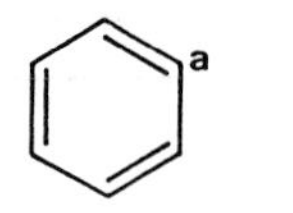

(a) 128.5

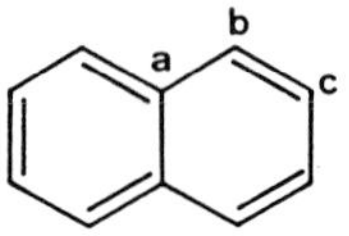

(a) 133.7
(b) 128.0
(c) 126.0

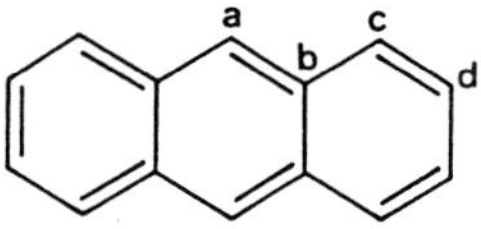

(a) 126.2
(b) 131.8
(c) 128.1
(d) 125.3

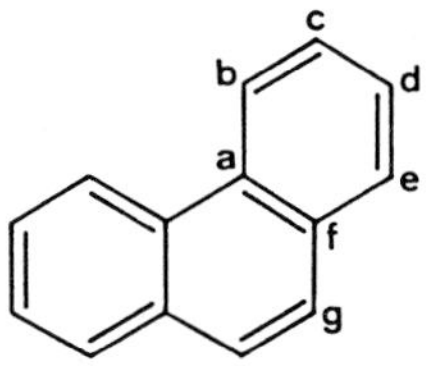

(a) 131.9
(b) 122.4
(c) 126.3
(d) 126.3
(e) 128.3
(f) 130.1
(g) 126.6

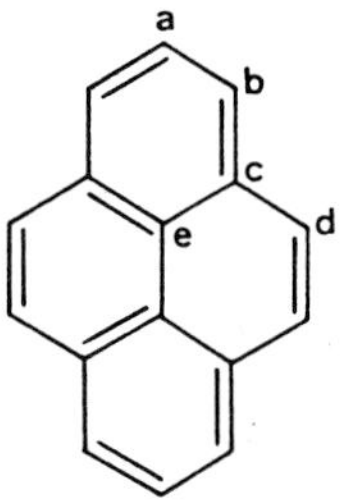

(a) 125.5
(b) 124.6
(c) 130.9
(d) 127.0
(e) 124.6

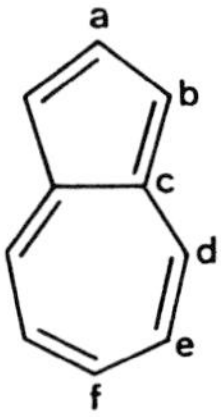

(a) 135.2
(b) 119.7
(c) 140.1
(d) 137.4
(e) 123.9
(f) 137.4

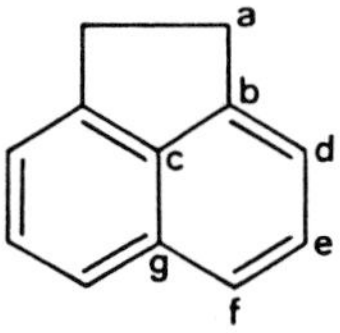

(a) 30.3
(b) 145.9
(c) 139.7
(d) 119.5
(e) 128.2
(f) 122.7
(g) 132.1

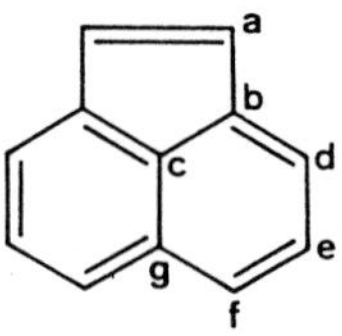

(a) 129.7
(b) 140.0
(c) 128.7
(d) 124.3
(e) 127.9
(f) 127.4
(g) 128.4

^{13}C-Chemical Shifts in Some Alkynes (δ in ppm relative to TMS)

$$\underset{\mathbf{a}\;\;\mathbf{b}}{X-C\equiv C-H}$$

X	a	b
-H	71.9	71.9
$-CH_3$	80.4	68.3
$-CH_2CH_3$	85.5	67.1
$-CH_2CH_2CH_3$	84.0	68.7
$-CH_2CH_2CH_2CH_3$	83.0	66.0
$-CH(CH_3)_2$	89.2	67.6
$-C(CH_3)_3$	92.6	66.8
-cyclohexyl	88.7	68.3
$-CH_2OH$	83.0	73.8
$-CH=CH_2$	82.8	80.0
$-C\equiv C-CH_3$	68.8	64.7
-phenyl	84.6	78.3
$-OCH_2CH_3$	88.2	22.0
$-SCH_2CH_3$	72.6	81.4
-CHO	81.8	83.1
$-COCH_3$	81.9	78.1
-COOH	74.0	78.6
$-COOCH_3$	74.8	75.6

Additivity rules for estimating the chemical shifts of sp-hybridized carbon atoms in alkynes: see W. Höbold, R. Radeglia, D. Klose, J. Prakt. Chem. 318, 519 (1976).

Effect of Substituents on the ^{13}C-Chemical Shifts in Monosubstituted Benzenes (δ in ppm relative to TMS, see also D.F. Ewing, Org. Magn. Res. 12, 499 (1979)

$$\delta_{C_i} = 128.5 + Z_i$$

(Benzene ring with substituent X at position 1; positions 1, 2, 3, 4 numbered.)

	Substituent X	Z_1	Z_2	Z_3	Z_4
	-H	0.0	0.0	0.0	0.0
	$-CH_3$	9.2	0.7	-0.1	-3.0
	$-CH_2CH_3$	15.7	-0.6	-0.1	-2.8
	$-CH(CH_3)_2$	20.2	-2.2	-0.3	-2.8
	$-CH_2CH_2CH_2CH_3$	14.2	-0.2	-0.2	-2.8
	$-C(CH_3)_3$	22.4	-3.3	-0.4	-3.1
	-cyclopropyl	15.1	-3.3	-0.6	-3.6
	$-CH_2Cl$	9.3	0.3	0.2	0.0
	$-CH_2Br$	9.5	0.7	0.3	0.2
C	$-CF_3$	2.5	-3.2	0.3	3.3
	$-CCl_3$	16.3	-1.7	-0.1	1.8
	$-CH_2OH$	12.4	-1.2	0.2	-1.1
	$-CHOCH_2$ (epoxide)	9.2	-3.1	-0.1	-0.5
	$-CH_2NH_2$	14.9	-1.4	-0.2	-2.0
	$-CH_2SCH_3$	9.8	0.4	-0.1	-1.6
	$-CH_2SOCH_3$	0.8	1.5	0.4	-0.2
	$-CH_2CN$	1.6	0.5	-0.8	-0.7
	$-CH{=}CH_2$	8.9	-2.3	-0.1	-0.8
	$-C{\equiv}CH$	-6.2	3.6	-0.4	-0.3
	-phenyl	13.1	-1.1	0.5	-1.1
H	-F	34.8	-13.0	1.6	-4.4
A	-Cl	6.3	0.4	1.4	-1.9
L	-Br	-5.8	3.2	1.6	-1.6
	-I	-34.1	8.9	1.6	-1.1
	-OH	26.9	-12.8	1.4	-7.4
	-ONa	39.6	-8.2	1.9	-13.6
	$-OCH_3$	31.4	-14.4	1.0	-7.7
O	$-OCH{=}CH_2$	28.2	-11.5	0.7	-5.8
	-Ophenyl	27.6	-11.2	-0.3	-6.9
	$-OCOCH_3$	22.4	-7.1	0.4	-3.2
	$-OSi(CH_3)_3$	26.8	-8.4	0.9	-7.1
	$-OPO(Ophenyl)_2$	21.9	-8.4	1.2	-3.0
	-OCN	25.0	-12.7	2.6	-1.0

	Substituent X	Z_1	Z_2	Z_3	Z_4
	$-NH_2$	18.2	-13.4	0.8	-10.0
	$-NHCH_3$	21.4	-16.2	0.8	-11.6
	$-N(CH_3)_2$	22.5	-15.4	0.9	-11.5
	-NHphenyl	14.7	-10.6	0.9	-10.5
	-N(phenyl)$_2$	19.8	-7.0	0.9	-5.6
	$-NH_3^+$	0.1	-5.8	2.2	2.2
N	$-N^+(CH_3)_3$	19.5	-7.3	2.5	2.4
	$-NHCOCH_3$	9.7	-8.1	0.2	-4.4
	$-NHNH_2$	22.8	-16.5	0.5	-9.6
	$-N(CH_3)NO$	13.7	-9.5	0.8	-1.4
	-N=N-phenyl	24.0	-5.8	0.3	2.2
	$-N^+\equiv N$	-12.7	6.0	5.7	16.0
	-NC	-1.8	-2.2	1.4	0.9
	-NCO	5.1	-3.7	1.1	-2.8
	-NCS	3.0	-2.7	1.3	-1.0
	-NO	37.4	-7.7	0.8	7.0
	$-NO_2$	19.9	-4.9	0.9	6.1
	-SH	2.1	0.7	0.3	-3.2
	$-SCH_3$	10.0	-1.9	0.2	-3.6
	$-SC(CH_3)_3$	4.5	9.0	-0.3	0.0
S	-Sphenyl	7.3	2.5	0.6	-1.5
	$-SOCH_3$	17.6	-5.0	1.1	2.4
	$-SO_2CH_3$	12.3	-1.4	0.8	5.1
	$-SO_2Cl$	15.6	-1.7	1.2	6.8
	$-SO_3H$	15.0	-2.2	1.3	3.8
	$-SO_2OCH_3$	6.4	-0.6	1.5	5.9
	-SCN	-3.7	2.5	2.2	2.2
	-CHO	8.2	1.2	0.5	5.8
	$-COCH_3$	8.9	0.1	-0.1	4.4
O	$-COCF_3$	-5.6	1.8	0.7	6.7
‖	-COphenyl	9.3	1.6	-0.3	3.7
C	-COOH	2.1	1.6	-0.1	5.2
/\	$-COO^-$	9.7	4.6	2.2	4.6
	$-COOCH_3$	2.0	1.2	-0.1	4.3
	$-CONH_2$	5.0	-1.2	0.1	3.4
	$-CON(CH_3)_2$	8.0	-1.5	-0.2	1.0
	-COCl	4.7	2.7	0.3	6.6
	-CSphenyl	18.7	1.0	-0.6	2.4
	-CN	-15.7	3.6	0.7	4.3
	$-P(CH_3)_2$	13.6	1.6	-0.6	-1.0
	-P(phenyl)$_2$	8.9	5.2	0.0	0.1
	$-PO(OCH_2CH_3)_2$	1.6	3.6	-0.2	3.4
	$-PS(OCH_2CH_3)_2$	6.1	2.8	-0.4	3.4
	$-SiH_3$	-0.5	7.3	-0.4	1.3
	$-Si(CH_3)_3$	11.6	4.9	-0.7	0.4
	$-Sn(CH_3)_3$	13.4	7.4	-0.2	-0.3
	$-Pb(CH_3)_3$	20.1	8.0	-0.1	-1.0

Effect of a Substituent in Position 1 of Monosubstituted Naphtalenes on the Chemical Shifts of the other Carbon Atoms (δ in ppm relative to TMS)

for X=H δ_{C_1} = 128.0, δ_{C_2} = 125.9, δ_{C_9} = 133.6

	Substituent X	C-1	C-2	C-3	C-4	C-5	C-6	C-7	C-8	C-9	C-10
C	$-CH_3$	6.0	0.5	0.6	-1.8	0.3	-0.7	-0.5	-4.1	-1.1	-0.2
	$-C(CH_3)_3$	17.9	-2.8	-0.9	-0.6	1.6	-1.4	-1.4	-1.2	-1.6	2.2
	$-CH_2Br$	4.0	1.1	-0.9	1.3	0.5	-0.1	0.3	-4.6	-2.8	0.1
	$-CH_2OH$	8.2	-0.9	-0.6	0.1	0.5	-0.3	0.1	-4.5	-2.6	0.0
	$-CF_3$		-1.3	-1.8	5.0	1.0	0.8	2.0	-3.4	1.0	-3.9
HAL	-F	31.5	-16.1	0.1	-3.8	0.1	1.4	0.7	-7.1	-9.3	2.1
	-Cl	3.9	0.2	-0.2	-0.9	0.2	3.1	0.8	-3.6	-2.8	1.0
	-Br	-5.4	3.6	-0.2	-0.5	-0.1	0.4	1.0	-1.3	-2.0	0.6
	-I	-28.4	12.3	1.7	1.7	1.4	1.6	2.6	4.4	1.3	1.3
O	-OH	23.5	-17.2	-0.1	-7.3	-0.4	0.5	0.3	-6.6	-9.3	1.0
	$-OCH_3$	27.3	-22.3	-0.2	-7.9	-0.7	0.3	-0.9	-6.1	-8.1	0.8
	$-OCOCH_3$	18.6	-7.9	-0.6	-2.1	0.0	0.4	0.4	-6.9	-6.9	0.9
N	$-NH_2$	14.0	-16.5	0.3	-9.3	0.3	-0.3	-1.3	-7.3	-10.2	0.6
	$-N(CH_3)_2$	23.7	-11.2	0.6	-4.6	1.0	0.4	-0.3	-3.2	-3.9	2.1
	$-NH_3^+$	-3.8	-4.6	-0.9	3.4	1.4	2.1	2.8	-9.0	-7.4	1.2
	$-NO_2$	18.5	-2.1	-2.0	6.5	0.5	1.3	3.4	-5.1	-8.7	0.6
O=C<	-CHO	2.9	10.8	-1.4	6.7	0.2	0.6	2.7	-3.5	-3.6	-0.3
	$-COCH_3$	6.9	2.9	-1.7	4.9	0.3	0.4	2.0	-2.0	-3.5	0.2
	-COOH	-1.5	3.6	-2.4	4.3	-0.6	-0.9	0.6	-3.2	-3.2	-0.8
	$-COOCH_3$	-0.9	4.5	-1.2	5.4	0.7	0.5	1.9	-1.8	-1.9	0.5
	$-CON(CH_3)_2$	6.8	-2.1	-0.8	0.9	0.4	0.4	1.0	0.1	-4.1	-0.2
	-COCl		10.6	-0.5	9.3	1.9	2.1	4.5	-2.1	-2.1	1.0
	-CN	-19.2	5.1	-2.4	3.8	-0.7	0.2	1.2	-4.5	-2.8	-2.2
	$-Si(CH_3)_3$	9.8	5.1	-0.4	1.7	1.2	-0.8	-0.7	0.1	3.8	0.2

Effect of a Substituent in Position 2 of Monosubstituted Naphthalenes on the Chemical Shifts of the other Carbon Atoms (δ in ppm relative to TMS)

for X=H δ_{C_1} = 128.0
δ_{C_2} = 125.9
δ_{C_9} = 133.6

Substituent X	C-1	C-2	C-3	C-4	C-5	C-6	C-7	C-8	C-9	C-10
C $-CH_3$	-1.3	9.3	2.0	-0.8	-0.5	-1.1	-0.2	-0.6	-0.1	-2.0
$-C(CH_3)_3$	-3.3	22.5	-3.0	-0.4	0.0	-0.7	-0.2	-0.6	0.4	-1.3
$-CH_2Br$	-1.7	9.0	1.9	-0.4	-0.5	0.7	0.3	0.6	-0.6	-0.7
$-CF_3$	-2.0		-4.2	1.1*	0.1*	2.4*	1.5	1.1	-1.1	1.3
HAL -F	-17.0	34.9	-9.6	2.4	0.0	-0.7	1.1	-0.6	0.7	-3.0
-Cl	-1.4	5.7	0.8	1.5	-0.2	0.2	1.1	-1.1	0.7	-1.9
-Br	1.8	-6.2	3.1	1.5	-0.3	0.2	0.8	-1.1	-2.0	0.7
-I	9.2	-34.1	9.0	2.3	0.5	1.3	1.5	-0.6	2.1	-0.8
O -OH	-18.6	27.3	-8.3	1.8	-0.3	-2.4	0.5	-1.7	0.9	-4.7
$-OCH_3$	-22.2	31.8	-7.1	1.5	-0.3	-2.2	0.5	-1.2	1.0	-4.3
$-OCOCH_3$	-9.5	22.5	-4.8	1.3	-0.4	-0.3	0.6	-0.4	0.1	-2.2
N $-NH_2$	-20.6	16.7	-8.9	-0.2	-1.6	-4.8	-0.9	-3.5	-0.1	-7.0
$-N(CH_3)_2$	-21.1	23.6	-8.8	1.2	0.0	-3.4	0.7	-1.1	2.4	-5.9
$-NH_3^+$	-5.9	-0.3	-6.5	3.2	0.2	2.3	2.0	0.2	0.1	-0.3
$-NO_2$	-3.4	20.0	-6.7	1.7	0.1	4.0	2.2	2.1	-1.1	2.4
O=C< -CHO	6.2	7.9	-3.6	0.8	-0.3	2.9	0.9	1.8	2.4	-1.4
$-COCH_3$	1.9	8.3	-2.2	0.2	-0.4	2.3	0.7	1.4	1.8	-1.3
-COOH	2.7	2.4	-0.6	0.2	-0.3	2.4	0.9	1.3	-1.3	1.5
$-COOCH_3$	3.0	1.8	-0.5	0.2	-0.1	2.4	0.9	1.4	-1.0	1.9
-CN	5.8	-16.7	0.1	1.0	-0.2	3.0	1.6	0.2	-1.6	0.7
$-Si(CH_3)_3$	5.8	11.9	3.9	-1.0	0.1	0.3	-0.2	0.1	-0.5	0.2

* assignment uncertain

MULTIPLY SUBSTITUTED BENZENES

The ^{13}C-chemical shifts in multiply substituted benzenes and naphthalenes can be estimated using the ^{13}C-chemical shifts listed on p. C120 and C125 for monosubstituted benzenes and on p. 126 and 127 for monosubstituted naphthalenes.

Example: Estimation of the ^{13}C-chemical shifts in 3,5-dimethylnitrobenzene:

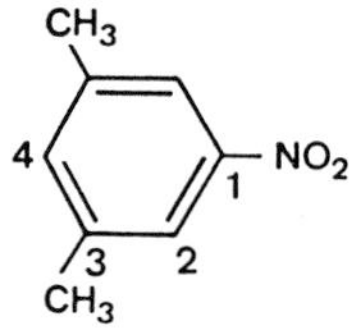

(C-1)		
	base value:	128.5
	$Z_1(NO_2)$	19.9
	$2Z_3(CH_3)$	-0.2
	estimated:	148.2
	determined:	148.5

(C-2)		
	base value:	128.5
	$Z_2(NO_2)$	-4.9
	$Z_2(CH_3)$	0.7
	$Z_4(CH_3)$	-3.0
	estimated:	121.3
	determined:	121.7

(C-3)		
	base value:	128.5
	$Z_1(CH_3)$	9.2
	$Z_3(CH_3)$	-0.1
	$Z_3(NO_2)$	0.9
	estimated:	138.5
	determined:	139.6

(C-4)		
	base value:	128.5
	$2Z_2(CH_3)$	1.4
	$Z_4(NO_2)$	6.1
	estimated:	136.0
	determined:	136.2

Larger discrepancies between estimated and experimentally determined values are to be expected if the substituents are ortho to each other.

^{13}C-Chemical Shifts in Five-Membered Heteroaromatic Rings

(δ in ppm relative to TMS)

(a) 109.9
(b) 143.0

(a) 126.4
(b) 124.9

(a) 107.7
(b) 118.0

(a) 104.7
(b) 133.3

(a) 103.4
(b) 138.5

(a) 109.0
(b) 135.0

(a) 136.2
(b) 122.3

(a) 145.1
(b) 126.8

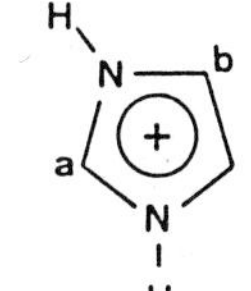

(a) 134.6
(b) 120.1

(a) 150.0
(b) 100.5
(c) 158.9

(a) 150.6
(b) 125.4
(c) 138.1

(a) 157.0
(b) 123.4
(c) 147.8

(a) 152.7
(b) 143.2
(c) 118.6

(a) 130.4

(a) 147.9

MONOSUBSTITUTED PYRIDINES

Effect of Substituents on the ^{13}C-Chemical Shifts in Monosubstituted Pyridines (δ in ppm relative to TMS)

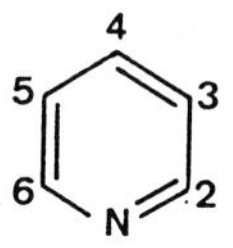

$$\delta_{C-2} = 149.8 + Z_{i2}$$
$$\delta_{C-3} = 123.6 + Z_{i3}$$
$$\delta_{C-4} = 135.7 + Z_{i4}$$
$$\delta_{C-5} = 123.6 + Z_{i5}$$
$$\delta_{C-6} = 149.8 + Z_{i6}$$

2- or 6-substituent (i = 2 or 6)	$Z_{22}=Z_{66}$	$Z_{23}=Z_{65}$	$Z_{24}=Z_{64}$	$Z_{25}=Z_{63}$	$Z_{26}=Z_{62}$
$-CH_3$	8.8	-0.6	0.2	-3.0	-0.4
$-CH_2CH_3$	13.6	-1.8	0.4	-2.9	-0.7
-F	14.4	-13.1	6.1	-1.5	-1.5
-Cl	2.3	0.7	3.3	-1.2	0.6
-Br	-6.7	4.8	3.3	-0.5	1.4
-OH	15.5	-3.5	-0.9	-16.9	-8.2
$-OCH_3$	15.3	-7.5	2.1	-13.1	-2.2
$-NH_2$	11.3	-14.7	2.3	-10.6	-0.9
$-NO_2$	8.0	-5.1	5.5	6.6	0.4
-CHO	3.5	-2.6	1.3	4.1	0.7
$-COCH_3$	4.3	-2.8	0.7	3.0	-0.2
$-CONH_2$	0.1	-1.2	1.5	2.8	-1.4
-CN	-15.9	5.0	1.6	3.6	1.4
$-Si(CH_3)_3$	17.6	3.8	-2.9	-2.0	-0.5
$-Sn(CH_3)_3$	22.5	6.8	-3.2	-2.5	-0.2

3- or 5-substituent (i = 3 or 5)	$Z_{32}=Z_{56}$	$Z_{33}=Z_{55}$	$Z_{34}=Z_{54}$	$Z_{35}=Z_{53}$	$Z_{36}=Z_{52}$
$-CH_3$	1.3	9.0	0.2	-0.8	-2.3
$-CH_2CH_3$	-0.4	15.5	-0.6	-0.4	-2.7
-F	-11.5	36.2	-13.0	0.9	-3.9
-Cl	-0.3	8.2	-0.2	0.7	-1.4
-Br	2.1	-2.6	2.9	1.2	-0.9
-I	7.1	-28.4	9.1	2.4	0.3
-OH	-10.7	31.4	-12.2	1.3	-8.6
$-OCH_3$	-12.5	31.6	-15.7	0.2	-8.4
$-NH_2$	-11.9	21.5	-14.2	0.9	-10.8
-CHO	2.4	7.9	0.0	0.6	5.4
$-COCH_3$	3.5	8.6	-0.5	-0.1	0.0
$-COOCH_3$	-0.6	1.1	-0.3	-1.7	1.8
$-CONH_2$	2.7	6.0	1.3	1.3	-1.5
-CN	3.6	-13.7	4.4	0.6	4.2
$-Si(CH_3)_3$	1.9	8.3	2.5	-3.1	-0.9
$-Sn(CH_3)_3$	5.1	12.2	6.6	-0.7	-1.2

4-substituent (i = 4)	$Z_{42}=Z_{46}$	$Z_{43}=Z_{45}$	Z_{44}
$-CH_3$	0.5	0.8	10.8
$-CH_2CH_3$	-0.1	-0.4	17.0
$-CH(CH_3)_2$	0.4	-1.8	21.4
$-C(CH_3)_3$	0.1	-3.4	23.4
$-CH=CH_2$	0.3	-2.9	8.6
-F	2.7	-11.8	33.0
-Br	3.0	3.4	-3.0
$-OCH_3$	0.9	-13.8	29.2
$-NH_2$	0.9	-13.8	19.6
-CHO	1.7	-0.6	5.5
$-COCH_3$	1.6	-2.6	6.8
$-COOCH_3$	1.0	-0.7	1.6
$-CONH_2$	0.4	-0.8	6.4
-CN	2.1	2.2	-15.7
$-Si(CH_3)_3$	-3.6	1.6	11.4
$-Sn(CH_3)_3$	-1.9	6.5	15.7

MULTIPLY SUBSTITUTED PYRIDINES

The ^{13}C-chemical shifts in multiply substituted pyridines can be estimated using the ^{13}C-chemical shifts listed on p. C140 and C145 for monosubstituted pyridines.

Example: Estimation of the ^{13}C-chemical shifts in 2,5-dimethylpyridine:

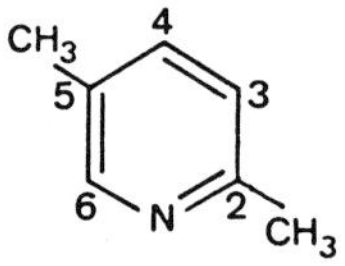

CH3, 4, 3, 5, 6, N, 2, CH3

(C-2)	base value:	149.8
	$Z_{22}(CH_3)$	8.8
	$Z_{52}(CH_3)$	-2.3
	estimated:	156.3
	determined:	155.2

(C-3)	base value:	123.6
	$Z_{23}(CH_3)$	-0.6
	$Z_{53}(CH_3)$	-0.8
	estimated:	122.2
	determined:	122.5

(C-4)	base value:	135.7
	$Z_{24}(CH_3)$	0.2
	$Z_{54}(CH_3)$	0.2
	estimated:	136.1
	determined:	136.7

(C-5)	base value:	123.6
	$Z_{55}(CH_3)$	9.0
	$Z_{25}(CH_3)$	-3.0
	estimated:	129.6
	determined:	129.6

(C-6)	base value:	149.8
	$Z_{56}(CH_3)$	1.3
	$Z_{26}(CH_3)$	-0.4
	estimated:	150.7
	determined:	149.4

^{13}C-Chemical Shifts in Six-Membered Heteroaromatic Rings

(δ in ppm relative to TMS)

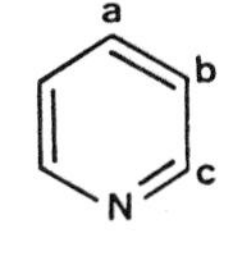

(a) 135.7
(b) 123.6
(c) 149.8

(a) 148.4
(b) 129.0
(c) 142.5

Substituted pyridines: see p. C140.

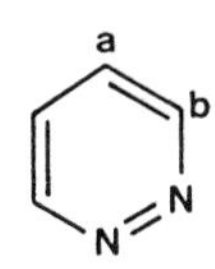

(a) 126.5
(b) 151.4

(a) 121.4
(b) 156.4
(c) 158.0

(a) 144.9

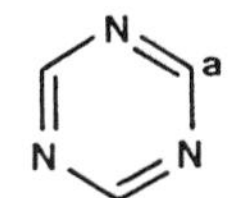

(a) 166.5

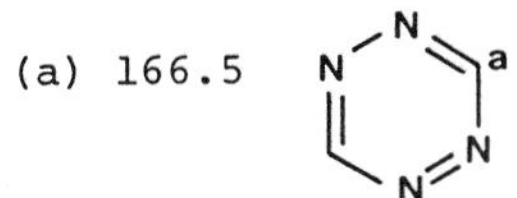

(a) 160.9

^{13}C-Chemical Shifts in Condensed Heteroaromatics

(δ in ppm relative to TMS)

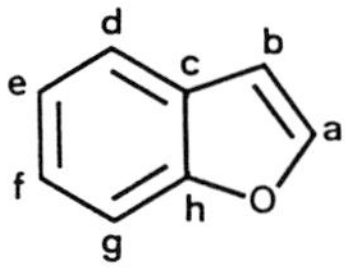

(a)	145.0	(e)	123.2
(b)	106.9	(f)	124.6
(c)	127.9	(g)	111.8
(d)	121.6	(h)	155.5

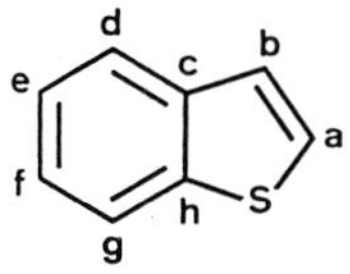

(a)	126.4	(e)	124.3
(b)	124.0	(f)	124.4
(c)	139.8	(g)	122.6
(d)	123.8	(h)	139.9

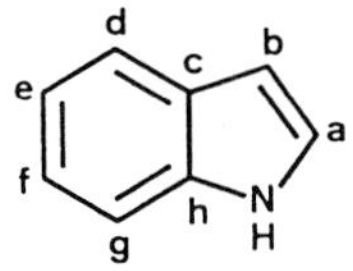

(a)	124.1	(e)	121.7
(b)	102.1	(f)	119.6
(c)	127.6	(g)	111.0
(d)	120.5	(h)	135.5

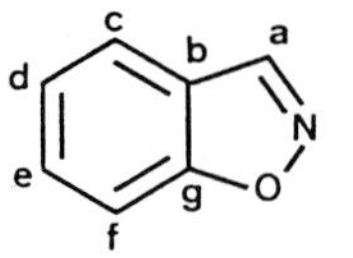

(a)	147.1	(e)	130.6
(b)	122.2	(f)	109.9
(c)	124.3	(g)	162.7
(d)	123.0		

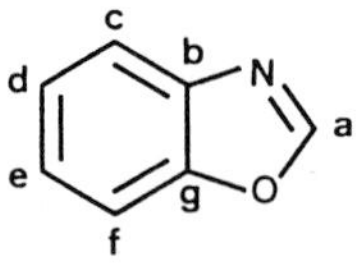

(a)	152.6	(e)	124.4
(b)	140.1	(f)	110.8
(c)	120.5	(g)	150.0
(d)	125.4		

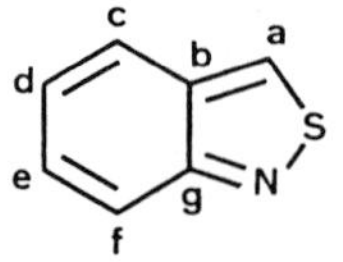

(a)	144.5	(e)	128.6
(b)	134.5	(f)	121.6
(c)	122.1	(g)	161.5
(d)	124.1		

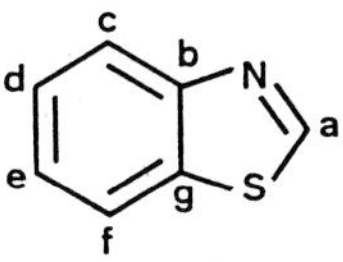

(a) 155.5 (e) 125.8*
(b) 152.6 (f) 122.7*
(c) 122.1* (g) 133.2
(d) 125.1*

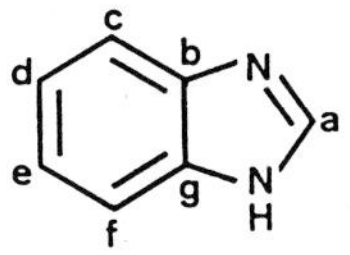

(a) 141.5
(b,g) 137.9
(c,f) 115.4
(d,e) 122.9

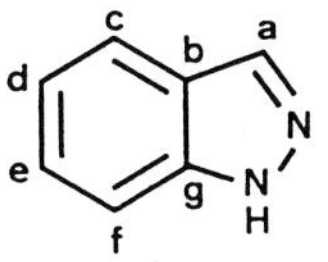

(a) 133.4 (e) 125.8
(b) 122.8 (f) 110.0
(c) 120.4 (g) 139.9
(d) 120.1

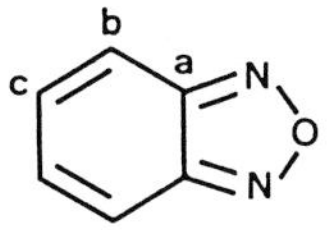

(a) 144.4
(b) 111.2
(c) 127.2

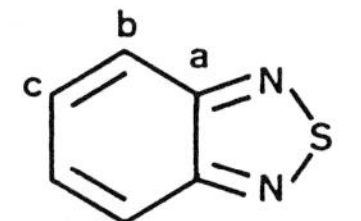

(a) 155.2
(b) 121.6
(c) 129.0

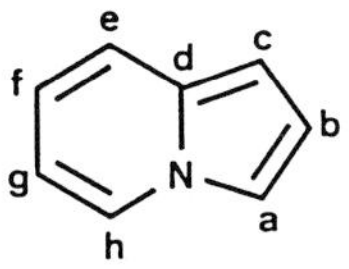

(a) 113.0 (e) 119.6
(b) 114.1 (f) 117.2
(c) 99.5 (g) 110.5
(d) 133.4 (h) 125.6

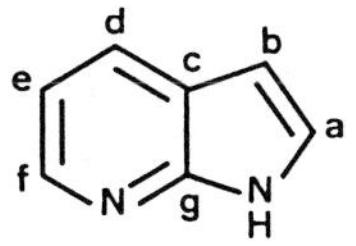

(a) 125.5 (e) 115.6
(b) 100.5 (f) 142.1
(c) 120.7 (g) 148.9
(d) 129.0

* assignment uncertain

^{13}C-Chemical Shifts in Condensed Heteroaromatics

(δ in ppm relative to TMS)

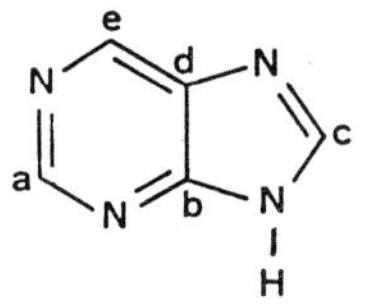

(a) 152.0 (b) 154.9 (c) 147.9 (d) 128.4 (e) 144.8

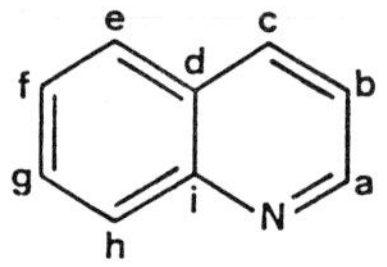

(a) 150.0 (b) 120.8 (c) 135.7 (d) 128.0 (e) 127.6 (f) 126.3 (g) 129.2 (h) 129.2 (i) 148.1

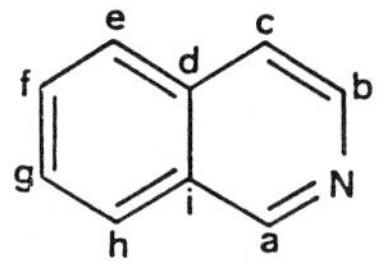

(a) 152.2 (b) 142.7 (c) 120.2 (d) 135.5 (e) 126.2 (f) 130.1 (g) 127.0 (h) 127.3 (i) 128.5

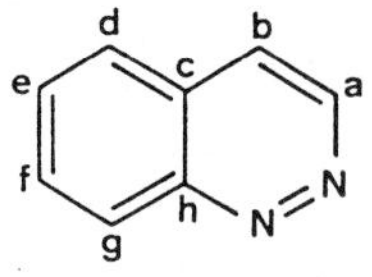

(a) 146.1 (b) 124.7 (c) 126.9 (d) 128.0 (e) 132.2 (f) 132.1 (g) 129.5 (h) 151.0

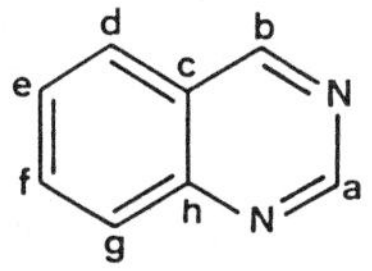

(a) 160.7 (b) 155.9 (c) 125.2 (d) 127.4 (e) 127.9 (f) 134.1 (g) 128.6 (h) 150.1

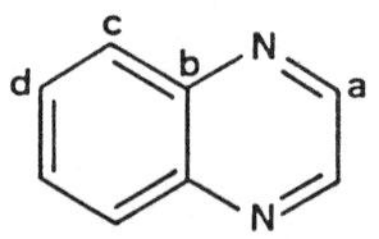

(a) 144.8 (b) 142.8 (c) 129.6 (d) 129.4

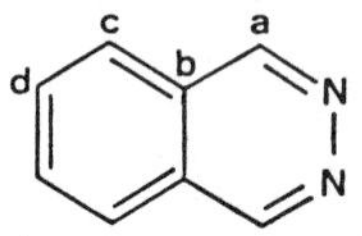

(a) 152.0 (b) 126.7 (c) 126.7 (d) 133.1

(a) 156.2 (d) 122.6
(b) 111.6 (e) 120.6
(c) 127.0 (f) 124.2

(a) 138.5 (d) 124.6
(b) 122.9 (e) 121.9
(c) 127.0 (f) 134.9

(a) 139.6 (d) 118.4
(b) 110.8 (e) 120.0
(c) 125.4 (f) 122.6

(a) 149.1 (e) 129.5
(b) 130.3 (f) 126.6
(c) 125.5 (g) 135.8
(d) 128.3

(a) 144.0
(b) 130.9
(c) 130.2

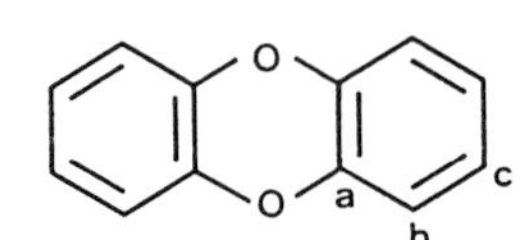

(a) 142.2
(b) 116.2
(c) 123.6

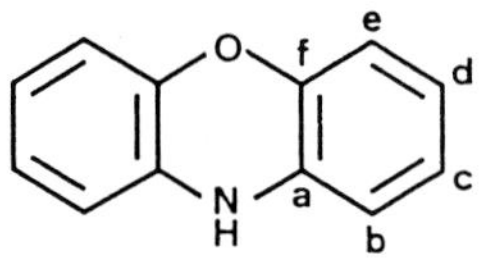

(a)	131.8	(d)	120.0
(b)	112.8	(e)	114.5
(c)	123.0	(f)	142.7

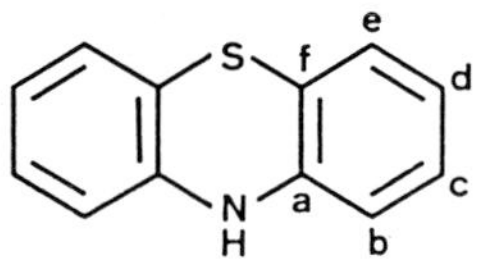

(a)	141.7	(d)	121.3
(b)	113.8	(e)	126.7*
(c)	125.6*	(f)	116.8

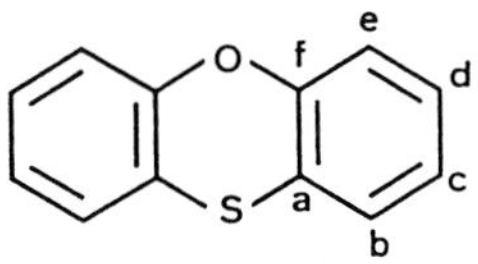

(a)	119.9	(d)	126.5*
(b)	127.4*	(e)	117.5
(c)	124.2	(f)	151.9

* assignment uncertain

^{13}C-Chemical Shifts in Halogenated Derivatives of Methane, Ethane and Ethylene (δ in ppm relative to TMS)

Compound type	Substituent X			
	F[1]	Cl	Br	I
CH_3X	71.6	25.6	9.6	-24.0
CH_2X_2	109.0	54.0	21.4	-54.0
CHX_3	116.4	77.2	12.1	-139.9
CX_4	118.5	96.1	-28.7	-292.5
CH_2X	80.1	39.9	27.6	-1.6
$\mid$ CH_3	15.8	18.9	19.4	20.6
CHX_2		69.3	40.1	
$\mid$ CH_3		31.6	31.8	
CX_3		96.2	31.5	
$\mid$ CH_3		46.3	49.4	
XCH_2CH_2X		51.7	32.4	3.0
CX_3CX_3	116.2	105.3	53.4	
CHX		126.1	114.7	
$\parallel$ CH_2		117.2	122.4	
CX_2		127.1	97.0	
$\parallel$ CH_2		113.3	127.2	
XCH=CHX cis		118.1	116.4	96.5
trans		119.9	109.4	79.4
CX_2		125.1	95.0	
$\parallel$ CHX		117.6	112.4	
$CX_2=CX_2$		121.3	93.7	

1) For ^{19}F-^{13}C-couplings see p. C245.

Polyhalogenated Alkanes

Additivity rules for estimating the chemical shifts in halomethanes: G.R. Somayajulu, J.R. Kennedy, T.M. Vickrey, B.J. Zwolinski, J. Magn. Res. 33, 559 (1979).

Additivity rules for estimating the chemical shifts in linear perfluoroalkanes

(see D.W. Ovenall, J.J. Chang, J. Magn. Res. 25, 361 (1977):

$$\delta = 124.8 + \sum_i Z_i$$

Increments for Z_i for the CF_2- or CF_3-substituents in position:		
α	β	γ
-8.6	1.8	0.5

Example: $CF_3CF_2CF_2CF_3$

CF_3: 124.8 − 8.6 + 1.8 + 0.5 = 118.5 (Exp. 118.5)

CF_2: 124.8 − 2·8.6 + 1.8 = 109.4 (Exp. 109.3)

Substituted Haloalkanes

The additivity rules for estimating the ^{13}C-chemical shifts for aliphatic compounds (see p. C10-C25) can be applied to those haloalkanes which do not have more than one halogen atom at a given carbon atom.

Compound		
CH_2F–COOH	CH_2F 78.9; COOH 173.5	$\lvert J\rvert_{CF}$ = 177 Hz; $\lvert J\rvert_{CCF}$ = 22 Hz
CHF_2–COOH	CHF_2 108.1; COOH 167.2	$\lvert J\rvert_{CF}$ = 239 Hz; $\lvert J\rvert_{CCF}$ = 28 Hz
CF_3–COOH	CF_3 115.0; COOH 163.0	$\lvert J\rvert_{CF}$ = 284 Hz; $\lvert J\rvert_{CCF}$ = 44 Hz
CH_2Cl–COOH	CH_2Cl 40.7; COOH 173.7	
$CHCl_2$–COOH	$CHCl_2$ 63.7; COOH 170.4	
CCl_3–COOH	CCl_3 88.9; COOH 167.0	
CH_2Br–COOH	CH_2Br 25.9; COOH 172.0	
$CHBr_2$–COOH	$CHBr_2$ 31.3; COOH 169.7	

Benzene derivatives C_6H_5–R (a = substituent carbon, b = ipso, c = ortho, d = meta, e = para):

Compound	(a)	(b)	(c)	(d)	(e)
$C_6H_5CH_2F$	84.9 $\lvert J\rvert$=166 Hz	137.0 $\lvert J\rvert$= 17 Hz	127.8 $\lvert J\rvert$= 6 Hz	128.9 $\lvert J\rvert$= 1 Hz	129.0 $\lvert J\rvert$= ~0 Hz
$C_6H_5CF_3$	124.5 $\lvert J\rvert$= 272 Hz	131.0 $\lvert J\rvert$= 32 Hz	125.3 $\lvert J\rvert$= 4 Hz	128.8 $\lvert J\rvert$= 1 Hz	131.8 $\lvert J\rvert$= ~0 Hz
$C_6H_5CH_2Cl$	46.2	137.8	128.8	128.7	128.5
$C_6H_5CHCl_2$	71.9	140.4	126.1	128.6	129.7
$C_6H_5CCl_3$	97.7	144.8	126.8	128.4	130.3
$C_6H_5CH_2Br$	33.4	138.0	129.2	128.8	128.7
$C_6H_5CH_2I$	5.9	138.5	128.5	128.5	127.6

^{13}C-Chemical Shifts in Alcohols

(δ in ppm relative to TMS)

OH	
CH_3	50.2

OH	
CH_2	57.8
CH_3	18.2

OH	
CH_2	64.2
CH_2	25.9
CH_3	10.3

OH	
CH_2	62.9
CH_2	36.0
CH_2	20.3
CH_3	15.2

OH	
CH_2	63.2
CH_2	33.6
CH_2	29.4
CH_2	23.8
CH_3	15.3

OH	
CH	64.0
$(CH_3)_2$	25.3

OH	
C	68.9
$(CH_3)_3$	31.2

OH	
CH_2	73.3
C	32.7
$(CH_3)_3$	26.2

OH	
CH_2	62.1
CH_2	32.9
CH_2	25.8
CH_2	31.9
CH_2	23.0
CH_3	14.2

CH_3	23.5
CH-OH	67.2
CH_2	39.2
CH_2	28.2
CH_2	23.2
CH_3	14.3

CH_3	10.1
CH_2	30.5
CH-OH	72.2
CH_2	39.4
CH_2	19.2
CH_3	14.3

^{13}C-Chemical Shifts in Glycols and Polyols

(δ in ppm relative to TMS)

CH_2OH	63.4
CH_2OH	

CH_2OH	60.2
CH_2	36.4
CH_2OH	

	in $CDCl_3$	in D_2O
CH_2OH	67.7	71.6
CHOH	68.2	72.7
CH_3	18.7	23.0

CH_2OH	64.5
CHOH	73.7
CH_2OH	

$(CH_2OH)_2$	64.3
C	48.3
$(CH_2OH)_2$	

CH_2OH	65.8
CHOH	74.5
CHOH	72.9
CHOH	74.3
CHOH	76.1
CH_2OH	66.1

D-sorbitol

^{13}C-Chemical Shifts in Substituted Alcohols

(δ in ppm relative to TMS

CH_2OH	63.4
CH	137.5
CH_2	114.9

CH_2OH	50.0
C	83.0
C	73.8
H	

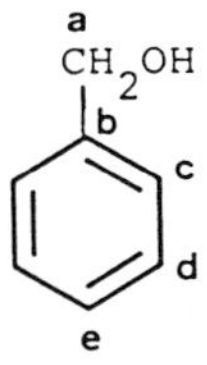

$C_6H_5CH_2OH$ (a = CH_2OH, b = ring C attached to CH_2OH, c, d, e = ring positions)

(a)	65.1
(b)	140.9
(c)	127.3
(d)	128.7
(e)	127.4

CH_2OH	61.4	$\|J\|_{CCF}$ = 35 Hz
CF_3	125.1	$\|J\|_{CF}$ = 278 Hz

CH_2OH	75.9
CCl_3	99.1

^{13}C-Chemical Shifts in Ethers

(δ in ppm relative to TMS)

CH_3	60.9
O	
CH_3	

CH_3	57.6
O	
CH_2	67.7
CH_3	14.7

CH_3	54.9
O	
CH	72.6
$(CH_3)_2$	21.4

CH_3	49.4
O	
C	72.7
$(CH_3)_3$	27.0

OCH_3	58.4
CH_2	72.3
CH_2	
OCH_3	

OCH_3	57.4
CH_2	73.1
CH	134.4
‖	
CH_2	116.4

OCH_3	52.5
CH	152.7
‖	
CH_2	84.4

$(OCH_3)_2$	53.7
CH_2	109.9

$\overset{a}{CH}(\overset{b}{OCH_3})_3$: (a) 115.0, (b) 51.1

$\overset{a}{CH}(\overset{b}{OCH_2}\overset{c}{CH_3})_3$: (a) 112.9, (b) 59.5, (c) 15.2

$(\overset{a}{CH_3})_2\overset{b}{C}(\overset{c}{OCH_3})_2$: (a) 24.0, (b) 99.9, (c) 48.1

Anisole ($^{a}CH_3$–O–C_6H_5; b = ipso, c = ortho, d = meta, e = para): (a) 54.8, (b) 159.9, (c) 114.1, (d) 129.5, (e) 120.8

1,3-Benzodioxole (a = OCH_2O, b = C–O, c, d = ring CH): (a) 100.7, (b) 147.8, (c) 108.8, (d) 121.8

^{13}C-Chemical Shifts in Cyclic Ethers

(δ in ppm relative to TMS)

(a) 39.5

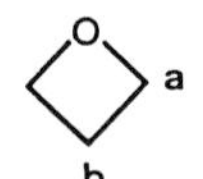

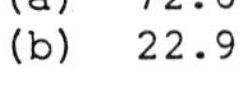

(a) 72.6
(b) 22.9

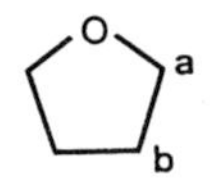

(a) 68.4
(b) 26.5

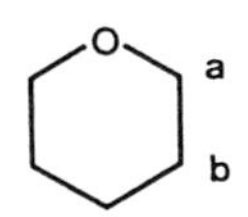

(a) 69.5
(b) 27.7
(c) 24.9

(a) 95.0
(b) 64.5

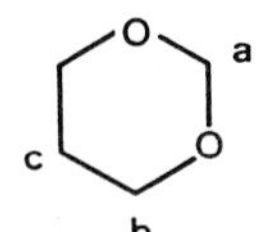

(a) 94.8
(b) 67.5
(c) 27.5

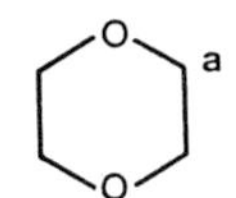

(a) 67.6

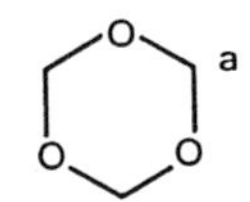

(a) 93.7

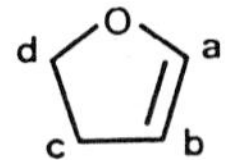

(a) 145.6
(b) 98.4
(c) 28.5
(d) 68.6

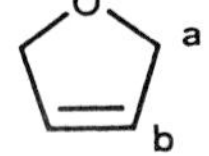

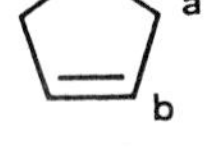

(a) 75.3
(b) 126.3

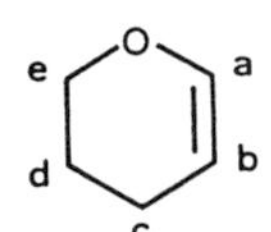

(a) 144.1
(b) 99.4
(c) 19.4
(d) 22.6
(e) 64.8

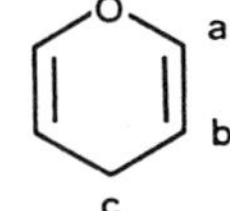

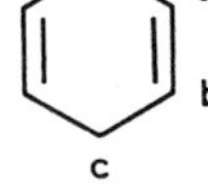

(a) 141.1
(b) 101.1

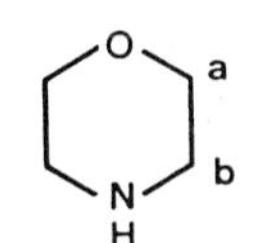

(a) 68.1
(b) 46.7

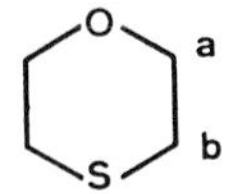

(a) 68.5
(b) 27.0

Protonation of Amines (see J.E. Sarneski, H.L. Surprenant, F.K. Molen, Ch.N. Reilley, Anal. Chem. 47, 2116 (1975)

The protonation of amines causes a shielding of the carbon atoms in the vicinity of the nitrogen. Such a shielding amounts to -2 ppm for an α-carbon atom, -3 to -4 ppm for a β-carbon and -0.5 to -1.0 ppm for a γ-carbon. The most frequent exceptions occur in branched systems: tertiary and quaternary carbon atoms in the α-position are generally deshielded by protonation of the nitrogen (+0.5 to +9 ppm).

^{13}C-Chemical Shifts in Amines (δ in ppm relative to TMS) as well as shifts induced by protonation (values in parentheses: $\delta_{amine\ hydrochloride} - \delta_{amine}$, measured in D_2O)

Compound	δ	Compound	δ
$\overset{a}{C}H_3NH_2$	(a) 28.3(-1.8)	$\overset{b}{C}H_3\overset{a}{C}H_2NH_2$	(a) 36.9(-0.2) (b) 19.0(-5.0)
$(\overset{a}{C}H_3)_2NH$	(a) 38.2(-2.0)	$(\overset{b}{C}H_3\overset{a}{C}H_2)_2NH$	(a) 44.5(-0.6) (b) 15.7(-3.2)
$(\overset{a}{C}H_3)_3N$	(a) 47.6(-1.2)	$(\overset{b}{C}H_3\overset{a}{C}H_2)_3N$	(a) 51.4(+1.3) (b) 12.9(-1.7)
$(\overset{a}{C}H_3)_4N^+I^-$	(a) 56.5	$(\overset{b}{C}H_3\overset{a}{C}H_2)_4N^+I^-$	(a) 54.4 (b) 9.5
$\overset{b}{C}H_3\overset{a}{C}H_2NH\overset{c}{C}H_3$	(a) 45.9(-0.4) (b) 14.3(-2.6) (c) 35.2(-1.8)	$\overset{b}{C}H_3\overset{a}{C}H_2N(\overset{c}{C}H_3)_2$	(a) 53.6(+0.5) (b) 12.8(-2.1) (c) 44.6(-1.3)

$\overset{b}{(CH_3)_2}\overset{a}{C}HNH_2$	(a) 43.0(+2.2) (b) 26.5(-4.9)	$\overset{b}{(CH_3)_3}\overset{a}{C}NH_2$	(a) 47.2(+5.7) (b) 32.9(-4.7)
$\overset{b}{(CH_3)_2}\overset{a}{C}HNH\overset{c}{CH_3}$	(a) 50.5(+1.9) (b) 22.5(-3.1) (c) 33.9(-2.5)	$\overset{b}{(CH_3)_3}\overset{a}{C}NH\overset{c}{CH_3}$	(a) 50.4(+6.6) (b) 28.2(-1.2) (c) 28.5(-2.7)
$\overset{b}{(CH_3)_2}\overset{a}{C}HN\overset{c}{(CH_3)_2}$	(a) 55.5(+3.8) (b) 18.7(-1.3) (c) 40.9(-0.8)	$\overset{b}{(CH_3)_3}\overset{a}{C}N\overset{c}{(CH_3)_2}$	(a) 53.6(+8.9) (b) 25.4(-0.8) (c) 38.7(+0.2)
$\overset{c}{CH_3}\overset{b}{CH_2}\overset{a}{CH_2}NH_2$	(a) 44.6(-1.8) (b) 27.4(-5.4) (c) 11.5(-0.4)	$(\overset{c}{CH_3}\overset{b}{CH_2}\overset{a}{CH_2})_2NH$	(a) 52.4(-1.4) (b) 24.0(-2.6) (c) 12.0(-0.5)
$(\overset{c}{CH_3}\overset{b}{CH_2}\overset{a}{CH_2})_3N$	(a) 56.8 (b) 21.3 (c) 12.0	$(\overset{c}{CH_3}\overset{b}{CH_2}\overset{a}{CH_2})_4N^+I^-$	(a) 60.4 (b) 16.0 (c) 10.9
$\overset{c}{CH_3}\overset{b}{CH_2}\overset{a}{CH_2}NH\overset{d}{CH_3}$	(a) 54.0(-2.1) (b) 23.2(-2.9) (c) 12.5(-0.9) (d) 36.1(-2.0)	$\overset{c}{CH_3}\overset{b}{CH_2}\overset{a}{CH_2}N\overset{d}{(CH_3)_2}$	(a) 61.8(-1.6) (b) 20.6(-2.0) (c) 11.9(-0.8) (d) 45.2(-1.2)

Cyclohexylamine (NH_2 at a; ring carbons b, c, d): (a) 51.1(+0.7) (b) 37.7(-5.5) (c) 25.8(-1.2) (d) 26.5(-1.3)

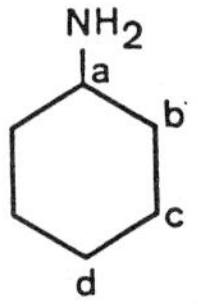

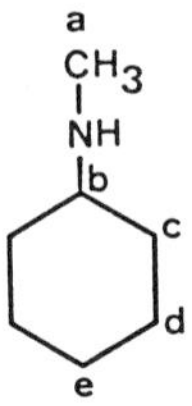

N-Methylcyclohexylamine (CH_3 a, NH, ring carbons b–e): (a) 33.5(-1.5) (b) 58.7(+0.6) (c) 32.7(-2.7) (d) 25.7(-0.3) (e) 26.8(-0.7)

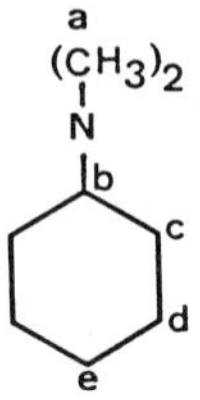

N,N-Dimethylcyclohexylamine ($(CH_3)_2$ a, N, ring carbons b–e): (a) 41.1(-0.7) (b) 64.3(+2.4) (c) 29.2(-1.6) (d) 26.5(-0.9) (e) 26.9(-1.2)

NH_2–CH_2 44.8
CH 139.9
$=CH_2$ 113.6

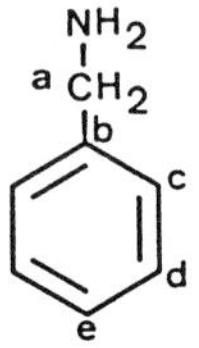

Benzylamine (NH_2, CH_2 a, ring carbons b–e): (a) 46.3 (b) 143.4 (c) 127.1 (d) 128.3 (e) 126.5

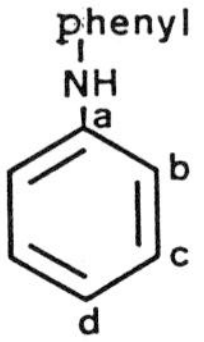

Diphenylamine (phenyl–NH, ring carbons a–d): (a) 143.2 (b) 117.9 (c) 129.4 (d) 118.0

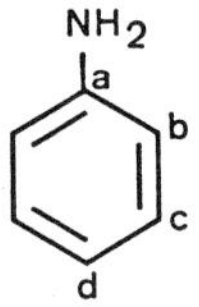

Aniline (NH_2, ring carbons a–d): (a) 146.7 (b) 115.1 (c) 129.3 (d) 118.5

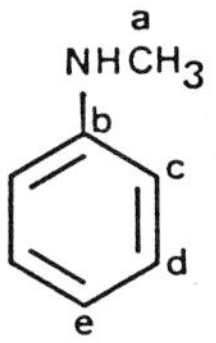

N-Methylaniline ($NHCH_3$ a, ring carbons b–e): (a) 30.2 (b) 149.9 (c) 112.3 (d) 129.3 (e) 116.9

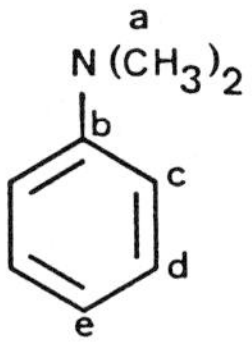

N,N-Dimethylaniline ($N(CH_3)_2$ a, ring carbons b–e): (a) 39.9 (b) 151.0 (c) 113.1 (d) 129.4 (e) 117.0

$NH_2CH_2CH_2OH$ (a, b): (a) 44.6(-1.9) (b) 64.2(-5.4)

$N(CH_2CH_2OH)_3$ (a, b): (a) 57.4(-1.0) (b) 60.3(-3.5)

$CH_3NHCH_2CH_2NHCH_3$ (a, b): (a) 36.6(-1.3/-0.5*) (b) 51.2(-3.0/-2.7*)

* values for the doubly protonated form

^{13}C-Chemical Shifts in Cyclic Amines

(δ in ppm relative to TMS)

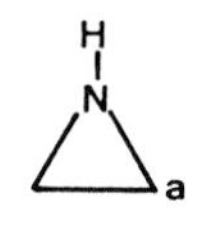

(a) 18.2

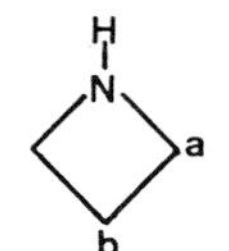

(a) 45.3
(b) 19.3

(a) 47.1
(b) 25.7

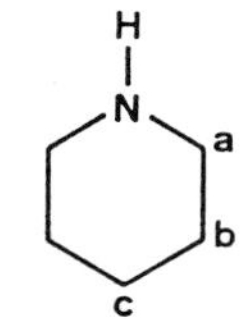

(a) 47.9
(b) 27.8
(c) 25.9

(a) 47.9

(a) 46.7
(b) 68.1

(a) 48.6
(b) 28.5

(a) 47.7
(b) 57.2
(c) 26.4
(d) 26.4

(a) 46.4
(b) 57.7
(c) 17.5

(a) 45.9
(b) 54.2
(c) 125.0
(d) 124.3
(e) 26.2
(f) 51.7

(a) 42.7
(b) 56.7
(c) 24.4

(a) 48.0
(b) 58.9
(c) 28.5
(d) 28.5

NITRO COMPOUNDS, NITROSO COMPOUNDS, NITROSAMINES

^{13}C-Chemical Shifts in Nitro and Nitroso Compounds

(δ in ppm relative to TMS)

		δ
CH_3NO_2	CH_3	61.2
$CH_3CH_2NO_2$	CH_2	70.8
	CH_3	12.3
$CH_3CH_2CH_2NO_2$	CH_2 (NO2)	77.4
	CH_2	21.2
	CH_3	10.8
$CH_3CH_2CH_2CH_2NO_2$	CH_2 (NO2)	75.6
	CH_2	29.6
	CH_2	19.8
	CH_3	13.3
$(CH_3)_2CHNO_2$	CH	78.8
	$(CH_3)_2$	20.8
$(CH_3)_3CNO_2$	C	85.2
	$(CH_3)_3$	26.9
$CH_3CH(NO_2)CH_2CH_3$	CH_3	18.7
	$CH-NO_2$	85.0
	CH_2	28.6
	CH_3	10.1
$CH_2=CHNO_2$	CH	145.6
	CH_2	122.4

Compound	(a)	(b)	(c)	(d)
Nitrocyclohexane	84.6	31.4	24.7	25.5
Nitrosobenzene	165.9	120.8	129.3	135.5
Nitrobenzene	148.4	123.6	129.4	134.6

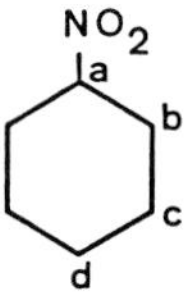

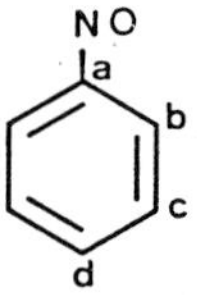

^{13}C-Chemical Shifts in Nitrosamines

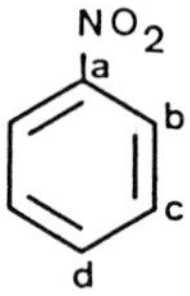

(δ in ppm relative to TMS)

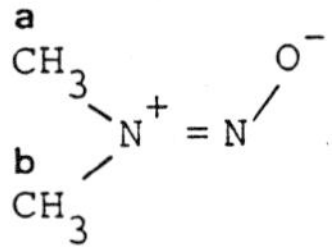

Compound	(a)	(b)	(c)	(d)	(e)	(f)
$(\overset{a}{CH_3})(\overset{b}{CH_3})N^+=N-O^-$	32.1	39.9				
$(\overset{b}{CH_3}\overset{a}{CH_2})(\overset{d}{CH_3}\overset{c}{CH_2})N^+=N-O^-$	38.4	11.5	47.0	14.5		
$(\overset{c}{CH_3}\overset{b}{CH_2}\overset{a}{CH_2})(\overset{f}{CH_3}\overset{e}{CH_2}\overset{d}{CH_2})N^+=N-O^-$	45.2	20.3	11.3	54.2	22.5	11.8
$(\overset{b}{(CH_3)_2}\overset{a}{CH})(\overset{d}{(CH_3)_2}\overset{c}{CH})N=N-O^-$	45.4	19.1	51.1	23.7		

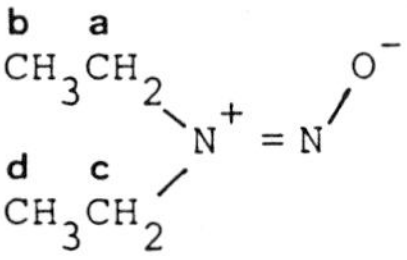

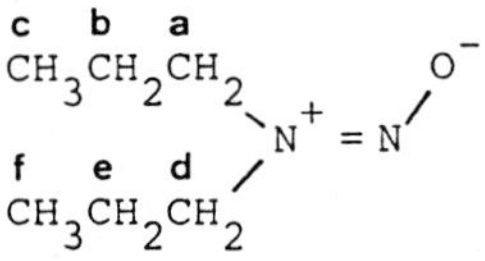

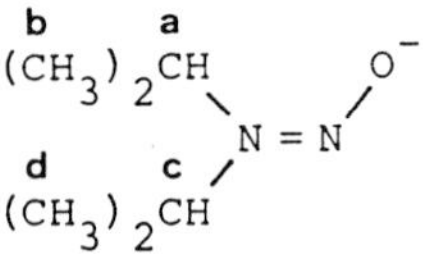

^{13}C-Chemical Shifts in Nitramines

(δ in ppm relative to TMS)

Compound	δ
CH_3NH-NO_2	32.5
$CH_3N(NO_2)_2$	41.5
$(CH_3)_2N-NO_2$	40.0

^{13}C-Chemical Shifts in Thiols

(δ in ppm relative to TMS)

Structure	δ
$SH-CH_3$	6.5

Structure	δ
$SH-CH_2-$	19.1
CH_3	19.7

Structure	δ
$SH-CH_2-$	26.4
CH_2	27.6
CH_3	12.6

Structure	δ
$SH-CH_2-$	23.7
CH_2	35.7
CH_2	21.0
CH_3	12.0

Structure	δ
$SH-CH$	29.9
$(CH_3)_2$	27.4

Structure	δ
$SH-C$	41.1
$(CH_3)_3$	35.0

Structure	δ
$SH-CH_2-$	38.8
C	31.8
$(CH_3)_3$	28.1

Structure	δ
$SH-CH_2-$	27.3
CH_2	64.2
OH	

Structure	δ
$SH-CH_2-CH_2-SH$	28.7

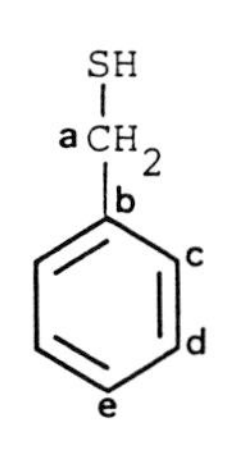

Carbon	δ
(a)	28.8
(b)	141.0
(c)	127.9
(d)	128.5
(e)	126.8

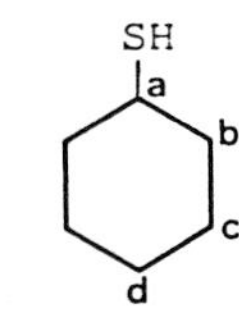

Carbon	δ
(a)	38.5
(b)	38.5
(c)	26.8
(d)	25.9

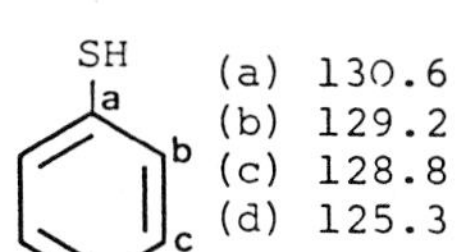

Carbon	δ
(a)	130.6
(b)	129.2
(c)	128.8
(d)	125.3

^{13}C-Chemical Shifts in Thioethers and Disulfides

(δ in ppm relative to TMS)

Compound	δ	Compound	δ
a CH_3SCH_3	(a) 19.3	b a $CH_3CH_2SCH_2CH_3$	(a) 25.5 (b) 14.8
c b a $(CH_3CH_2CH_2)_2S$	(a) 34.3 (b) 23.2 (c) 13.7	b a $(CH_3)_2CHSCH(CH_3)_2$	(a) 33.4 (b) 23.6
b a $(CH_3)_3CSC(CH_3)_3$	(a) 45.6 (b) 33.2	d c b a $(CH_3CH_2CH_2CH_2)_2S$	(a) 34.1 (b) 31.4 (c) 22.0 (d) 13.7
a b c d e $CH_3SCH_2CH_2CH_2CH_3$	(a) 15.5 (b) 34.1 (c) 31.4 (d) 22.0 (e) 13.7	c ba $((CH_3)_3CCH_2)_2S$	(a) 50.1 (b) 32.6 (c) 28.9
a $CH_3S\text{-}SCH_3$	(a) 22.0	b a $CH_3CH_2SSCH_2CH_3$	(a) 32.8 (b) 14.5
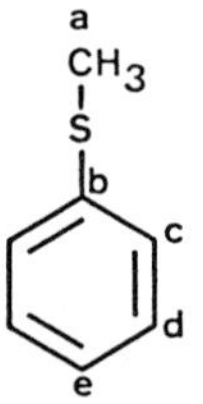 a CH_3 S b, c, d, e	(a) 15.6 (b) 138.5 (c) 126.6 (d) 128.7 (e) 124.9	a b c $(CH_3)_2C(SCH_2CH_3)_2$	(a) 30.4 (b) 54.8 (c) 23.2

^{13}C-Chemical Shifts in some Sulfonium Salts, Sulfoxides, Sulfones, Sulfonic Acids, and Sulfates (δ in ppm relative to TMS)

Compound	δ	Compound	δ	Compound	δ
$(CH_3)_3S^+I^-$	27.5	CH_3SOCH_3	40.1	$CH_3SO_2CH_3$	42.6
$\overset{a}{CH_3}SO\text{-}S\overset{b}{CH_3}$	(a) 42.7 (b) 13.7	$\overset{a}{CH_3}SO_2\text{-}S\overset{b}{CH_3}$	(a) 48.7 (b) 18.2	CH_3SO_3H	39.6
$\overset{b}{CH_3}\overset{a}{CH_2}SO_2\overset{c}{CH_3}$	(a) 48.2 (b) 6.7 (c) 39.3	$\overset{b}{CH_3}\overset{a}{CH_2}SO_2Cl$	(a) 60.2 (b) 9.1	$\overset{b}{CH_3}\overset{a}{CH_2}SO_3H$	(a) 46.7 (b) 8.0

CH_3	40.3	Cl		OH	
SO_2		SO_2		SO_2	
CH_2	56.3	CH_2	67.1	CH_2	53.7
CH_2	16.3	CH_2	18.4	CH_2	18.8
CH_3	13.0	CH_3	12.1	CH_3	13.7

CH_3	37.1	Cl		OH	
SO_2		SO_2		SO_2	
CH	53.5	CH	67.6	CH	52.9
$(CH_3)_2$	15.2	$(CH_3)_2$	17.1	$(CH_3)_2$	16.8

For aliphatic sulfinic acids and derivatives see: F. Freeman, Ch.N. Angeletakis, Org. Magn. Res. 21, 86 (1983)

SULFONES, SULFONIC ACIDS, SULFATES, CYCLIC THIOETHERS

Structure	δ	Structure	δ	Structure	δ
CH_3	34.2	Cl		OH	
SO_2		SO_2		SO_2	
C	57.6	C	74.2	C	55.9
$(CH_3)_3$	22.7	$(CH_3)_3$	24.5	$(CH_3)_3$	25.0

Structure	δ	Structure	δ	Structure	δ
CH_3	59.1	CH_3	15.4	CH_3	14.5
O		CH_2	58.3	CH_2	69.6
SO_2		O		O	
O		SO		SO_2	
CH_3		O		O	
		CH_2		CH_2	
		CH_3		CH_3	

^{13}C-Chemical Shifts in Cyclic Thioethers, Sulfoxides and Sulfones

(δ in ppm relative to TMS)

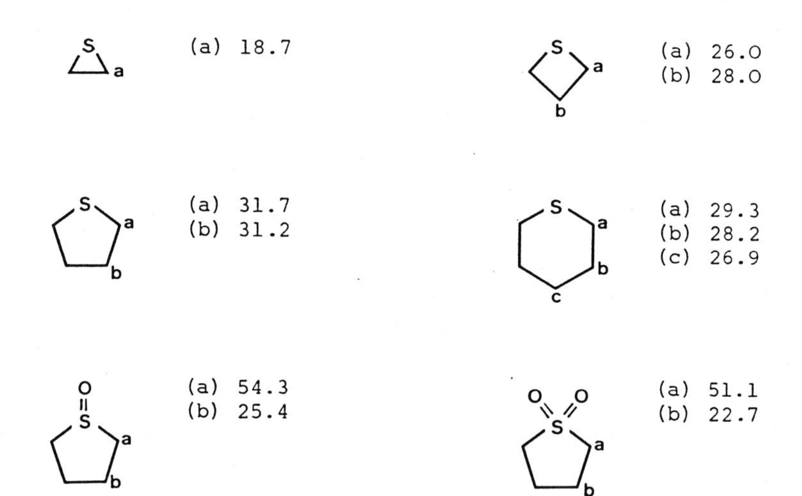

^{13}C-Chemical Shifts in Acetyl Derivatives CH_3COX

(δ in ppm relative to TMS)

Substituent X	δ_{CO}	δ_{CH_3}
-H	200.5	31.2
$-CH_3$	206.7	30.7
$-CH_2CH_3$	207.6	27.5
$-CH(CH_3)_2$	211.8	27.5
$-C(CH_3)_3$	213.5	24.5
$-CH_2CH_2CH_3$	206.6	29.3
$-CH_2CH_2CH_2CH_3$	206.8	29.4
$-nC_8H_{17}$	207.9	29.5
$-CH_2Cl$	200.1	27.2
$-CHCl_2$	193.6	22.1
$-CCl_3$	186.3	21.1
-cyclohexyl	209.4	27.6
-cyclopropyl	207.9	29.9
$-CH=CH_2$	197.5	25.7
-phenyl	196.9	25.7
-Cl	170.4	33.6
-Br	165.7	39.1
-I	158.9	
-OH	176.9	21.7
$-O^-$	182.6	24.4
$-OCH_3$	171.3	20.6
-Ophenyl	169.2	20.8
$-OCOCH_3$	167.4	21.8
$-NH_2$	173.4	22.3
$-NHCH_3$	171.7	22.7
$-N(CH_3)_2$	170.0	21.5
-NHcyclohexyl	168.1	22.7
-NHphenyl	169.5	24.1
-SH	194.5	32.6
$-SCH_3$	195.4	30.2
$-COCH_3$	197.7	23.2
$-Si(CH_3)_3$	248.8	35.7

Additivity Rules for Estimating the Chemical Shift for Carbonyl Groups

(δ in ppm relative to TMS)

Aldehydes and Ketones

$$\delta_{C=O} = 193.0 + \sum_i Z_i$$

$$-C_\beta-C_\alpha-\overset{\overset{\displaystyle O}{\|}}{C}-C_\alpha-C_\beta-$$

Substituent i	Z_α	Z_β
$-C\lessgtr$	6.5	2.6
$-CH=CH_2$	-0.8	0.0
$-CH=CH-CH_3$	0.2	0.0
-phenyl	-1.2	0.0

Carboxylic Acids and Esters

$$\delta_{C=O} = 166.0 + \sum_i Z_i$$

$$-C_\gamma-C_\beta-C_\alpha-\overset{\overset{\displaystyle O}{\|}}{C}-O-C_{\alpha'}$$

Substituent i	Z_α	Z_β	Z_γ	$Z_{\alpha'}$
$-C\lessgtr$	11.0	3.0	-1.0	-5.0
$-CH=CH_2$	5.0			-9.0
-phenyl	6 0	1.0		-8.0

Amides

$$\delta_{C=O} = 165.0 + \sum_i Z_i$$

$$C_\gamma - C_\beta - C_\alpha - \overset{O}{\overset{\|}{C}} - N \begin{matrix} C_{\alpha'} - C_{\beta'} \\ C_{\alpha'} - C_{\beta'} \end{matrix}$$

Substituent i	Z_α	Z_β	Z_γ	$Z_{\alpha'}$	$Z_{\beta'}$
-C$\lessgtr$	7.7	4.5	-0.7	-1.5	-0.3
$-CH=CH_2$	3.3				
-phenyl	4.7			-4.5	

^{13}C-Chemical Shifts for Carbonyl Groups (δ in ppm relative to TMS)

R	R-CHO	R-$COCH_3$	R-COOH	R-COO^-	R-$COOCH_3$	R-$CONH_2$	R-COOCOR	R-COCl
-H	197.0	199.7	166.3	171.3	161.6	167.6	158.5	
-CH_3	200.5	206.7	176.9	182.6	171.3	173.4	167.4	170.4
-CH_2CH_3	202.7	207.6	180.4	185.1	173.3	177.2	170.3	174.7
-$CH(CH_3)_2$	204.6	211.8	184.1		177.4		172.8	178.0
-$C(CH_3)_3$	205.6	213.5	185.9	188.6	178.8	180.9	173.9	180.3
-n-C_8H_{17}	202.6	207.9	180.7	183.1	174.4	176.3	169.4	173.8
-CH_2Cl	193.3	200.1	173.8	175.9	167.8	168.3	162.1	167.7
-$CHCl_2$		193.6	170.3	171.8	165.1		157.6	165.5
-CCl_3	176.9	186.3	167.1	167.6	162.5		154.1	
-cyclohexyl	204.7	209.4	182.1	185.4	175.3	177.3		176.3
-CH=CH_2	194.4	197.5	171.7	174.5	166.5	168.3		165.6
-C≡CH	176.8		156.5		153.4			
-phenyl	192.0	196.9	172.6	177.6	166.8	169.7	162.8	168.0

^{13}C-Chemical Shifts in Aldehydes

(δ in ppm relative to TMS)

Compound	δ
CH_2O	197.0

Compound	δ
CHO	200.5
CH_3	31.3

Compound	δ
CHO	202.7
CH_2	36.7
CH_3	5.2

Compound	δ
CHO	201.6
CH_2	45.7
CH_2	15.7
CH_3	13.3

Compound	δ
CHO	201.3
CH_2	43.6
CH_2	24.3
CH_2	22.4
CH_3	13.8

Compound	δ
CHO	204.6
CH	41.1
$(CH_3)_2$	15.5

Compound	δ
CHO	205.6
C	42.4
$(CH_3)_3$	23.4

Compound	δ
CHO	194.4
CH	138.6
CH_2 (=)	137.8

Compound	δ
CHO	176.8
C	81.8
CH (≡)	83.1

Cyclohexanecarbaldehyde (a = CHO, b = ring C-1, c, d, e):

(a)	204.7
(b)	50.1
(c)	26.1
(d)	25.2
(e)	25.2

Benzaldehyde (a = CHO, b = ring C-1, c, d, e):

(a)	192.0
(b)	136.7
(c)	129.7
(d)	129.0
(e)	134.3

Compound	δ
CHO	176.9
CCl_3	95.3

^{13}C-Chemical Shifts in Ketones

(δ in ppm relative to TMS)

CH_3	30.7
C=O	206.7
CH_3	

CH_3	27.5
C=O	207.6
CH_2	35.2
CH_3	7.0

CH_3	29.3
C=O	206.6
CH_2	45.2
CH_2	17.5
CH_3	13.5

CH_3	29.4
C=O	206.8
CH_2	43.5
CH_2	31.9
CH_2	23.8
CH_3	14.0

CH_3	27.5
C=O	211.8
CH	41.6
$(CH_3)_2$	18.2

CH_3	24.5
C=O	213.5
C	44.3
$(CH_3)_3$	26.5

CH_3	25.7
C=O	197.5
CH	137.1
‖ CH_2	128.0

CH_3	
C=O	
C	81.9
≡ CH	78.1

$\overset{}{CH_3}\overset{a\,b\,c}{CH_2COCH_2}CH_3$ (a: CO, b: CH_2, c: CH_3)

(a) 210.7
(b) 35.5
(c) 8.0

$(CH_3)_2CHCOCH(CH_3)_2$ (a: CO, b: CH, c: CH_3)

(a) 215.1
(b) 38.0
(c) 17.8

$(CH_3)_3CCOC(CH_3)_3$ (a: CO, b: C, c: CH_3)

(a) 218.0
(b) 45.6
(c) 28.6

CH_3	27.2
CO	200.1
CH_2Cl	49.4

CH_3	22.1
CO	193.6
$CHCl_2$	70.2

CH_3	21.1
CO	186.3
CCl_3	96.5

CCl_3	90.2
CO	175.5
CCl_3	

CH_3	27.0
CO	199.0
CH_2Br	35.5

CH_3	25.1
CO	203.5
CH_2F	84.9

CH_3	23.1
CO	187.5
CF_3	115.6

CH_3	23.2
CO	197.7
CO	
CH_3	

CH_3 *	28.5
CO	201.1
CH_2	56.6
CO	
CH_3	

CH_3	29.6
CO	206.9
CH_2	37.0
CH_2	
CO	
CH_3	

* for the enol form see p. C97

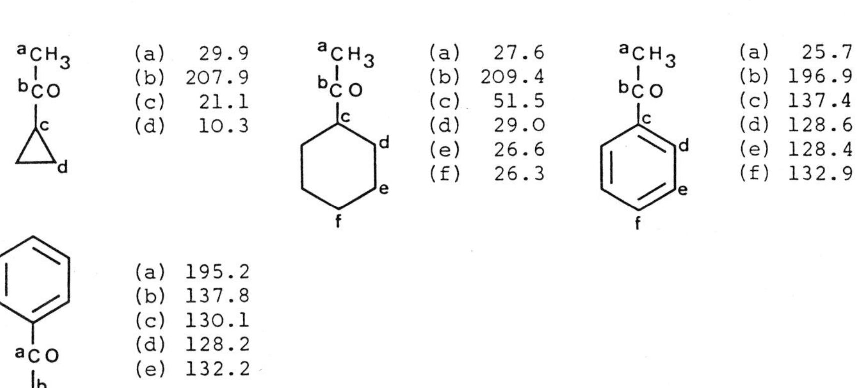

^{13}C-NMR

CYCLIC KETONES, QUINONES

^{13}C-Chemical Shifts in Cyclic Ketones and in Quinones

(δ in ppm relative to TMS)

(a) 208.9
(b) 47.8
(c) 9.9

(a) 219.1
(b) 38.2
(c) 23.4

(a) 209.7
(b) 41.5
(c) 26.6
(d) 24.6

(a) 214.9
(b) 43.9
(c) 24.5
(d) 30.6

(a) 155.1
(b) 158.3

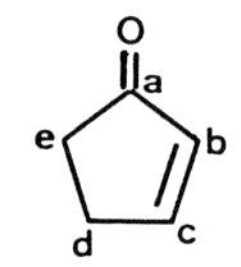

(a) 209.8
(b) 134.2
(c) 165.3
(d) 29.1
(e) 34.0

(a) 199.0
(b) 129.9
(c) 150.6
(d) 25.8
(e) 22.9
(f) 38.2

(a) 185.8
(b) 127.3
(c) 156.7
(d) 37.9
(e) 26.7

(a) 187.0
(b) 136.4

(a) 180.4
(b) 130.8
(c) 139.7

(a) 184.7
(b) 138.5
(c) 131.8
(d) 126.2
(e) 133.7

^{13}C-Chemical Shifts in Carboxylic Acids

(δ in ppm relative to TMS)

Carbon	δ
HCOOH	166.3

Carbon	δ
COOH	176.9
CH_3	21.7

Carbon	δ
COOH	180.4
CH_2	28.5
CH_3	9.6

Carbon	δ
COOH	179.4
CH_2	36.2
CH_2	18.7
CH_3	13.7

Carbon	δ
COOH	180.6
CH_2	34.8
CH_2	27.7
CH_2	22.7
CH_3	14.2

Carbon	δ
COOH	184.1
CH	34.1
$(CH_3)_2$	18.8

Carbon	δ
COOH	185.9
C	38.7
$(CH_3)_3$	27.1

Carbon	δ
COOH	171.7
CH	128.3
$=CH_2$	133.1

Carbon	δ
COOH	156.5
C	74.0
$\equiv$CH	78.6

Cyclohexanecarboxylic acid (a = COOH, b = ring C bearing COOH, c, d, e = ring positions)

Carbon	δ
(a)	182.1
(b)	43.7
(c)	29.6
(d)	26.2
(e)	26.6

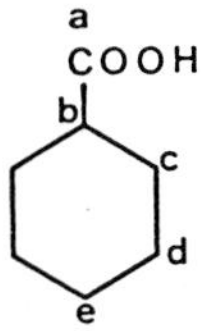

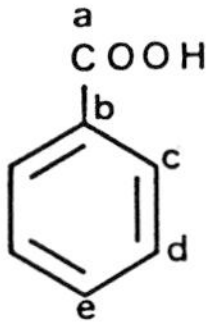

Benzoic acid (a = COOH, b = ring C bearing COOH, c, d, e = ring positions)

Carbon	δ
(a)	172.6
(b)	130.6
(c)	130.1
(d)	128.4
(e)	133.7

Carbon	δ
COOH	163.0
CF_3	115.0

Carbon	δ
COOH	173.7
CH_2Cl	40.7

Carbon	δ
COOH	170.4
$CHCl_2$	63.7

Carbon	δ
COOH	167.1
CCl_3	88.9

Dicarboxylic acids

Carbon	δ
COOH	160.1
COOH	

Carbon	δ
COOH	169.2
CH_2	40.9
COOH	

Carbon	δ
COOH	173.9
CH_2	28.9
CH_2	
COOH	

	cis	trans
COOH	166.1	166.6
CH	130.4	134.2
=CH		
COOH		

^{13}C-Chemical Shifts in Carboxylate Anions

(δ in ppm relative to TMS; measured in water unless indicated otherwise)

Compound	Carbon	δ	δ
$HCOO^-$	$HCOO^-$	171.3	
CH_3COO^-	COO^-	182.6	177.6[a]
	CH_3	24.4	20.8[a]
$CH_3CH_2COO^-$	COO^-	185.1	181.3[b]
	CH_2	31.5	28.4[b]
	CH_3	11.1	10.6[b]
$(CH_3)_3CCOO^-$	COO^-	188.6	
$CH_2=CHCOO^-$	COO^-	174.5	
	CH	134.3	
	CH_2	126.7	
CH_2ClCOO^-	COO^-	175.9	
	CH_2Cl	45.0	
$CHCl_2COO^-$	COO^-	171.8	
	$CHCl_2$	65.6	
CCl_3COO^-	COO^-	167.6	
	CCl_3	96.2	
Cyclohexanecarboxylate	(a) COO^-	185.4	
	(b)	47.2	
	(c)	30.9	
	(d)	26.9	
	(e)	26.9	
Benzoate	(a) COO^-	177.6	
	(b)	138.2	
	(c)	133.1	
	(d)	130.7	
	(e)	133.1	

[a] solvent: $CDCl_3$

[b] solvent: $CDCl_3/CD_3SOCD_3$

^{13}C-Chemical Shifts in Esters of Carboxylic Acids

(δ in ppm relative to TMS)

Methyl Esters

OCH_3	49.1
CO	161.6
H	

OCH_3	51.5
CO	171.3
CH_3	20.6

OCH_3	51.0
CO	173.3
CH_2	27.2
CH_3	9.2

OCH_3	51.9
CO	172.2
CH_2	35.6
CH_2	18.9
CH_3	13.8

OCH_3	51.6
CO	173.4
CH_2	34.9
CH_2	28.5
CH_2	23.9
CH_3	15.1

OCH_3	51.5
CO	177.4
CH	34.1
$(CH_3)_2$	19.1

OCH_3	51.5
CO	178.8
C	38.7
$(CH_3)_3$	27.3

OCH_3	51.5
CO	166.5
CH	128.8
CH_2 (=)	130.4

OCH_3	
CO	153.4
C	74.8
CH (≡)	75.6

Methyl cyclohexanecarboxylate ($^{a}OCH_3$, ^{b}CO, ring carbons c–f):

(a)	51.2
(b)	175.3
(c)	43.4
(d)	29.6
(e)	26.0
(f)	26.4

Methyl benzoate ($^{a}OCH_3$, ^{b}CO, ring carbons c–f):

(a)	51.8
(b)	166.8
(c)	130.5
(d)	129.7
(e)	128.4
(f)	132.8

OCH_3	53.0
CO	167.8
CH_2Cl	40.7

OCH_3	54.2
CO	165.1
$CHCl_2$	64.1

OCH_3	55.7
CO	162.5
CCl_3	89.6

^{13}C-Chemical Shifts in Esters of Carboxylic Acids

(δ in ppm relative to TMS)

Acetates

CH_3	20.6
CO	171.3
O	
CH_3	51.5

CH_3	20.9
CO	170.7
O	
CH_2	60.4
CH_3	14.4

CH_3	21.3
CO	170.3
O	
CH	67.5
$(CH_3)_2$	21.9

CH_3	22.3
CO	170.2
O	
C	79.9
$(CH_3)_3$	28.1

CH_3	20.5
CO	167.9
O	
CH	141.7
‖	
CH_2	96.6

Cyclohexyl acetate: $^aCH_3$–^bCO–O–^{c}(cyclohexyl ring, positions d, e, f)

(a)	21.0
(b)	169.2
(c)	72.3
(d)	32.2
(e)	24.4
(f)	26.1

Phenyl acetate: $^aCH_3$–^bCO–O–^{c}(benzene ring, positions d, e, f)

(a)	20.8
(b)	169.2
(c)	150.9
(d)	121.4
(e)	128.9
(f)	125.3

Esters of Dicarboxylic Acids

OCH_3	53.1
CO	158.4
CO	
OCH_3	

OCH_3	52.3
CO	167.6
CH_2	41.2
CO	
OCH_3	

OCH_3	51.3
CO	173.1
CH_2	29.1
CH_2	
CO	
CH_3	

	cis	trans
OCH_3	52.1	52.2
CO	165.8	165.3
CH	130.1	133.5
‖		
CH		
CO		
OCH_3		

OCH_3	53.6
CO	152.3
C	74.6
≡	
C	
CO	
OCH_3	

^{13}C-Chemical Shifts in Lactones

(δ in ppm relative to TMS)

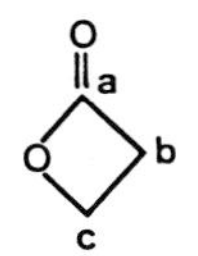

(a) 168.6
(b) 39.1
(c) 58.7

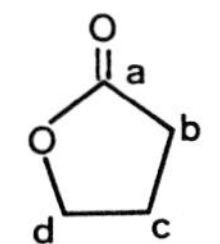

(a) 178.1
(b) 27.8
(c) 22.3
(d) 68.8

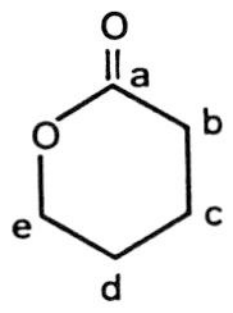

(a) 171.2
(b) 29.2
(c) 19.1
(d) 22.3
(e) 69.3

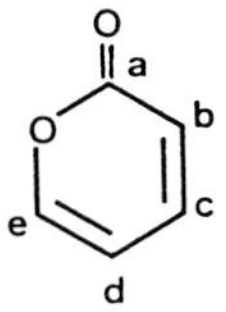

(a) 161.6
(b) 117.0
(c) 142.9
(d) 106.0
(e) 152.1

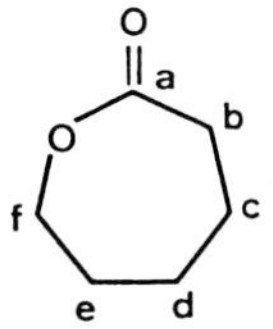

(a) 176.0
(b) 34.6
(c) 23.1*
(d) 29.5*
(e) 28.9*
(f) 69.2

*assignment uncertain

AMIDES

^{13}C-Chemical Shifts in Amides

(δ in ppm relative to TMS)

Compound	Carbon	δ
H–C(a)(=O)–NH_2	(a)	167.6
H–C(a)(=O)–NH–CH_3(b) (cis)	(a)	165.5
	(b)	25.4
H–C(a)(=O)–NH–CH_3(b) (trans)	(a)	168.7
	(b)	29.0
H–C(a)(=O)–N(CH_3(b))(CH_3(c))	(a)	162.6
	(b)	31.5
	(c)	36.5
H–C(a)(=O)–N(CH_2(b)CH_3(c))(CH_2(d)CH_3(e))	(a)	162.2
	(b)	36.7
	(c)	12.8
	(d)	41.9
	(e)	14.9

CH_3CONH_2	CH_3	22.3
	CO	173.4*
$CH_3CONHCH_3$	CH_3	22.7
	CO	171.7
	N–CH_3	26.1
$CH_3CONHCH_2CH_3$	CH_3	22.8
	CO	171.0
	CH_2	34.4
	CH_3	14.6
$CH_3CONHCH_2CH_2CH_3$	CH_3	22.5
	CO	169.8
	CH_2	40.7
	CH_2	22.5
	CH_3	11.1
$CH_3CONHCH(CH_3)_2$	CH_3	22.6
	CO	168.6
	CH	40.5
	$(CH_3)_2$	22.3
$CH_3CONHC(CH_3)_3$	CH_3	23.6
	CO	169.0
	C	49.9
	$(CH_3)_3$	28.6

*in water: 177.0

Compound	Carbon	δ
CH_3(a)–CO(b)–NH–C(c) cyclohexyl (d, e, f)	(a)	22.7
	(b)	168.1
	(c)	47.5
	(d)	32.6
	(e)	24.7
	(f)	25.4
CH_3(a)–CO(b)–NH–C(c) phenyl (d, e, f)	(a)	24.1
	(b)	169.5
	(c)	138.2
	(d)	120.4
	(e)	128.7
	(f)	124.1

(a) 21.5
(b) 170.0
(c) 35.0
(d) 38.0

(a) 21.4
(b) 169.6
(c) 40.0
(d) 13.1
(e) 42.9
(f) 14.2

(a) 21.5
(b) 170.1
(c) 47.4
(d) 21.0
(e) 11.4
(f) 50.6
(g) 22.2
(h) 11.2

(a) 14.0
(b) 18.9
(c) 35.1
(d) 172.2
(e) 40.1
(f) 13.2
(g) 42.0
(h) 14.4

^{13}C-Chemical Shifts in Lactams

(δ in ppm relative to TMS)

(a) 179.4
(b) 30.3
(c) 20.8
(d) 42.4

(a) 175.5
(b) 127.6
(c) 147.6
(d) 49.5

(a) 171.9
(b) 31.5
(c) 20.9
(d) 22.3
(e) 42.0

(a) 169.4
(b) 32.3
(c) 21.6
(d) 23.3
(e) 49.9
(f) 34.4

(a) 165.0
(b) 120.8
(c) 142.0
(d) 106.6
(e) 136.1

(a) 177.5
(b) 36.5
(c) 23.2
(d) 30.7*
(e) 29.9*
(f) 42.0

* assignment uncertain

^{13}C-Chemical Shifts in Carboxylic Acid Anhydrides

(δ in ppm relative to TMS)

H–CO–O–CO–H: CO 158.5

CH_3–CO–O–CO–CH_3: CH_3 21.8; CO 167.4

CH_3–CH_2–CO–O–CO–CH_2CH_3: CH_3 8.5; CH_2 27.4; CO 170.9

CH_3–CH_2–CH_2–CO–O–CO–$CH_2CH_2CH_3$: CH_3 13.4; CH_2 18.2; CH_2 37.2; CO 169.6

Succinic anhydride: (a) 172.5 (b) 28.2

Maleic anhydride: (a) 165.9 (b) 137.4

Phthalic anhydride: (a) 163.1 (b) 131.1 (c) 125.3 (d) 136.1

^{13}C-Chemical Shifts in Carboxylic Acid Halides

(δ in ppm relative to TMS)

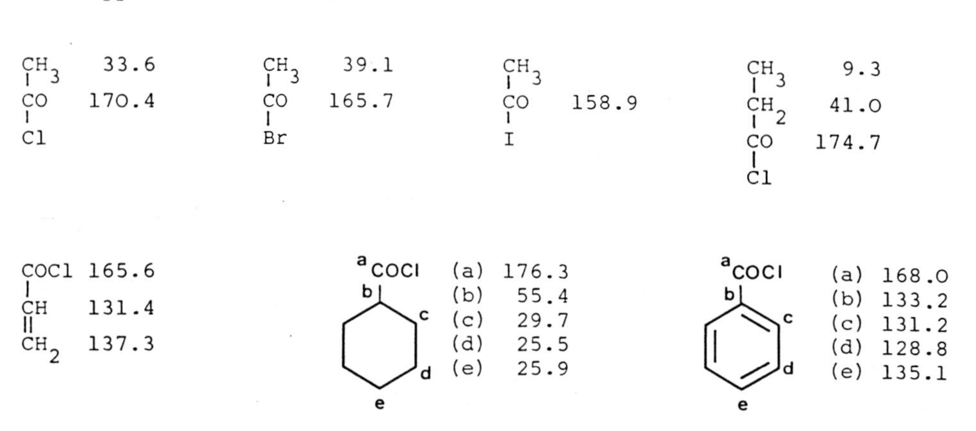

^{13}C-Chemical Shifts in Carbonic Acid Derivatives

(δ in ppm relative to TMS)

CH_3	54.9
O	
CO	156.5
O	
CH_3	

CH_3	20.2
S	
CS	226.2
S	
CH_3	

CH_3	13.6
CH_2	19.1
CH_2	30.9
CH_2	67.3
O	
CO	155.9
OC_4H_9	

CH_3	14.7
CH_2	60.7
O	
CO	157.8
NH	
CH_3	27.4

CO_2 124.2 CO_3^{--} 168.5 CS_2 192.8

NH_2	
CO	163.5
NH_2	

$(CH_3)_2N-CO-N(CH_3)_2$: CH_3 38.5; CO 165.4

$(CH_3)_2N-CS-N(CH_3)_2$: CH_3 43.0; CS 193.9

Ethylene carbonate (1,3-dioxolan-2-one): (a) 156.7, (b) 65.1

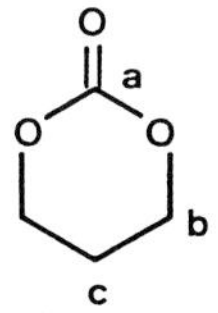

Trimethylene carbonate (1,3-dioxan-2-one): (a) 148.7, (b) 68.1, (c) 21.7

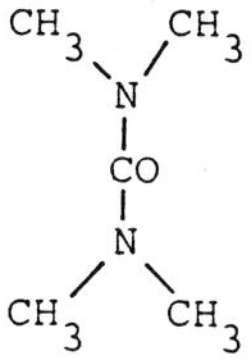

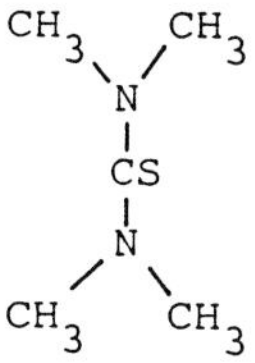

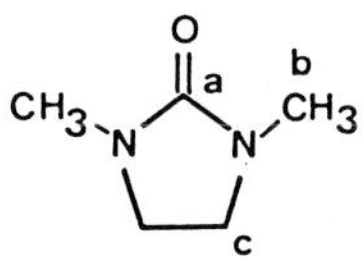

1,3-Dimethyl-2-imidazolidinone: (a) 161.3, (b) 31.2, (c) 45.0

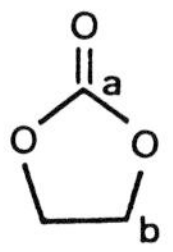

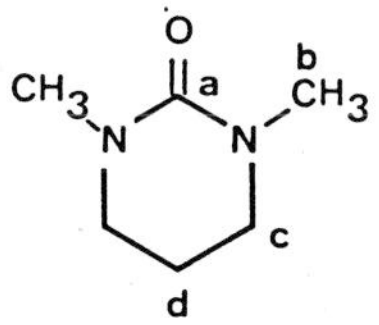

1,3-Dimethyl-tetrahydro-2-pyrimidinone: (a) 156.7, (b) 35.6, (c) 48.1, (d) 22.5

NITRILES, ISONITRILES

^{13}C-Chemical Shifts in Nitriles

(δ in ppm relative to TMS)

Compound	Carbon	δ
CH_3-CN	CN	117.4
	CH_3	1.7
CH_3-CH_2-CN	CN	120.8
	CH_2	10.8
	CH_3	10.6
$(CH_3)_2CH-CN$	CN	123.7
	CH	19.8
	$(CH_3)_2$	19.9
$(CH_3)_3C-CN$	CN	125.1
	C	28.1
	$(CH_3)_3$	28.5
$CH_2=CH-CN$	CN	117.2
	CH	107.8
	CH_2	137.5
Cyclohexyl-CN	(a)	122.4
	(b)	28.3
	(c)	30.1
	(d)	24.6
	(e)	25.8
Phenyl-CN	(a)	118.7
	(b)	112.5
	(c)	132.0
	(d)	129.2
	(e)	132.8
$NC-CH_2-CN$ (a b)	(a)	110.5
	(b)	8.6
$NC-CH_2-CH_2-CN$ (a b)	(a)	118.0
	(b)	14.6

^{13}C-Chemical Shifts and ^{13}C-^{14}N-Couplings in Isonitriles

(δ in ppm relative to TMS)

Because of the symmetrical electron distribution around the nitrogen atom the ^{13}C-^{14}N-coupling can be observed in the ^{13}C-NMR spectra of isonitriles.

Compound	Carbon	δ	J_{CN}
$C^-\equiv N^+-CH_3$	C^-	158.2	5.8 Hz
	CH_3	26.8	7.5 Hz
$C^-\equiv N^+-CH_2-CH_3$	C^-	156.8	5.3 Hz
	CH_2	36.4	6.5 Hz
	CH_3	15.3	∿0 Hz
$C^-\equiv N^+-CH=CH_2$	C^-	165.7	5.0 Hz
	CH	119.4	11.7 Hz
	CH_2	120.6	∿0 Hz
$C^-\equiv N^+$-Phenyl	(a)	165.7	5.2 Hz
	(b)	126.7	13.2 Hz
	(c)	126.3	∿0 Hz
	(d)	129.9	∿0 Hz
	(e)	129.4	∿0 Hz

^{13}C-Chemical Shifts in Imines (δ in ppm relative to TMS)

$H(C_bH\text{-}...)$: H–bC(aCH$_3$)=N–cC(dCH$_3$)$_3$

(a)	22.6
(b)	154.2
(c)	56.6
(d)	29.7

(aCH$_3$)(bCH$_3$)cC=N–dCH(eCH$_3$)$_2$

(a)	17.8
(b)	29.3
(c)	163.4
(d)	50.6
(e)	23.6

^{13}C-Chemical Shifts in Oximes (δ in ppm relative to TMS)

H–aC(=N–OH)–bCH$_2$–cCH$_2$–dCH$_3$

(a)	152.3
(b)	31.5
(c)	20.2
(d)	13.9

H–aC(=N\OH)–bCH$_2$–cCH$_2$–dCH$_3$

(a)	151.9
(b)	27.1
(c)	19.6
(d)	13.6

HO–N=aC(cCH$_3$)(CH$_3$b)

(a)	155.5
(b)	20.9
(c)	14.3

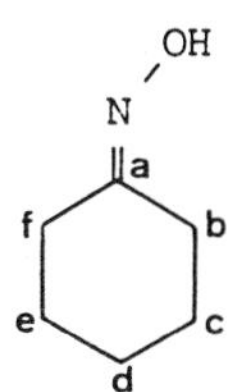

(a)	159.4
(b)	27.5
(c)	26.1
(d)	24.6
(e)	26.3
(f)	32.3

^{13}C-Chemical Shifts in Isocyanates (δ in ppm relative to TMS)

NCO	121.5
CH_3	26.3

NCO	125.0 (broad)
CH_2	43.3
CH_2	34.2
CH_2	20.4
CH_3	13.6

^{13}C-Chemical Shifts in Hydrazones and Carboiimides

(δ in ppm relative to TMS)

a CH_3 \ b C = N \ N(d CH_3)(d CH_3); c CH_3 /

(a)	18.0
(b)	164.6
(c)	25.1
(d)	47.1

a,b $(CH_3)_3C$ \ c C = N \ NH_2; e,d $(CH_3)_3C$ /

(a)	30.6
(b)	40.9
(c)	159.6
(d)	37.2
(e)	29.3

a CH_3 \ b C = N \ N(f CH_3)(f CH_3); c CH_2 – d CH_2 – e CH_3 /

(a)	22.6
(b)	167.2
(c)	33.1
(d)	19.7
(e)	14.2
(f)	47.0

a CH_3 \ b C = N / N(f CH_3)(f CH_3); c CH_2 – d CH_2 – e CH_3 /

(a)	16.2
(b)	165.5
(c)	40.5
(d)	20.1
(e)	13.7
(f)	46.5

$(CH_3)_2$	24.8
CH	49.0
N	
C	140.2
N	
CH	
$(CH_3)_2$	

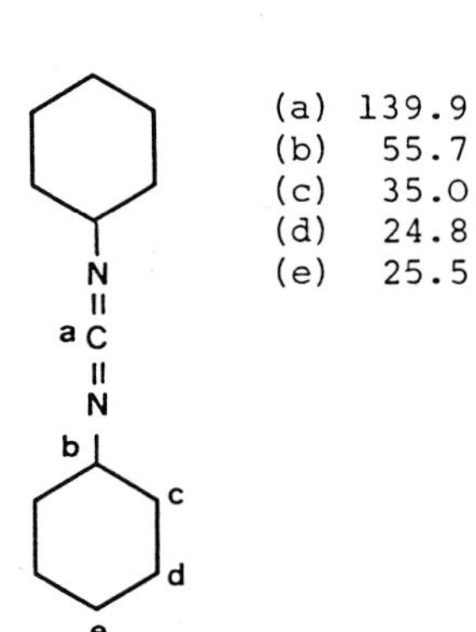

(a)	139.9
(b)	55.7
(c)	35.0
(d)	24.8
(e)	25.5

^{13}C-Chemical Shifts in Sulfur-containing Carbonyl Derivatives

(δ in ppm relative to TMS)

Thiocarbonyl groups cause an increase in ^{13}C-chemical shift by ∿30 ppm vs. the corresponding carbonyl groups ($\delta_{C=S} = 1.5 \cdot \delta_{C=O} - 57.5$). Carbonyl groups of thiolic acids and their esters are more strongly deshielded by ca. 20 ppm compared to the corresponding carboxylic acids and esters.

CH_3	32.6
CO	194.5
SH	

CH_3	30.2
CO	195.4
S	
CH_3	11.3

CH_3	39.2
CS	234.1
S	
CH_3	20.6

CH_3	30.1
CO	194.1
S	
CH_2	28.4
CH_2	32.1
CH_2	22.2
CH_3	13.6

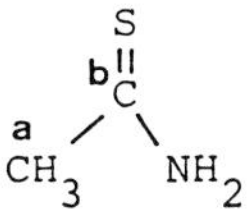

Thioacetamide, CH_3(a)–C(b)(=S)–NH_2:

(a) 33.3
(b) 205.6

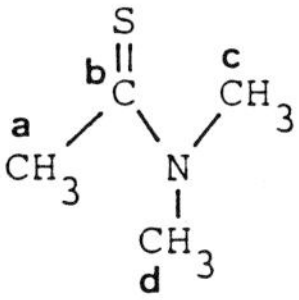

N,N-Dimethylthioacetamide, CH_3(a)–C(b)(=S)–N(CH_3(c))(CH_3(d)):

(a) 32.7
(b) 199.4
(c) 44.3*
(d) 42.3*

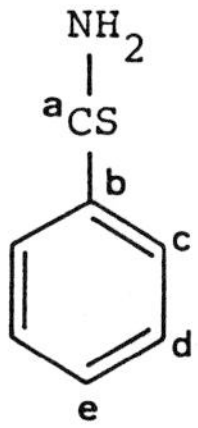

Thiobenzamide, C_6H_5–C(a)S–NH_2 (ring carbons b ipso, c ortho, d meta, e para):

(a) 202.1
(b) 140.1
(c) 128.1
(d) 128.8
(e) 132.1

* assignment uncertain

^{13}C-Chemical Shifts in Thiocyanates and Isothiocyanates

(δ in ppm relative to TMS)

SCN	111.8	SCN^-	133.3	NCS	128.7	NCS	131 (broad)
|				|		|	
CH_2	28.7			CH_3	29.3	CH_2	45.0
|						|	
CH_3	15.4					CH_2	32.3
						|	
						CH_2	20.0
						|	
						CH_3	13.3

^{13}C-Chemical Shifts in Amino Acids

(δ in ppm relative to TMS, solvent: water)

NH_2-CH_2-COOH (b: CH_2, a: COOH)

	pH = 0.45	pH = 4.53	pH = 12.01
(a)	171.2	173.6	182.7
(b)	41.5	42.8	46.0

$NH_2CH_2CH_2CH_2COOH$ (c, b, a)

	pH = 0.49	pH = 5.03	pH = 12.56
(a)	175.7	179.4	182.7
(b)	36.6	38.0	39.3
(c)	32.2	34.8	41.6

NH_2-CH(CH_3)-COOH (c: CH_3, b: CH, a: COOH)

	pH = 0.43	pH = 4.96	pH = 12.52
(a)	174.0	177.0	185.7
(b)	50.1	51.9	52.7
(c)	16.5	17.5	21.7

NH_2-CH(CH(CH_3)$_2$)-COOH (d,e: $(CH_3)_2$, c: CH, b: CH, a: COOH)

	pH = 0.30	pH = 5.64	pH = 12.60
(a)	172.9	175.4	184.1
(b)	59.8	61.8	63.2
(c)	30.3	30.3	32.9
(d)	18.5	19.2	20.3
(e)	18.0	17.8	17.9

NH_2-CH(CH_2CH(CH_3)$_2$)-COOH (e,f: $(CH_3)_2$, d: CH, c: CH_2, b: CH, a: COOH)

	pH = 0.37	pH = 7.00	pH = 13.00
(a)	174.0	176.3	185.4
(b)	52.8	54.4	55.9
(c)	40.1	40.7	45.5
(d)	25.1	25.1	25.6
(e)	22.7	22.9	23.7
(f)	22.1	21.8	22.5

^{13}C-NMR

AMINO ACIDS

```
e CH3
   |
d CH2
   |  2 f
c CH-CH3
   |
NH2-CH-COOH
    b    a
```

	pH = 0.28	pH = 6.04	pH = 12.84
(a)	172.8	175.0	184.1
(b)	58.7	60.9	62.3
(c)	37.1	37.1	39.8
(d)	25.9	25.6	25.2
(e)	12.1	12.4	12.3
(f)	15.3	15.9	16.7

```
   OH
   |
c CH2
   |
NH2-CH-COOH
    b    a
```

	zwitterion
(a)	173.1
(b)	57.4
(c)	61.3

```
d CH3
   |
c CH-OH
   |
NH2-CH-COOH
    b    a
```

	zwitterion
(a)	173.8
(b)	61.7
(c)	67.1
(d)	20.8

```
   SH
   |
c CH2
   |
NH2-CH-COOH
    b    a
```

	HCl salt	zwitterion
(a)	172.0	173.6
(b)	57.5	57.0
(c)	27.4	25.8

```
    e
   SCH3
   |
d CH2
   |
c CH2
   |
NH2-CH-COOH
    b    a
```

	zwitterion
(a)	175.3
(b)	55.3
(c)	31.0
(d)	30.1
(e)	15.2

NH_2-CH-COOH / CH_2(c) / S / S / CH_2 / NH_2-CH(b)-COOH(a)

	$(HCl)_2$ salt
(a)	175.6
(b)	54.7
(c)	39.0

Phenyl (d, e, f, g) – c CH_2 – NH_2-CH(b)-COOH(a)

	zwitterion
(a)	175.0
(b)	57.3
(c)	37.5
(d)	(~145)
(e)	131.1
(f)	130.7
(g)	129.5

4-Hydroxyphenyl (OH at g; d, e, f, g) – c CH_2 – NH_2-CH(b)-COOH(a)

	zwitterion
(a)	175.0
(b)	57.3
(c)	37.5
(d)	(~138)
(e)	130.5
(f)	117.5
(g)	156.3

d COOH / c CH_2 / NH_2-CH(b)-COOH(a)

	pH = 0.41	pH = 6.73	pH = 12.73
(a)	172.0	175.5	183.4
(b)	50.6	53.5	55.3
(c)	35.0	37.8	44.5
(d)	174.4	178.7	181.3

e COOH / d CH_2 / c CH_2 / NH_2-CH(b)-COOH(a)

	pH = 0.32	pH = 6.95	pH = 12.51
(a)	172.6	175.8	183.9
(b)	53.4	56.0	57.2
(c)	26.1	28.2	33.0
(d)	30.7	34.7	35.3
(e)	172.4	182.4	184.0

NH_2 – e CH_2 – d CH_2 – c CH_2 – NH_2-CH(b)-COOH(a)

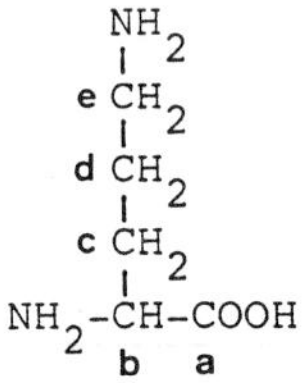

	pH = 0.46	pH = 5.02	pH = 13.53
(a)	172.8	175.3	184.6
(b)	53.7	55.5	57.2
(c)	28.1	28.7	33.3
(d)	24.0	24.0	29.4
(e)	40.2	40.3	41.9

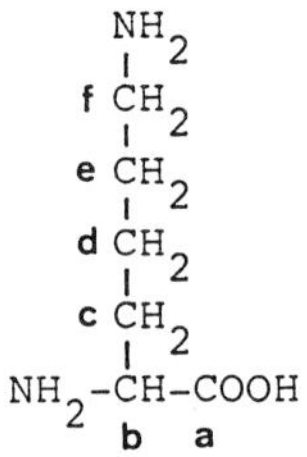

NH_2 – f CH_2 – e CH_2 – d CH_2 – c CH_2 – NH_2-CH(b)-COOH(a)

	pH = 0.50	pH = 6.03	pH = 13.85
(a)	173.2	175.8	184.8
(b)	54.0	55.9	57.3
(c)	30.5	31.2	35.7
(d)	22.6	22.6	23.6
(e)	27.6	27.7	33.0
(f)	40.5	40.5	41.8

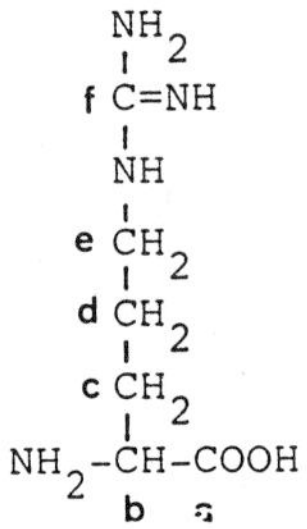

NH_2 – f C=NH – NH – e CH_2 – d CH_2 – c CH_2 – NH_2-CH(b)-COOH(a)

	zwitterion
(a)	175.2
(b)	55.6
(c)	28.8
(d)	25.2
(e)	41.7
(f)	157.8

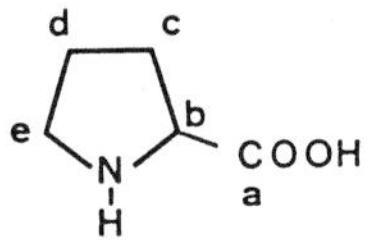

Pyrrolidine ring (N-H; e, d, c, b) – COOH (a)

	zwitterion
(a)	175.2
(b)	62.4
(c)	30.3
(d)	25.0
(e)	47.4

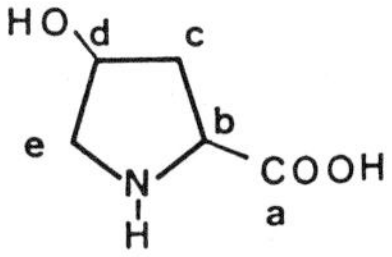

HO-pyrrolidine ring (N-H; e, HO d, c, b) – COOH (a)

	zwitterion
(a)	174.9
(b)	60.7
(c)	38.2
(d)	70.9
(e)	53.9

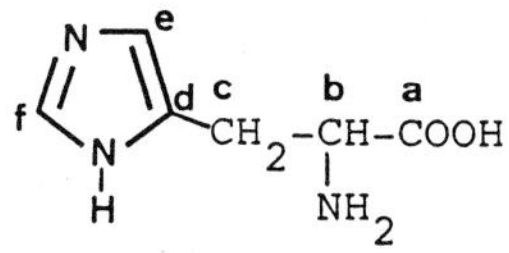

Histidine structure (imidazole ring: N, e, d, f, N-H; side chain d–CH₂(c)–CH(b)(NH₂)–COOH(a))

	after addition of HCl	zwitterion
(a)	174.0	174.9
(b)	55.1	58.8
(c)	27.6	29.0
(d)	130.0	
(e)	118.5	117.9
(f)	135.8	136.9

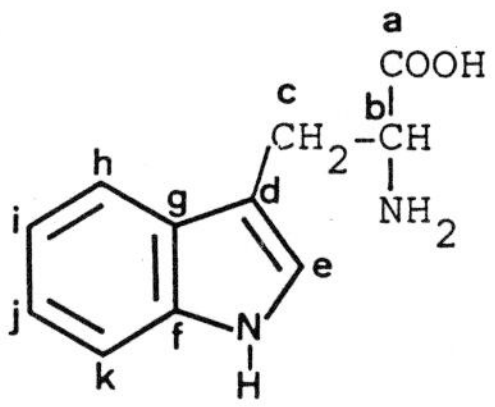

	after addition of HCl	zwitterion
(a)	174.4	175.5
(b)	55.5	56.4
(c)	27.3	27.6
(d)	108.2	108.8
(e)	126.5	126.1
(f)		127.8
(g)		137.8
(h)		123.4
(i)		119.6
(j)		120.8
(k)		113.3

^{13}C-Chemical Shifts in a few Monosaccharides

(δ in ppm relative to TMS)

D-ribofuranose

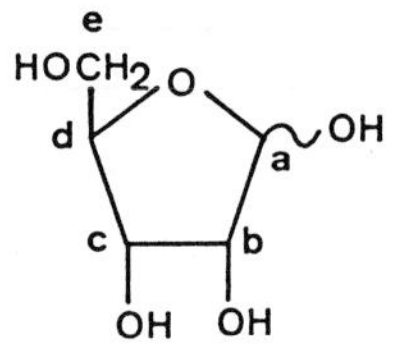

	α-anomer	β-anomer
(a)	99.9	103.6
(b)	73.7	78.0
(c)	72.8	73.2
(d)	85.8	85.2
(e)	64.1	65.2

D-ribopyranose

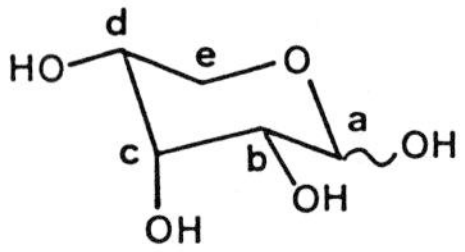

	α-anomer	β-anomer
(a)	96.2	96.6
(b)	72.8	73.8
(c)	72.0	71.7
(d)	70.0	70.0
(e)	65.7	65.7

D-glucopyranose

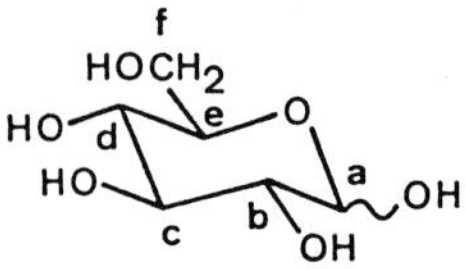

	α-anomer	β-anomer
(a)	93.3	97.1
(b)	73.1	75.6
(c)	74.4	77.3
(d)	71.2	71.2
(e)	72.9	77.3
(f)	62.4	62.4

D-fructofuranose

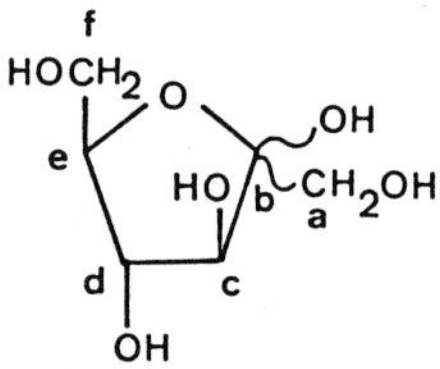

	α-anomer	β-anomer
(a)	62.1	63.9
(b)	105.3	102.4
(c)	83.0	76.5
(d)	77.0	75.5
(e)	82.2	81.5
(f)	62.1	63.3

^{13}C-Chemical Shifts in the most important Pyrimidine Bases and Nucleosides (δ in ppm relative to TMS; in CD_3SOCD_3 unless indicated otherwise)

Cytosine

(a)	171.0
(b)	93.9
(c)	158.1
(d)	166.9

(in: CD_3SOCD_3/ water 1:2

Uracil

(a)	164.4
(b)	100.3
(c)	142.2
(d)	151.5

Thymine

(a)	165.0
(b)	107.8
(c)	137.8
(d)	151.6
(e)	11.9

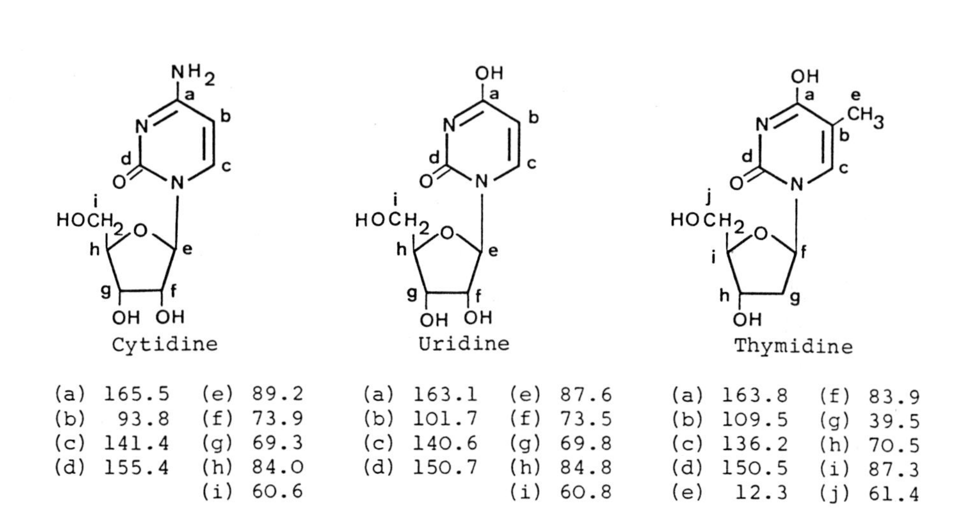

Cytidine

(a)	165.5	(e)	89.2
(b)	93.8	(f)	73.9
(c)	141.4	(g)	69.3
(d)	155.4	(h)	84.0
		(i)	60.6

Uridine

(a)	163.1	(e)	87.6
(b)	101.7	(f)	73.5
(c)	140.6	(g)	69.8
(d)	150.7	(h)	84.8
		(i)	60.8

Thymidine

(a)	163.8	(f)	83.9
(b)	109.5	(g)	39.5
(c)	136.2	(h)	70.5
(d)	150.5	(i)	87.3
(e)	12.3	(j)	61.4

PURINE BASES AND NUCLEOSIDES

^{13}C-Chemical Shifts in the most important Purine Bases and Nucleosides

(δ in ppm relative to TMS; in CD_3SOCD_3 unless indicated otherwise)

CD_3SOCD_3)

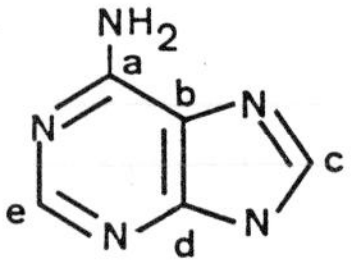

Adenine

(a)	155.4
(b)	117.5
(c)	139.3
(d)	151.3
(e)	152.4

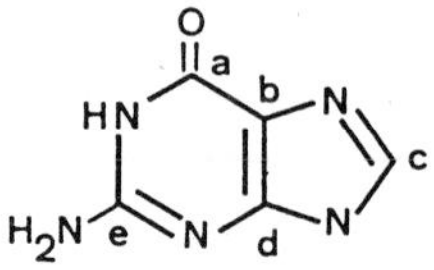

Guanine

(a)	168.8
(b)	119.6
(c)	150.1
(d)	162.2
(e)	160.0

(in water)

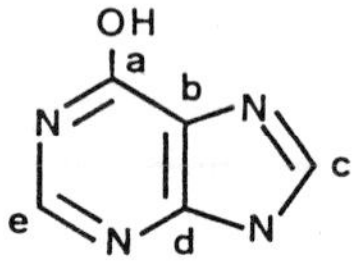

Hypoxanthine

(a)	155.4
(b)	119.2
(c)	140.2
(d)	153.2
(e)	144.6

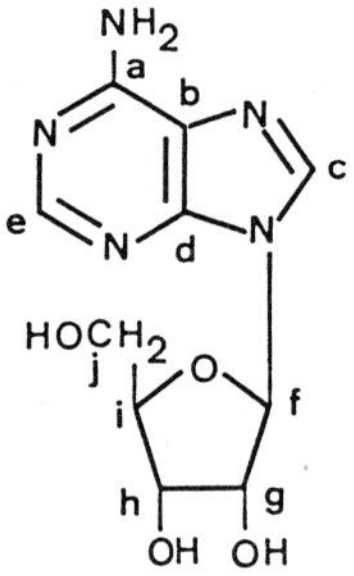

Adenosine

(a)	156.3	(f)	88.2
(b)	119.6	(g)	73.8
(c)	140.2	(h)	70.9
(d)	149.3	(i)	86.2
(e)	152.6	(j)	61.9

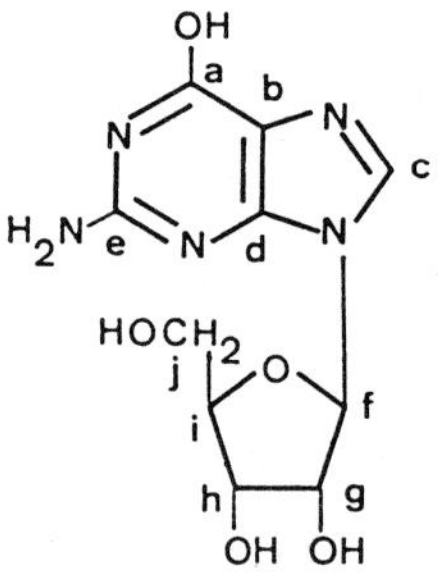

Guanosine

(a)	157.1	(f)	86.7
(b)	116.7	(g)	73.9
(c)	136.1	(h)	70.6
(d)	151.5	(i)	85.5
(e)	153.8	(j)	61.6

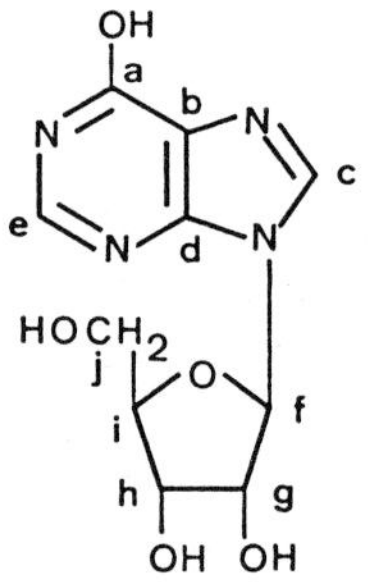

Inosine

(a)	156.9	(f)	87.8
(b)	124.6	(g)	74.4
(c)	139.1	(h)	70.6
(d)	148.5	(i)	85.9
(e)	146.2	(j)	61.6

^{13}C-Chemical Shifts (δ in ppm relative to TMS) and ^{31}P-^{13}C-Coupling Constants (|J| in Hz) in Some Phosphorous Compounds

$X-CH_2CH_2CH_2CH_3$

Substituent X	X-CH_2-		-CH_2-		-CH_2-		CH_3		P-CH_3	
	δ	|J|	δ	|J|	δ	|J|	δ	|J|	δ	|J|
-PCl_2	42.9	44	25.1	14	23.4	11	13.7	0		
-$P(CH_3)_2$	32.6	12	28.3	14	24.5	11	13.9	0	14.4	16
-$P(n-C_4H_9)_2$	28.6	14	27.9	15	24.8	10	14.0	0		
-$P(OCH_3)_2$	33.6	20	24.8	16	24.7	11	14.0	0		
-$P^+(n-C_4H_9)_3$	18.7	48	23.7	4	24.1	15	13.3	0		
-$PO(n-C_4H_9)_2$	27.8	66	24.0	5	24.4	13	13.6	0		
-$PS(CH_3)_2$	34.6	54	24.8	4	23.9	16	13.6	0	20.8	54
-$PS(n-C_4H_9)_2$	30.9	51	24.6	4	24.0	16	13.6	0		
-$OP(O-n-C_4H_9)_2$	61.9	11	33.4	5	19.1	0	13.7	0		
-$OPO(O-n-C_4H_9)_2$	67.2	6	32.6	7	18.9	0	13.6	0		

Substituent X	a		b		c		d	
	δ	|J|	δ	|J|	δ	|J|	δ	|J|
-P(phenyl)$_2$	137.2	12	133.6	20	128.4	7	128.5	0
-O-P(O-phenyl)$_2$	151.5	4	124.1	7	129.5	0	124.1	0
-O-PO(O-phenyl)$_2$	150.4	8	120.1	5	129.7	0	125.5	0

^{13}C-^{1}H-Coupling Constants

see J.L. Marshall, Carbon-Carbon and Carbon-Proton NMR Couplings, Verlag Chemie International, Deerfield Beach, Florida, 1983.

$$J_{C-H} \approx 0.62\ J_{H-H}$$

Couplings Through One Bond ($^{1}J_{C-H}$ in Hz)

CH_4 125

CH_3Cl 151

cyclopropane 160

cyclobutane 134

$CH_2=CH_2$ 156

benzene 159

cyclopentane 128

cyclohexane 125

$HC\equiv CH$ 249

furan: (a) 175, (b) 202

pyrrole: (a) 169, (b) 183

imidazole: (a) 206, (b) 189

pyrazole: (a) 186, (b) 177

2H-1,2,3-triazole: (a) 194

4H-1,2,4-triazole: (a) 209

pyridine: (a) 161, (b) 163, (c) 178

Additivity Rule for Estimating the ^{13}C-^{1}H-Coupling Constants Across a Bond in Aliphatic Compounds ($^{1}J_{C-H}$ in Hz)

$$J_{CH\ Z_1Z_2Z_3} = 125.0 + \sum_i Z_i$$

Substituent	Increments Z_i
-H	0.0
$-CH_3$	1.0
$-C(CH_3)_3$	-3.0
$-CH_2Cl$	3.0
$-CH_2Br$	3.0
$-CH_2I$	7.0
$-CHCl_2$	6.0
$-CCl_3$	9.0
-C≡CH	7.0
-phenyl	1.0
-F	24.0
-Cl	27.0
-Br	27.0
-I	26.0
-OH	18.0
-O-phenyl	18.0
$-NH_2$	8.0
$-NHCH_3$	7.0
$-N(CH_3)_2$	6.0
$-SOCH_3$	13.0
-CHO	2.0
$-COCH_3$	-1.0
-COOH	5.5
-CN	11.0

Example: Estimating the ^{13}C-^{1}H-coupling constant in $CHCl_3$

$J = 125.0 + 3 \cdot 27.0 = 206.0$ Hz (determined: 209.0 Hz)

^{13}C-NMR

^{13}C-^{1}H-COUPLING CONSTANTS

Coupling Through Two Bonds ($|^{2}J_{C-H}|$ in Hz)

Typical ranges:		Examples:	
$H-C-^{13}C-$	1 - 6	$H-CH_2-^{13}CH_3$	4.5
$H-C=^{13}C-$	0 - 16	$H-CH=^{13}CH_2$	2.4
		$H-CCl=^{13}CHCl$	16.0
$H-C\equiv^{13}C-$	40 - 60	$H-C\equiv^{13}CH$	49.3
$H-C-^{13}CO-$	5 - 8	$H-CH_2-^{13}CO-CH_3$	5.5
$H-CO-^{13}C-$	20 - 50	$H-CO-^{13}CH_3$	26.7
H–C(ring)–^{13}C (benzene ring, substituent R)	1 - 4	H–C(ring)–^{13}C (benzene)	1.0

R: any substituent

The ^{13}C-^{1}H-coupling constants through two bonds with an aldehyde or acetylene proton are so large that they can be observed as split peaks in "off resonance" decoupled spectra.

Coupling Through Three Bonds ($|^{3}J_{C-H}|$ in Hz)

$$|^{3}J_{C-H}| = 0.....10$$

These coupling constants depend on the dihedral angle in the same way as the vicinal ^{1}H-^{1}H-coupling constants (see p. H20).

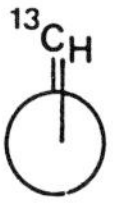

~ 6

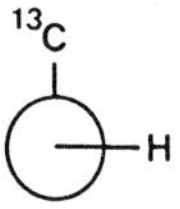

~ 0

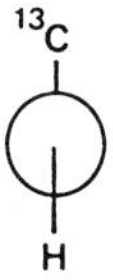

~ 9

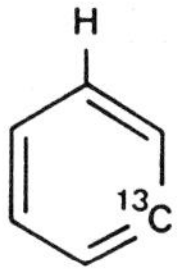

7 - 10

(in benzene: 7.6)

The trans-H-C-C-^{13}C-coupling constant in alkenes is always larger than the corresponding cis-coupling constant (see U. Vögeli, W. v. Philipsborn, Org. Magn. Res. 7, 617 (1975):

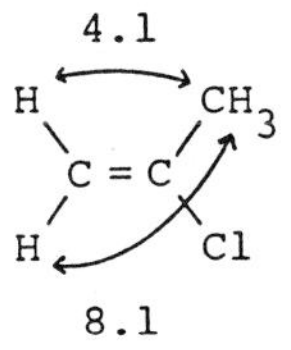

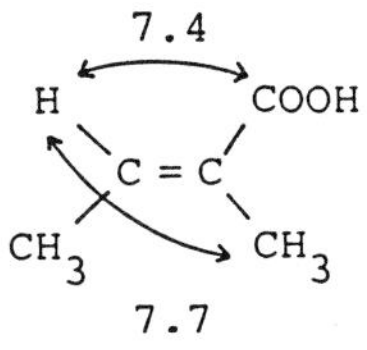

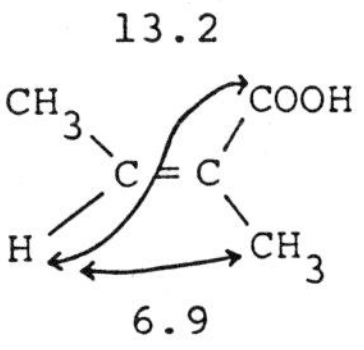

4.1 8.1 7.4 7.7 13.2 6.9

^{13}C-NMR

^{13}C-^{13}C-COUPLING CONSTANTS

^{13}C-^{13}C-Couplings Through One Bond ($|^{1}J_{C-C}|$ in Hz)

CH_3-CH_3	34.6
$CH_2=CH_2$	67.6
$\overset{a}{C}H_2=\overset{b}{C}H-\overset{c}{C}H_3$	(a,b) 70.0 (b,c) 41.9
H-C≡C-H	171.5

12.4

32.7

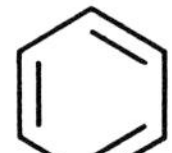

56.0

^{13}C-^{13}C-Couplings Through Two or Three Bonds ($|^{2}J_{C-C}|$ and $|^{3}J_{C-C}|$ in Hz)

$\overset{c}{C}H_3\overset{b}{C}H_2CH_2\overset{a}{C}H_2OH$	(a,b) < 1 (a,c) 4.6
$CH_3\overset{b}{C}H_2CH_2\overset{a}{C}OOH$	(a,b) 1.8
$\overset{a}{C}H_3CO\overset{b}{C}H_3$	(a,b) 16.1
$\overset{a}{C}H_3O\overset{b}{C}H_3$	(a,b) 2.4

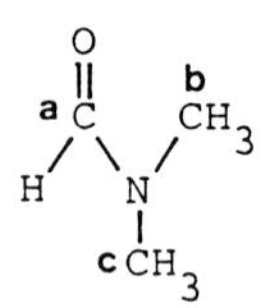

(a,b) 0.5
(a,c) 4.9

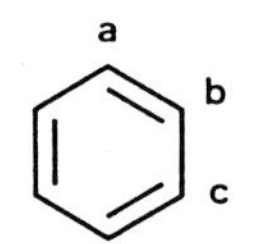

(a,b) 56.0
(a,c) 2.5
(a,d) 10.0

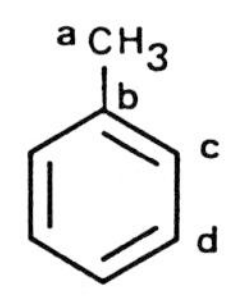

(a,c) 3.1
(a,d) 3.8
(a,e) 0.9

Coupling Between ^{13}C and ^{19}F (spin quantum number I = 1/2, natural abundance 100 %) (|J| in Hz), see also p. C169)

F–CH_2–CH_2–CH_2–CH_2–C_4H_9

	|J|
CH_2	164.8
CH_2	18.3
CH_2	6.2
CH_2	~ 0

CF_3–COOH

	|J|
CF_3	284
COOH	44

Fluorocyclohexane

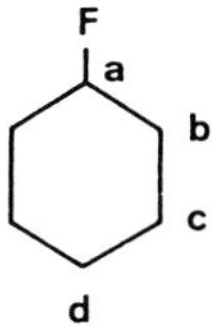

	|J|
(a)	171
(b)	19
(c)	5
(d)	0

CH_3F	161.9
CH_2F_2	234.8
CHF_3	274.3
CF_4	259.2

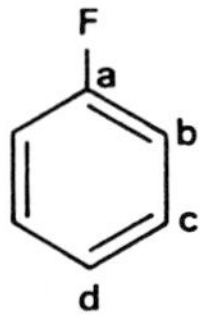

Fluorobenzene

	|J|
(a)	245.1
(b)	21.0
(c)	7.8
(d)	3.2

Benzotrifluoride (CF_3 = a)

	|J|
(a)	271.7
(b)	32.3
(c)	3.9
(d)	1.3
(e)	~ 0

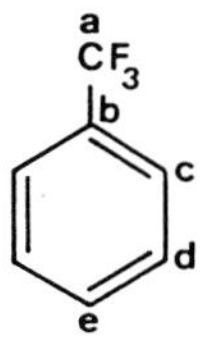

2-Fluoropyridine

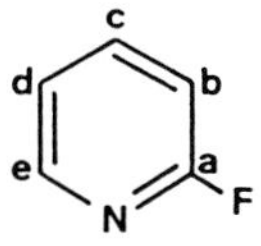

	|J|
(a)	236.3
(b)	37.6
(c)	7.5
(d)	4.2
(e)	14.9

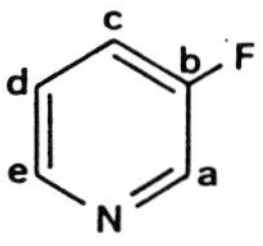

3-Fluoropyridine

	|J|
(a)	22.6
(b)	255.1
(c)	17.7
(d)	3.7
(e)	3.7

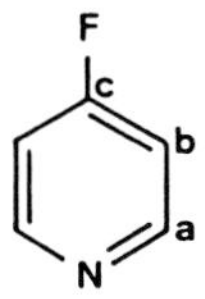

4-Fluoropyridine

	|J|
(a)	6.4
(b)	16.1
(c)	261.8

^{13}C-NMR Spectra of Common Solvents

(δ in ppm relative to TMS, idealized line intensities)

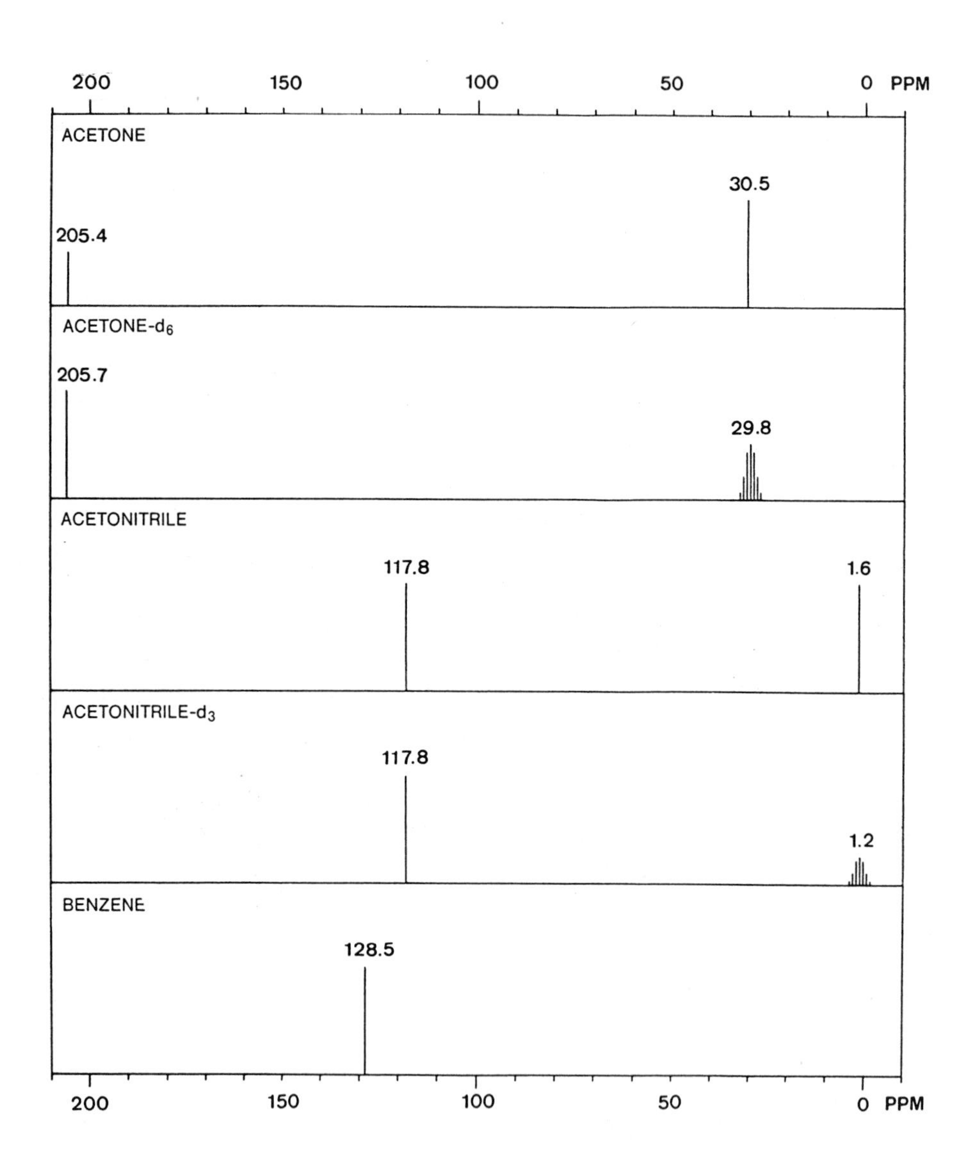

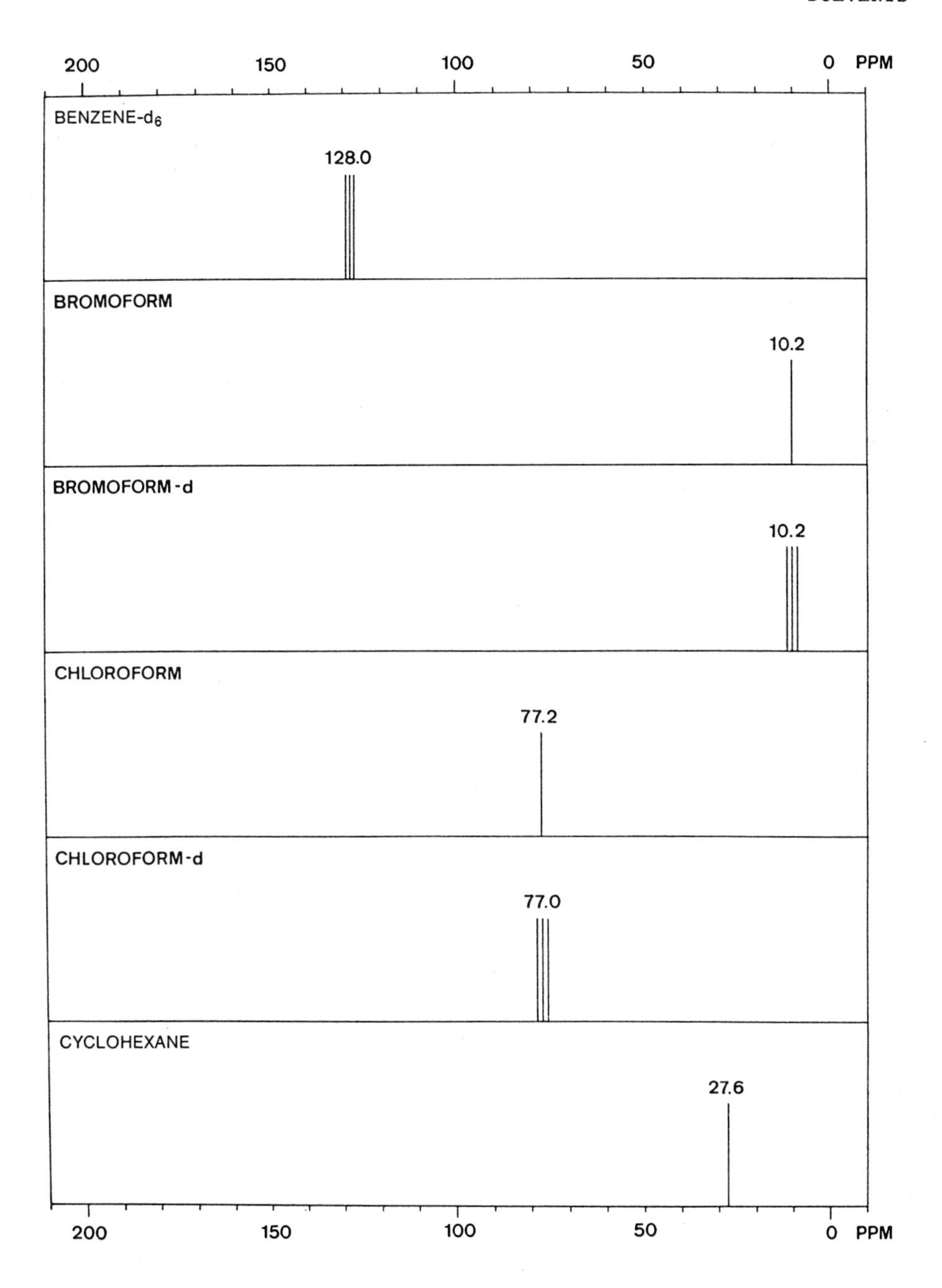
200
150
100
50
0 PPM
BENZENE-d_6
128.0
BROMOFORM
10.2
BROMOFORM-d
10.2
CHLOROFORM
77.2
CHLOROFORM-d
77.0
CYCLOHEXANE
27.6
200
150
100
50
0 PPM

^{13}C-NMR

SOLVENTS

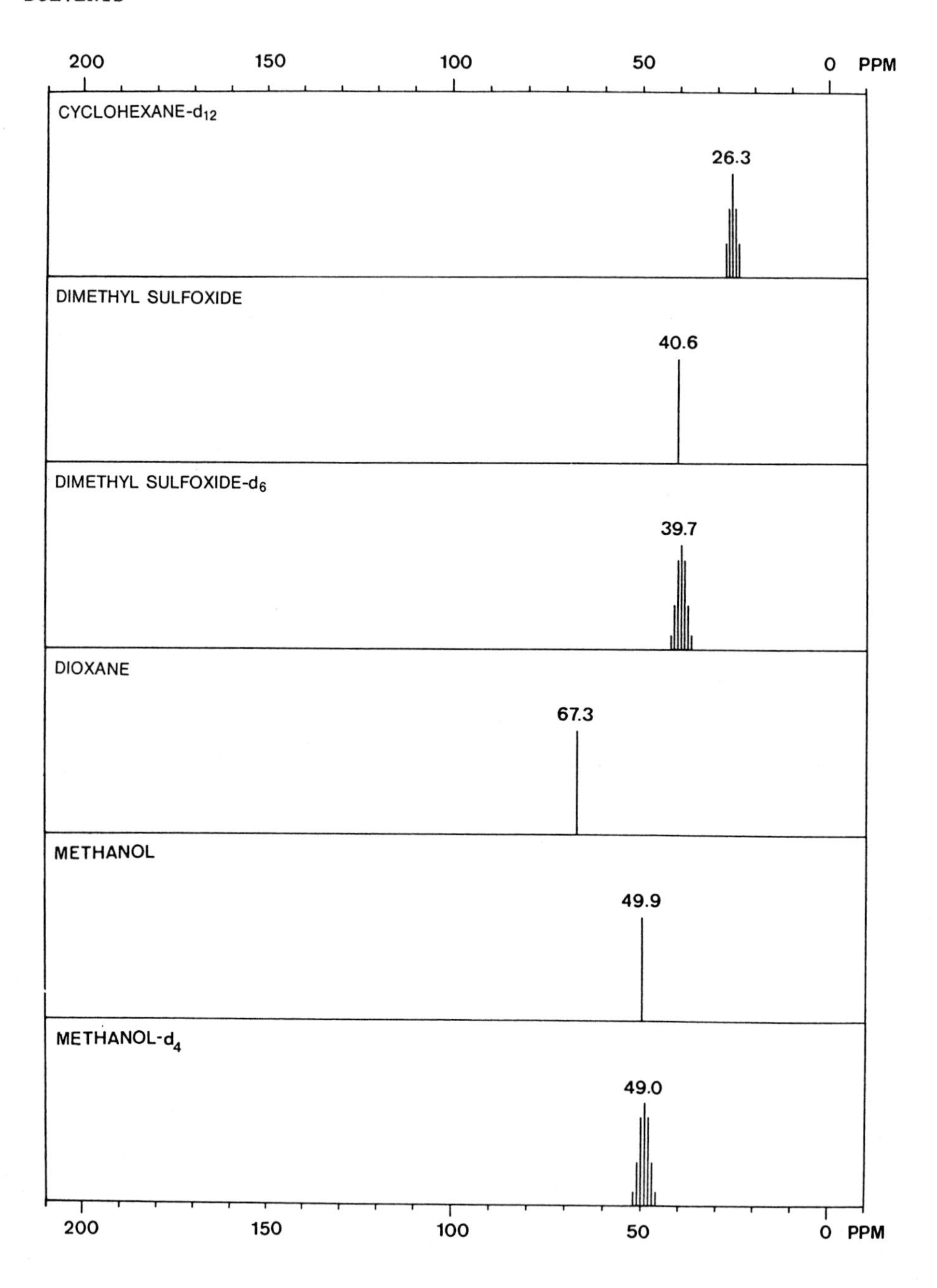

^{13}C-NMR

SOLVENTS

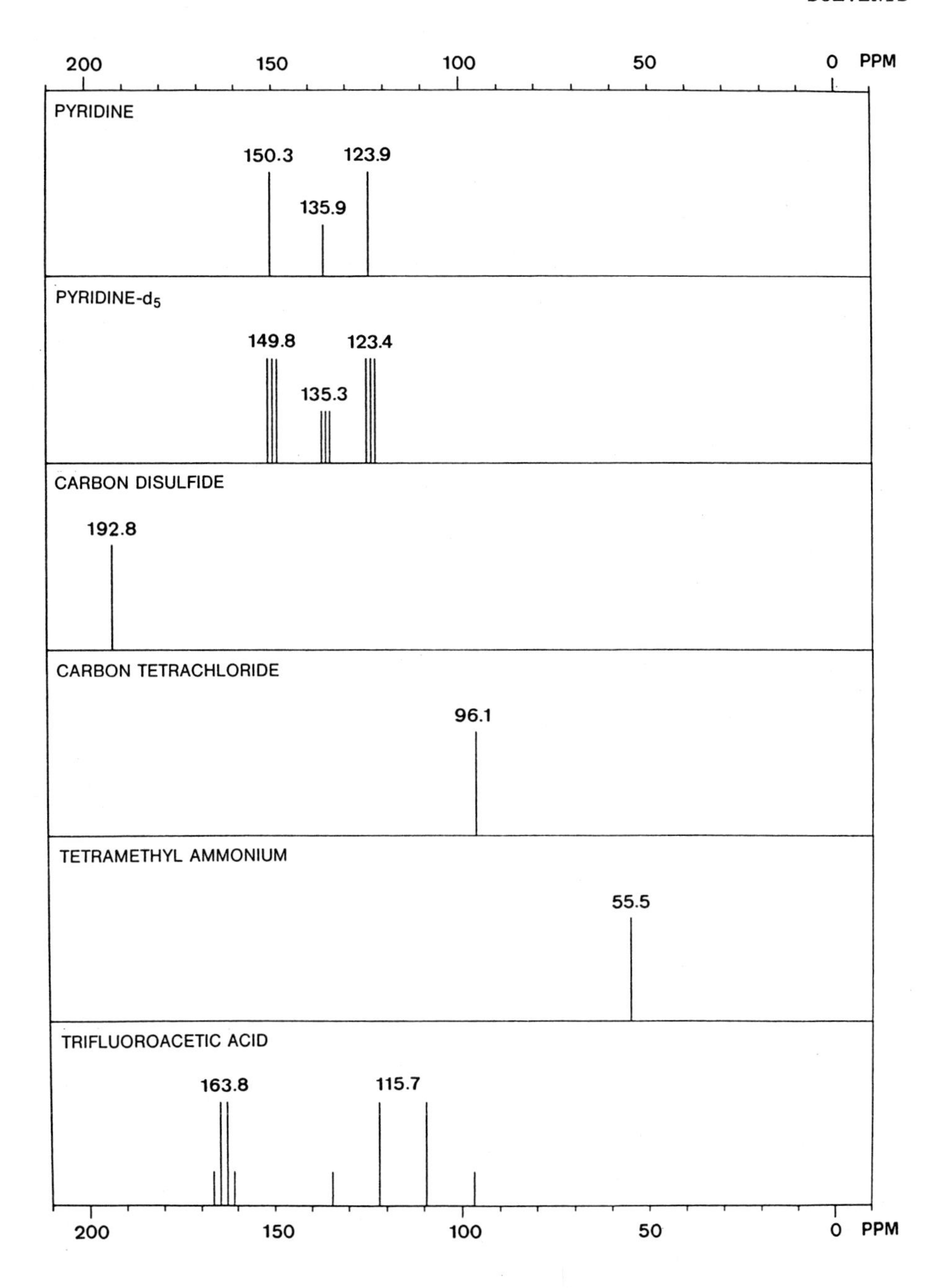

^{1}H-Chemical Shifts in Monosubstituted Alkanes

(δ in ppm relative to TMS)

	Substituent	Methyl	Ethyl		n-Propyl			Isopropyl		t-Butyl	Additional Data
		$-CH_3$	$-CH_2$	$-CH_3$	$-CH_2$	$-CH_2$	$-CH_3$	$-CH$	$-CH_3$	$-CH_3$	
	-H	0.23	0.86	0.86	0.91	1.33	0.91	1.33	0.91	0.89	H20
C	$-CH=CH_2$	1.71	2.00	1.00						1.02	H205
C	-C≡CH	1.80	2.16	1.15	2.10	1.50	0.97	2.59	1.15	1.22	H225
C	-phenyl	2.35	2.63	1.21	2.59	1.65	0.95	2.89	1.25	1.32	H30
HAL	-F [1]	4.27	4.36	1.24						1.34	H45
HAL	-Cl	3.06	3.47	1.33	3.47	1.81	1.06	4.14	1.55	1.60	H45
HAL	-Br	2.69	3.37	1.66	3.35	1.89	1.06	4.21	1.73	1.76	H45
HAL	-I	2.16	3.16	1.88	3.16	1.88	1.03	4.24	1.89	1.95	H45
O	-OH	3.39	3.59	1.18	3.49	1.53	0.93	3.94	1.16	1.22	H50
O	-O-alkyl	3.24	3.37	1.15	3.27	1.55	0.93	3.55	1.08	1.24	H60
O	$-OCH=CH_2$	3.5	3.66	1.21							H60
O	-O-phenyl	3.73	3.98	1.38	3.86	1.70	1.05	4.51	1.31		H60
O	$-OCOCH_3$	3.67	4.05	1.21	3.98	1.56	0.97	4.94	1.22	1.45	H140
O	-OCO-phenyl	3.88	4.37	1.38	4.25	1.76	1.07	5.22	1.37	1.58	H140
O	$-OSO_2$-p-toluyl	3.70	4.07	1.30	3.94	1.60	0.95	4.70	1.25		H140
N	$-NH_2$	2.47	2.74	1.10	2.61	1.43	0.93	3.07	1.03	1.15	H75
N	$-NHCOCH_3$	2.71	3.21	1.12	3.18	1.55	0.96	4.01	1.13	1.28	H150
N	$-NO_2$	4.29	4.37	1.58	4.28	2.01	1.03	4.44	1.53	1.59	H90

1) For ^{19}C-^{1}H-coupling see p. H357.

Substituent	Methyl	Ethyl		n-Propyl			Isopropyl		t-Butyl	Additional Data
	$-CH_3$	$-CH_2$	$-CH_3$	$-CH_2$	$-CH_2$	$-CH_3$	$-CH$	$-CH_3$	$-CH_3$	
-SH	2.00	2.44	1.31	2.46	1.57	1.02	3.16	1.34	1.43	H95
S -S-alkyl	2.09	2.49	1.25	2.43	1.59	0.98	2.93	1.25	1.39	
-S-S-alkyl	2.30	2.67	1.35	2.63	1.71	1.03			1.32	H110
-CHO	2.20	2.46	1.13	2.42	1.67	0.97	2.39	1.13	1.07	H120
$-COCH_3$	2.09	2.47	1.05	2.32	1.56	0.93	2.54	1.08	1.12	H125
-CO-phenyl	2.55	2.92	1.18	2.86	1.72	1.02	3.58	1.22		
C=O -COOH	2.08	2.36	1.16	2.31	1.68	1.00	2.56	1.21	1.23	H135
$-COOCH_3$	2.01	2.28	1.12	2.22	1.65	0.98	2.48	1.15	1.16	H140
$-CONH_2$	2.02	2.23	1.13	2.19	1.68	0.99	2.44	1.18	1.22	H150
-C=N-OH	1.9									H175
-CN	1.98	2.35	1.31	2.29	1.71	1.11	2.67	1.35	1.37	H180

Estimation of the Chemical Shift in Polysubstituted Alkanes

(δ in ppm relative to TMS)

$$\delta_{CH_2R_1R_2} = 1.25 + \sum_{1}^{2} Z_i \qquad \delta_{CHR_1R_2R_3} = 1.50 + \sum_{1}^{3} Z_i$$

	Substituent	Z_i
C	-alkyl	0.0
	-C=C-	0.8
	-C≡C-	0.9
	-phenyl	1.3
HAL	-Cl	2.0
	-Br	1.9
	-I	1.4
O	-OH	1.7
	-O-alkyl	1.5
	-O-phenyl	2.3
	-OCO-alkyl	2.7
	-OCO-phenyl	2.9
N	$-NH_2$	1.0
	$-N\text{-alkyl}_2$	1.0
	$-NO_2$	3.0
S	-S-alkyl	1.0
O=C	-CHO	1.2
	-CO-alkyl	1.2
	-COOH	0.8
	-COO-alkyl	0.7
	-CN	1.2

Example:

OCH_3

CH

COOH

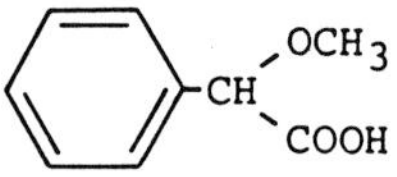

base value:	1.5
-O-alkyl	1.5
-COOH	0.8
-phenyl	1.3
estimated:	5.1
determined:	4.8

Estimation of the ^{1}H-Chemical Shift of Methylene Groups $R_1CH_2R_1$ (δ in ppm relative to TMS); see H.M. Bell, L.K. Berry, E.A. Madigan, Org. Magn. Res. 22, 693 (1984).

$$\delta_{CH_2R_1R_2} = 0.23 + Z_{R_1} + Z_{R_2}$$

	Substituent R	Z_R
	-alkyl	see p. H17
	-C=C-	1.33
	-C≡CH	1.52
C	-C≡Calkyl	1.52
	-C≡Cphenyl	1.77
	-phenyl	1.85
H	-F	3.15
A	-Cl	2.48
L	-Br	2.29
	-OH	2.46
	-Oalkyl	2.27
O	-Ophenyl	2.89
	-OCOalkyl	2.98
	-OCOphenyl	3.23
	$-NH_2$	1.69
	-NHalkyl	1.60
	$-N(alkyl)_2$	1.41
	-NHphenyl	2.15
N	-N(alkyl)phenyl	2.39
	$-NH_3^+$	2.31 (for D_2O [a])
	$-NH_2^+alkyl$	2.31 (for D_2O [a])
	$-NH^+(alkyl)_2$	2.46 (for D_2O [a])
	$-N^+(alkyl)_3$	2.56 (for D_2O [b])
	-NHCOalkyl	2.23
	-N(alkyl)COalkyl	2.23
	-NHCOphenyl	2.33
	-SH	1.63
S	-Salkyl	1.63
	-Sphenyl	1.92
	-COalkyl	1.58
	-COphenyl	2.08
O	-COOH	1.49
‖	-COOalkyl	1.49
C	-COOphenyl	1.74
/ \	$-CONH_2$	1.39
	$-CON(alkyl)_2$	1.39
	-CONHphenyl	1.59
	-C≡N	1.73

[a] for CD_3SOCD_3 or $CD_3SOCD_3/CDCl_3$ as solvent 0.2-0.3 ppm lower values must be used.

[b] for CDCl as solvent 0.3-0.5 ppm higher values must be used.

Estimation of the ^{1}H-Chemical Shift of Methylene Groups

$R_1CH_2-CR'_1R'_2R'_3$ (δ in ppm relative to TMS); see H.M. Bell, L.K. Berry, E.A. Madigan, Org. Magn. Res. 22, 693 (1984).

$$\delta_{R_1CH_2-CR'_1R'_2R'_3} = 1.20 + Z_{R_1} + Z_{R'_1} + Z_{R'_2} + Z_{R'_3}$$

	Substituent R	Z_R (α-Effect)	$Z_{R'}$ (β-Effect)
C	-alkyl	0.00	0.00
	-C=C-	0.78	0.16
	-C≡CH; -C≡Calkyl	0.98	0.21
	-C≡Cphenyl	1.23	0.21
	-phenyl	1.31	0.37
HAL	-F	2.76	0.24
	-Cl	2.04	0.36
	-Br	1.87	0.52
O	-OH	2.07	0.23
	-Oalkyl	1.91	0.28 a)
	-Ophenyl	2.48	0.58
	-OCOalkyl	2.61	0.43
	-OCOphenyl	2.91	0.53
N	$-NH_2$	1.27	0.25
	-NHalkyl	1.23	0.25
	$-N(alkyl)_2$	1.00	0.25
	-NHphenyl	1.74	0.31
	-N(alkyl)phenyl	1.97	0.31
	$-NH_3^+$; $-NH_2^+alkyl$	1.77 } for D_2O b)	0.60 } for D_2O c)
	$-NH^+(alkyl)_2$	1.95 } for D_2O b)	0.60 } for D_2O c)
	$-N^+(alkyl)_3$	2.07 for D_2O d)	0.70
	-NHCOalkyl; -N(alkyl)COalkyl	1.79	0.26
	-NHCOphenyl	1.89	0.26
S	-SH; -Salkyl	1.15	0.30
	-Sphenyl	1.44	0.28
O=C	-CO alkyl	1.11	0.38
	-CO phenyl	1.61	0.58
	-COOH; -COOalkyl	1.02	0.34
	-COO phenyl	1.27	0.34
	$-CONH_2$; $-CON(alkyl)_2$	0.89	0.40
	-CONH phenyl	1.09	0.40
	-C≡N	1.15	0.42

a) $Z_{R'_2}$ = 0 for the second O-alkyl group (in acetals and ketals).

b) for DC_3SOCD_3 and $CD_3SOCD_3/CDCl_3$ as solvent 0.2-0.3 ppm lower values must be used.

c) for CD_3SOCD_3 and $CD_3SOCD_3/CDCl_3$ as solvent 0.05-0.1 ppm lower values must be used.

d) for $CDCl_3$ as solvent 0.3-0.5 ppm higher values must be used.

Coupling in Aliphatic Compounds

for coupling in alicyclics see p. H185

Geminal coupling

$$J_{HCH} = -8 \cdots\cdots -18 \text{ Hz}$$

Electronegative substituents cause a decrease in $|J_{gem}|$ while an sp^2- or sp-hybridized carbon atom as the substituent causes an increase:

Compound	J_{gem} [Hz]	Compound	J_{gem} [Hz]
CH_4	-12.4	CH_3COCH_3	-14.9
CH_3Cl	-10.8	CH_3-phenyl	∿ -14.3
CH_2Cl_2	-7.5	CH_3CN	-16.9
CH_3OH	-10.8	$CH_2(CN)_2$	-20.3

Vicinal coupling

H–C–C–H

- for free rotation $J_{HCCH} = \sim 7$ Hz
- for fixed conformation $J_{HCCH} = 0\text{-}18$ Hz

The nature of the substituent has little influence on the vicinal coupling constants if free rotation is possible:

CH_3CH_2-Li $J_{vic} = 8.4$ Hz

CH_3CH_2-F $J_{vic} = 6.9$ Hz

The vicinal coupling constants are mainly dependent on the dihedral angle ϕ:

$$J = J^{0} \cos^2 \phi - 0.3 \qquad 0^\circ < \phi < 90^\circ$$

$$J = J^{180} \cos^2 \phi - 0.3 \qquad 90^\circ < \phi < 180^\circ$$

The same relationship between dihedral angle and vicinal coupling constant is frequently observed in substituted alkanes. Both J^0 and J^{180} can be affected by the substituents. The following figure gives the values calculated for various values of J^0 and J^{180} as a function of the dihedral angle.

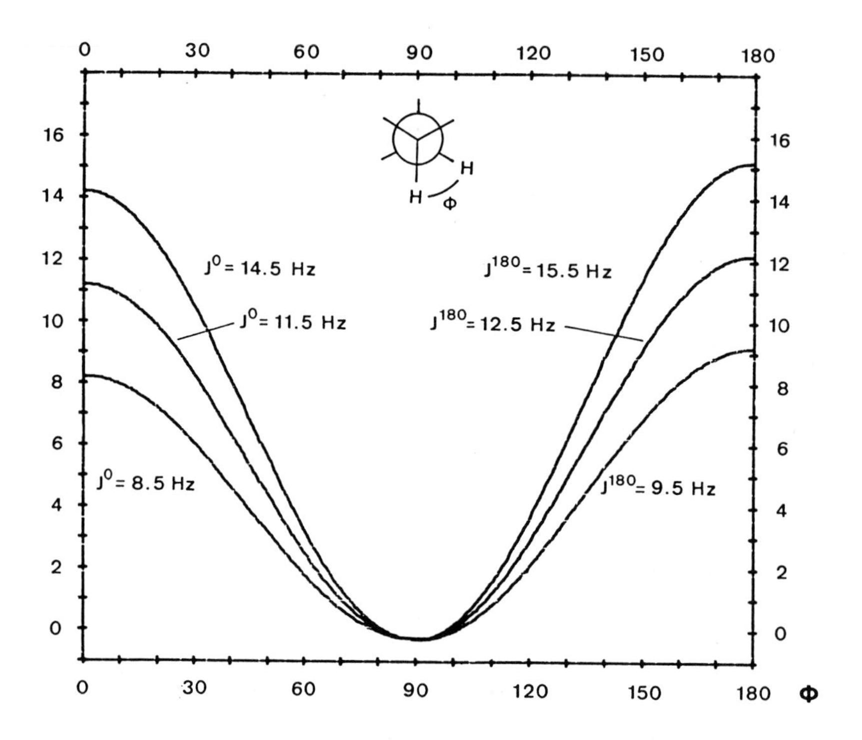

An excellent discussion of the scope and limitations of conformational analysis based on such relationships can be found on p. 281 of Jackman and Sternhell (see p. A15).

Coupling constants across more than three bonds ("long range" coupling) in saturated hydrocarbons are generally in the order of $<\sim 1$ Hz. However, they are of significance for fixed conformations of condensed, alicyclic systems (see p. H190) and in unsaturated compounds (see p. H205, H210).

^{1}H-Chemical Shifts in Aromatic Substituted Alkanes

(δ in ppm relative to TMS)

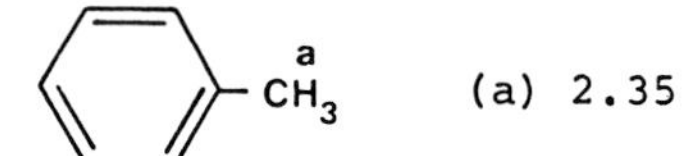

a b
CH_2-CH_3 (a) 2.63
(b) 1.21

a
CH_3 (a) 2.46

a
CH_3 (a) 2.65

O a
CH_3 (a) 2.17

a
CH_3
O (a) 1.94

N
CH_3
a (a) 3.50

N
H CH_3
a (a) 2.16

a
CH_3
N
H (a) 2.05

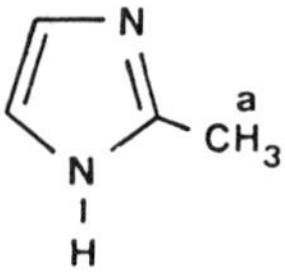

(a) 2.42

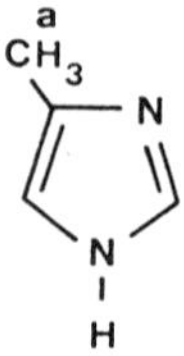

(a) 2.27

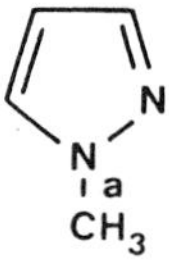

(a) 3.80

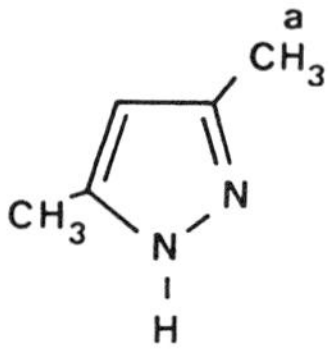

(a) 2.79

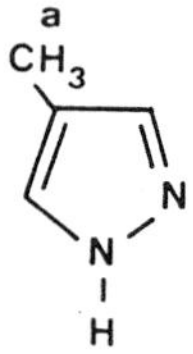

(a) 2.05

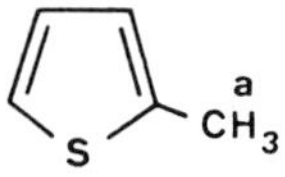

(a) 2.41

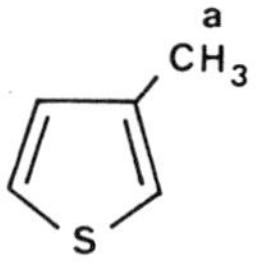

(a) 2.21

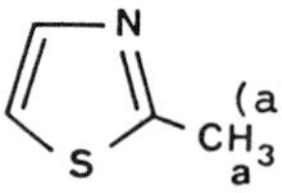

(a) 2.74

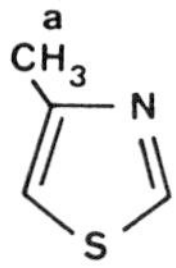

(a) 2.47

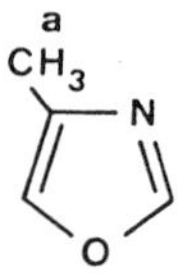

(a) 2.18

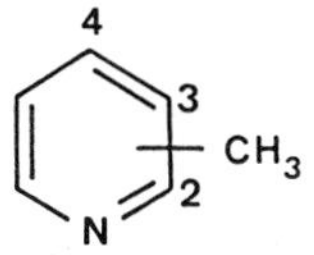

position	δ_{CH_3}
2	2.55
3	2.32
4	2.37

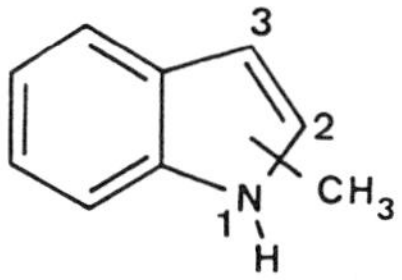

position	δ_{CH_3}
1	3.60
2	2.30
3	2.30

^{1}H-Chemical Shifts in Halogenated Methane and Ethane Derivatives

(δ in ppm relative to TMS)

Compound type	Chemical shift for X =			
	F[1)]	Cl	Br	I
CH_3X	4.27	3.06	2.69	2.16
CH_2X_2	5.45	5.33	4.94	3.90
CHX_3	6.49	7.24	6.82	4.91
CH_2-X	4.36	3.47	3.37	3.16
CH_3	1.24	1.33	1.66	1.88
CHX_2		5.89	5.86	
CH_3		2.07	2.47	
XCH_2CH_2X		3.67	3,63	3.70

1) For ^{19}F-^1H-coupling see p. H357.

Chemical Shift for Hydroxyl Protons
(δ in ppm relative to TMS)

Aliphatic (and alicyclic) alcohols: 0.5 - 5.0
Phenols: 5.0 - 8.0

Hydrogen bonds strongly deshield hydroxyl protons (in the enol form of β-dicarbonyl compounds to over 15). The position of the signal may be strongly dependent on the experimental conditions. If the substance contains, in addition, acidic protons (-OH, -COOH, H_2O) only one signal in the average position is generally observed because of rapid exchange.

In dimethyl sulfoxide as solvent this exchange is often so slow that isolated signals are observed (most important exceptions are substances containing strong acids or amines). In this solvent the chemical shifts of hydroxyl protons are characteristic (see p. H55).

Chemical Shifts of Hydroxyl Protons in Dimethyl Sulfoxide as the Solvent (δ in ppm relative to TMS)

Compound	δ	Compound	δ
CH_3CH_2OH	4.5	OH	4.0 ··· 4.5
CCl_3CH_2OH	6.8	OH	3.8 ··· 4.2
$-CH_2OH$	5.2	$-OH$	9.3
$( )_2CHOH$	5.6	NO_2- $-OH$	11.0
$( )_3COH$	6.4	OH, N	9.8

Chemical Shifts for Protons in the Vicinity of Hydroxyl Groups (δ in ppm relative to TMS)

For alicyclic alcohols see p. H195.
For phenols see p. H255.

Compound	δ
$\overset{a}{CH_3}OH$	(a) 3.39
$\overset{b}{CH_3}\overset{a}{CH_2}OH$	(a) 3.59 (b) 1.18
$\overset{c}{CH_3}\overset{b}{CH_2}\overset{a}{CH_2}OH$	(a) 3.49 (b) 1.53 (c) 0.93
$(\overset{b}{CH_3})_2\overset{a}{CH}OH$	(a) 3.94 (b) 1.16
$(\overset{a}{CH_3})_3COH$	(a) 1.22

Coupling of Hydroxyl Protons (J in Hz)

The coupling between hydroxyl protons and the protons on the adjacent carbon atoms are normally not observable in the spectrum because of the fast exchange of the hydroxyl protons. However, in very pure (acid-free) solutions or in dimethyl sulfoxide as the solvent the exchange is sufficiently slow that the H-O-C-H couplings become visible. They exhibit a dependence on the conformation analogous to that shown by the H-C-C-H coupling (see p. H20). In case of free rotation:

$$J_{HOCH} = \sim 5$$

In cyclohexanols the coupling constants for axial and equatorial hydroxyls measured in dimethyl sulfoxide differ significantly:

cyclohexane-OH (equatorial OH)	$J_{HOCH} = 4.2 \cdots 5.7$
cyclohexane-OH (axial OH)	$J_{HOCH} = 3.0 \cdots 4.2$

^{1}H-Chemical Shifts in Aliphatic Ethers (δ in ppm relative to TMS)

Compound	δ	Compound	δ
$\overset{a}{CH_3}$O-alkyl	(a) 3.24	$\overset{b}{CH_3}\overset{a}{CH_2}$O-alkyl	(a) 3.37 (b) 1.15
$\overset{a}{CH_3}$O-C=C	(a) 3.5	$\overset{b}{CH_3}\overset{a}{CH_2}$O-C=C	(a) 3.7 (b) 1.3
$\overset{a}{CH_3}$O-C_6H_4(H b)	(a) 3.73 $\lvert J \rvert_{ab}$ = 0 - 0.8	$\overset{b}{CH_3}\overset{a}{CH_2}$O-$C_6H_5$	(a) 3.98 (b) 1.38
$CH_3O\overset{a}{CH_3}$	(a) 3.3	$\overset{b}{CH_2}(O\overset{a}{CH_3})_2$	(a) 3.3 (b) 4.5
$\overset{b}{CH}(O\overset{a}{CH_3})_3$	(a) 3.3 (b) 5.1	$(\overset{b}{CH_3})_2\overset{a}{CH}$O-alkyl	(a) 3.55 (b) 1.08
$C(O\overset{a}{CH_3})_4$	(a) 3.29		

^{1}H-NMR

CYCLIC ETHERS

^{1}H-Chemical Shift (δ in ppm relative to TMS) and Coupling Constants (J in Hz) in Non-Aromatic Cyclic Ethers

For heteroaromatic systems see p. H265 - H350.

(a) 2.54

geminal coupling: J_{gem} = 6

vicinal couplings: J_{cis} = 4.5

J_{trans} = 3.1

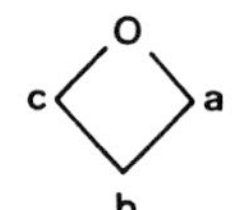

(a,c) 4.73
(b) 2.72

geminal couplings: (a) J_{gem} = -5.8

(b) J_{gem} = -11.0

vicinal couplings: J_{cis} = 8.7

J_{trans} = 6.6

"long-range" coupling: $|J|_{ac}$ < 0.3

(a) 3.75
(b) 1.85

(a) 4.20
(b) 2.53
(c) 4.82
(d) 6.22

J_{ab} = 9.3 J_{bd} = 2.6

J_{bc} = 2.5 J_{cd} = 2.6

(a,d) 4.43

(b,c) 5.78

J_{ab} = ∿ 1

J_{ac} = ∿-1

J_{bc} = ∿ 6

(a) 3.9 geminal coupling (a) J_{gem} = ∿-7.5
(b) 4.9 vicinal couplings J_{cis} = 7.3
J_{trans} = 6.0

(a) 5.90 (a) $|J|_{gem}$ = ∿1.5

(a) 3.6
(b,c) ∿1.6

(a)	4.0	(c)	∿1.9	(e)	6.4
(b)	∿1.9	(d)	4.6		

(a,e)	6.17	J_{ab} = 7.0		J_{ae} = 1.5	
(b,d)	4.63	J_{ac} = 1.7		J_{bc} = 3.4	
(c)	2.66				

(a,d) 6.35 J_{ab} = 5.9
(b,c) 7.71 J_{ac} = 0.4
J_{ad} = 2.7
J_{bc} = 1.1

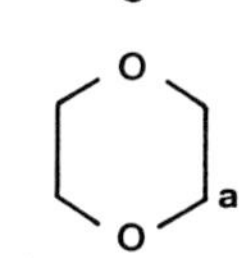

(a) 3.71

(a) 4.70
(b) 3.80
(c) 1.68

(a) 5.00

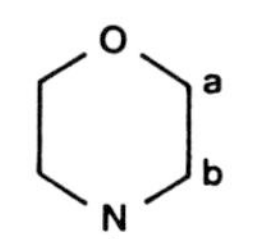

(a) 3.67
(b) 2.87
(c) 1.92

(a) 3.88
(b) 2.57

^{1}H-Chemical Shifts in Aliphatic Amines (δ in ppm relative to TMS)

alkyl-NH_2 (a); $(alkyl)_2$-NH (a)	(a) 0.5-4.0
C_6H_5-NH_2 (a); C_6H_5-NH-alkyl (a); C_6H_5-NH-C_6H_5 (a)	(a) 2.5-5.0
CH_3NH_2 (a = CH_3)	(a) 2.47
$CH_3CH_2NH_2$ (b = CH_3, a = CH_2)	(a) 2.74 (b) 1.10
$(CH_3)_2CHNH_2$ (b = CH_3, a = CH)	(a) 3.07 (b) 1.03
$(CH_3)_2$-N-C_6H_{11} (a = CH_3)	(a) 2.25
CH_3-NC_5H_{10} (N-methylpiperidine, a = CH_3)	(a) 2.21
$(CH_3)_2$-N-C_6H_5 (a = CH_3)	(a) 2.85
$(CH_3)_2$-N-C_6H_4-NO_2 (a = CH_3)	(a) 3.09

alkyl-NH_3^+ (a); $(alkyl)_2$-NH_2^+ (a); $(alkyl)_3$-NH^+ (a)	(a) ∿6-9
C_6H_5-NH_3^+ (a); C_6H_5-NH_2^+-alkyl (a); C_6H_5-NH^+-$(alkyl)_2$ (a)	(a) ∿6-9
$(CH_3)_2NH$ (a = CH_3)	(a) ∿2.3
CH_3-NH-C_6H_{11} (a = CH_3)	(a) 2.44
CH_3-NH-C_6H_5 (a = CH_3, b = NH)	(a) 2.78 (b) 3.55
CH_3NH-C_6H_4-NO_2 (a = CH_3, b = NH)	(a) 2.93 (b) 4.62
$(CH_3)_4N^+$ (a = CH_3)	(a) 3.3
$(CH_3CH_2)_4N^+$ (b = CH_3, a = CH_2)	(a) 3.25 (b) 1.25
$(CH_3)_3N^+$-C_6H_5 (a = CH_3)	(a) 3.72

Coupling in Amines (J in Hz)

Coupling with Amine Protons

The H-N-C-H coupling is not observable because of the rapid exchange of the NH-protons. However, for a CH-NH-C= moiety (enamines, aromatic amines, amides, etc.) splitting is often observed. The coupling constant exhibits a dependence on the conformation analogous to that shown by vicinal H-C-C-H couplings (see p. H20). In case of free rotation:

$$J_{H-C-N-H} = 5 - 6$$

In trifluoroacetic acid as the solvent the exchange of the ammonium protons is slowed to such an extent that the coupling of $H-N^{+}-C-H$ generally becomes observable; in case of free rotation: $J_{H-N^{+}-C-H} = 5 - 6$.

Coupling with ^{14}N

Generally $^{14}N-^{1}H$-couplings are not observable because of the quadrupole relaxation of the nitrogen nucleus, even if the exchange of the amino protons is slow (as in weakly basic amines, amides and ammonium ions). For the same reason ^{14}N-CH-CH couplings are generally not observable. In ammonium compounds of high symmetry the contribution of the quadrupole relaxation is, however, so small that the $^{14}N-^{1}H$ and the ^{14}N-CH-CH couplings become observable:

$^{+}NH_4$	$\lvert J\rvert_{^{14}N-H}$	= 52.8
$^{+}N(CH_2CH_2CH_3)_4$	$\lvert J\rvert_{^{14}N-CH}$	= 0.5
	$\lvert J\rvert_{^{14}N-C-CH}$	= 1.6
	$\lvert J\rvert_{^{14}N-C-C-CH}$	= 0

In many cases the combination of quadrupole relaxation and exchange of NH-protons is not sufficiently large to completely eliminate the large $^{14}N-^{1}H$ coupling across one bond. In such a case the signal for the NH-protons is often broad (for example in ammonium compounds, amides, NH-protons of heteroaromatics, etc.). However, such a line broadening has no effect on the $\underline{H}$-N-C-$\underline{H}$ coupling.

^{1}H-Chemical Shifts (δ in ppm relative to TMS) and Coupling Constants (J in Hz) in Non-Aromatic Cyclic Amines

For heteroaromatic systems see p. H265 - H350.

Aziridine (N–H a; ring CH$_2$ b):
(a) 0.9
(b) 1.61

geminal coupling: $J_{gem} = \sim 1$
vicinal couplings: $J_{cis} = \sim 6$, $J_{trans} = \sim 4$

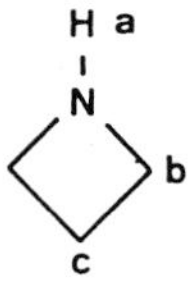

Azetidine (N–H a; b; c):
(a) 2.38
(b) 3.54
(c) 2.23

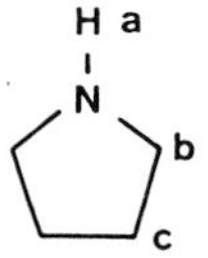

Pyrrolidine (N–H a; b; c):
(a) 2.01
(b) 2.75
(c) 1.59

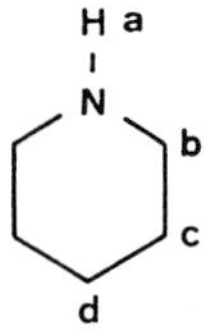

Piperidine (N–H a; b; c; d):
(a) 1.84
(b) 2.74
(c) 1.5
(d) 1.5

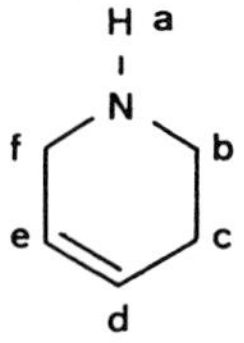

1,2,3,6-Tetrahydropyridine (N–H a; b; c; d; e; f):
(a) 1.63
(b) 2.95
(c) 2.07
(d), (e) 5.72 & 5.77
(f) 3.33

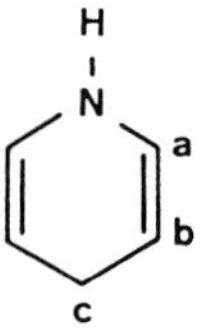

1,4-Dihydropyridine (N–H; a; b; c):
(a) 5.73
(b) 4.42
(c) 3.15

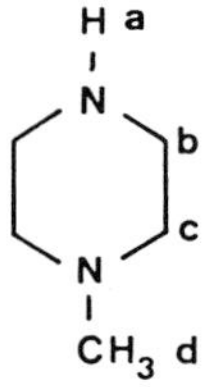

N-Methylpiperazine (N–H a; b; c; CH_3 d):
(a) 2.12
(b) 2.88
(c) 2.37
(d) 2.27

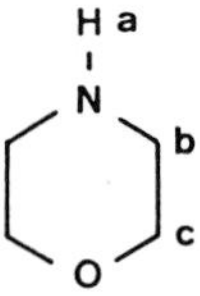

Morpholine (N–H a; b; c):
(a) 1.92
(b) 2.87
(c) 3.67

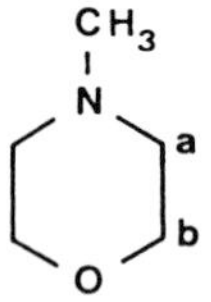

N-Methylmorpholine (CH_3–N; a; b):
(a) 2.32
(b) 3.62

^{1}H-Chemical Shifts in Aliphatic Nitro Compounds

(δ in ppm relative to TMS)

$\overset{a}{CH_3}NO_2$	(a) 4.29
$\overset{b}{CH_3}\overset{a}{CH_2}NO_2$	(a) 4.37 (b) 1.58
$\overset{c}{CH_3}\overset{b}{CH_2}\overset{a}{CH_2}NO_2$	(a) 4.28 (b) 2.01 (c) 1.03
$(\overset{b}{CH_3})_2\overset{a}{CH}NO_2$	(a) 4.44 (b) 1.53
$(\overset{a}{CH_3})_3CNO_2$	(a) 1.59

$NO_2-CH_2-CH_2-CH_2-CH_2-CH_3$	
NO_2	
CH_2	4.3
CH_2	2.0
CH_2	~ 1.4
CH_2	~ 1.4
CH_3	0.9

^{1}H-Chemical Shifts in Aliphatic N-Nitroso, Azo and Azoxy Compounds

(δ in ppm relative to TMS)

$(\overset{a}{CH_3})(\overset{b}{CH_3})N^+=N-O^-$ (a) 2.96 (b) 3.76

$[(\overset{b}{CH_3})_2\overset{a}{CH}][(\overset{d}{CH_3})_2\overset{c}{CH}]N^+=N-O^-$ (a) 4.89 (b) 1.15 (c) 4.26 (d) 1.52

$\delta_{cis} < \delta_{trans}$ for αCH_3-, αCH_2- and βCH_3-protons

$\delta_{cis} > \delta_{trans}$ for αCH-protons

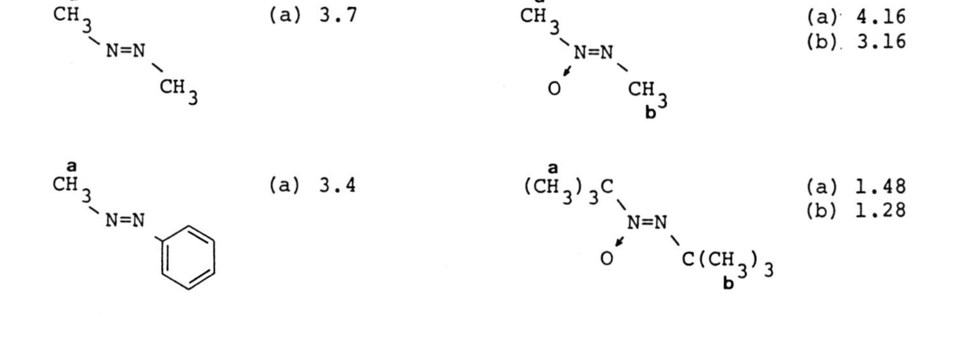

ALIPHATIC THIOLS,
THIOETHERS

^{1}H-Chemical Shifts (δ in ppm relative to TMS) and Coupling Constants (J in Hz) in Aliphatic Thiols

alkyl-SH (a)	(a) 1-2
C_6H_5-SH (a)	(a) 2-4

The exchange with other -SH, -COOH or -OH protons is generally so slow that the chemical shift is characteristic and the coupling with the -SH protons becomes observable (J = 5 - 9 Hz for free rotation).

CH_3SH (a: CH_3)	(a) 2.00
CH_3CH_2SH (b: CH_3, a: CH_2)	(a) 2.44 (b) 1.31
$CH_3CH_2CH_2SH$ (c: CH_3, b: CH_2, a: CH_2)	(a) 2.46 (b) 1.57 (c) 1.02
$(CH_3)_2CHSH$ (c: CH_3, b: CH, a: SH)	(a) 1.56 (b) 3.16 (c) 1.34
$(CH_3)_3CSH$ (a: SH)	(a) 1.43
$HSCH_2CH_2CH_2SH$ (c: CH_2, b: CH_2, a: SH)	(a) 1.35 (b) 2.68 (c) 1.88

^{1}H-Chemical Shifts in Aliphatic Thioethers

(δ in ppm relative to TMS)

CH_3S-alkyl (a: CH_3)	(a) 2.09
CH_3S-C=C (a: CH_3)	(a) 2.25
CH_3S-C_6H_5 (a: CH_3)	(a) 2.47
CH_3CH_2S-alkyl (b: CH_3, a: CH_2)	(a) 2.49 (b) 1.25
$CH_3CH_2CH_2S$-alkyl (c: CH_3, b: CH_2, a: CH_2)	(a) 2.43 (b) 1.59 (c) 0.98
$(CH_3)_2CHS$-alkyl (b: CH_3, a: CH)	(a) 2.93 (b) 1.25
$(CH_3)_3CS$-alkyl (a: CH_3)	(a) 1.39

^{1}H-Chemical Shifts (δ in ppm relative to TMS) and Coupling Constants (J in Hz) in Non-Aromatic Cyclic Thioethers and Sulfones

For heteroaromatic systems see p. H265 - H350.

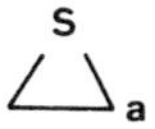

(a) 2.27

geminal coupling: J_{gem} = 0

vicinal couplings: J_{cis} = 7.2; J_{trans} = 5.7

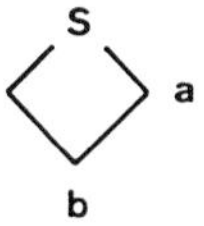

(a) 3.21
(b) 2.94

geminal couplings: (a) J_{gem} = -8.7; (b) J_{gem} = -11.7

vicinal couplings: J_{cis} = 8.9; J_{trans} = 6.3

"long-range" couplings: J_{cis} = 1.2; J_{trans} = -0.2

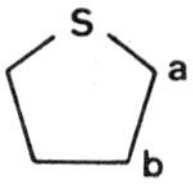

(a) 2.75
(b) 1.88

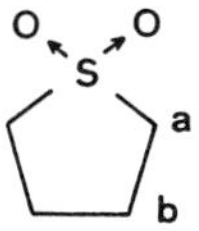

(a) 3.00
(b) 2.23

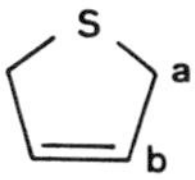

(a) 3.67
(b) 5.81

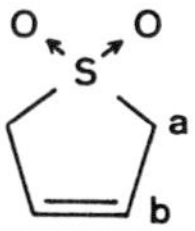

(a) 3.74
(b) 6.08

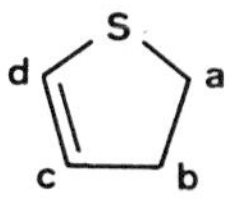

(a) 3.08
(b) 2.62
(c) 5.48
(d) 6.06

J_{ab} = 9.2
J_{bc} = 2.5
J_{cd} = 6.1
J_{bd} = 2.2

^{1}H-NMR

CYCLIC THIOETHERS

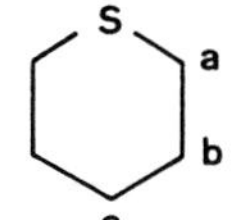

(a) 2.52
(b), (c) 1.6-1.8

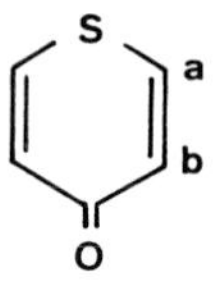

(a) 7.89
(b) 7.09

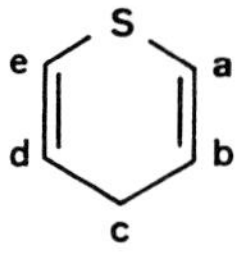

(a,e) 5.97
(b,d) 5.55
(c) 2.84

J_{ab} = 10.0
J_{bc} = 3.9
J_{ac} = 1.1

J_{bd} = 0.0
J_{ae} = 2.9
J_{ad} = 0.0

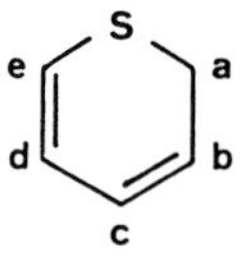

(a) 3.19
(b), (c), (d), (e) 5.5-6.2

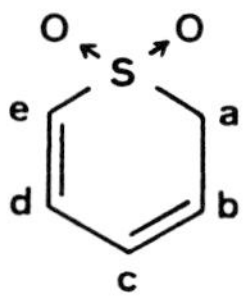

(a) 4.00
(b), (c), (d), (e) 6.3-6.7

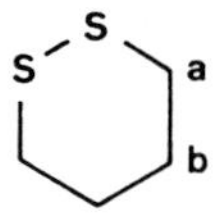

(a) 2.7
(b) 1.9

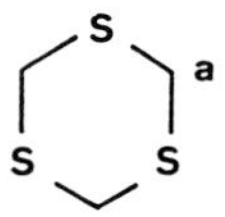

(a) 4.2

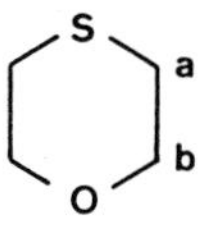

(a) 2.57
(b) 3.88

^{1}H-Chemical Shifts in Aliphatic S-Compounds

(δ in ppm relative to TMS)

For thiols and thioethers see p. H95 - H105.

Disulfides

$\overset{a}{CH_3}SS$-alkyl	(a) 2.30
$\overset{b}{CH_3}\overset{a}{CH_2}SS$-alkyl	(a) 2.67 (b) 1.35
$\overset{c}{CH_3}\overset{b}{CH_2}\overset{a}{CH_2}SS$-alkyl	(a) 2.63 (b) 1.71 (c) 1.03
$(\overset{a}{CH_3})_3CSS$-alkyl	(a) 1.32

Sulfoxides

Compound	δ
$\overset{a}{CH_3}SOCH_3$	(a) 2.50

Sulfones

Compound	δ
$\overset{a}{CH_3}SO_2CH_3$	(a) 2.84
$\overset{a}{CH_3}SO_2\overset{b}{CH_2}\overset{c}{CH_3}$	(a) 2.80 (b) 2.94 (c) 1.47
$\overset{a}{CH_3}SO_2\overset{b}{CH}=\overset{c}{CH_2}$	(a) 2.62 (b) 6.70 (c) cis 6.13 trans 5.95
$(\overset{a}{CH_3})_3CSO_2C(CH_3)_3$	(a) 1.44

Derivatives of Sulfinic and Sulfonic Acids

Compound	δ
$\overset{a}{CH_3}SON(\overset{b}{CH_3})_2$	(a) 2.50 (b) 2.68
$\overset{a}{CH_3}SO_2$-O-alkyl	(a) 3.0
$\overset{a}{CH_3}SO_2\overset{b}{NH}-C_6H_5$	(a) 2.82 (b) ~7.2
$\overset{a}{CH_3}SO_2Cl$	(a) 3.6
alkyl-$SO_2\overset{a}{OH}$ $C_6H_5-SO_2\overset{a}{OH}$	(a) 11-12
Cyclic: O←S(→O) in ring O–S–$\overset{a}{CH_2}$–$\overset{b}{CH_2}$–$\overset{c}{CH_2}$–O	(a) 3.19 (b) 2.44 (c) 4.27
$CH_3-C_6H_4-SO_2\overset{a}{NH_2}$	(a) ~5

For esters of sulfonic acids see also p. H140, H145.

Esters of Sulfuric Acid and Sulfurous Acid

$\overset{a}{CH_3}OSO-OCH_3$	(a) 3.58	cyclic ethylene sulfate (O–S(=O)(=O)–O ring, CH₂ = a)	(a) 4.68
$\overset{a}{CH_3}OSO_2-OCH_3$	(a) 3.94		

Thiocarboxylic Acids and Derivatives

$\overset{a}{CH_3}CO\overset{b}{S}H$	(a) 2.4 (b) 4.7	2H-pyran-2-thione (ring H a, b, c, d)	(a) 7.1 (b) 6.9 (c) 6.4 (d) 7.7
$\overset{a}{CH_3}COS\overset{b}{CH_3}$	(a) 2.30 (b) 2.27		
		thiophen-2(5H)-one (ring H a, b, c)	(a) 6.47 J_{ab} = 5.9 (b) 7.84 J_{ac} = 2.0 (c) 4.29 J_{bc} = 2.8

Thiocyanates, Isothiocyanates

$\overset{a}{CH_3}SCN$	(a) 2.61	$\overset{a}{CH_3}NCS$	(a) 3.37
$\overset{b}{CH_3}\overset{a}{CH_2}SCN$	(a) 2.98 (b) 1.52	$\overset{b}{CH_3}\overset{a}{CH_2}NCS$	(a) 3.64 (b) 1.40
$(CH_3)_2\overset{a}{CH}SCN$	(a) 3.48	$(CH_3)_2\overset{a}{CH}NCS$	(a) 3.98

^{1}H-Chemical Shifts (δ in ppm relative to TMS) and Coupling Constants (J in Hz) in Aliphatic Aldehydes

alkyl-CHO (a); alkenyl-CHO (a)	(a) 9.0-10.1
C_6H_5-CHO (a)	unsubstituted in the ortho-position: (a) 9.6-10.2 substituted in the ortho-position: (a) 10.2-10.5

Couplings:

>C(H)-C(H)=O	$\lvert J\rvert_{HC-CHO} = 0...3$ (depending on conformation)
$H^c H^d C=C(H^b)-C(H^a)=O$	$\lvert J\rvert_{ab} = \sim 8$ $\lvert J\rvert_{ac} = \sim 0.3$ $\lvert J\rvert_{ad} = \sim 0.1$
benzaldehyde (H^a on CHO; ring H^b, H^c, $H^{c'}$, H^d)	$\lvert J\rvert_{ab} = \sim 0.2$ $\lvert J\rvert_{ac} = 0$ $\lvert J\rvert_{ac'} = 0.4 - 1.0$ $\lvert J\rvert_{ad} = 0.1$

H-CHO	δ = 9.60 $\lvert J\rvert$ = 42.4
CH_3CHO (b a)	(a) 9.80 (b) 2.20 J_{ab} = 3
CH_3CH_2CHO (b a)	(a) 2.46 (b) 1.13
$CH_3CH_2CH_2CHO$ (d c b a)	(a) 9.74 (b) 2.42 (c) 1.67 (d) 0.97 J_{ab} = 2
$(CH_3)_2CHCHO$ (b a)	(a) 2.39 (b) 1.13
$(CH_3)_3CCHO$ (a)	(a) 1.07

^{1}H-NMR

ALIPHATIC KETONES

[1]H-Chemical Shifts (δ in ppm relative to TMS) and Coupling Constants (J in Hz) in Aliphatic Ketones

Compound	δ	Compound	δ
$\overset{a}{CH_3}COCH_3$	(a) 2.09	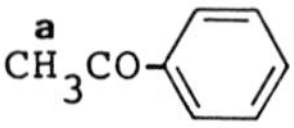 $\overset{a}{CH_3}CO-C_6H_5$	(a) 2.55
$\overset{a}{CH_3}COC{=}C$	(a) 2.3		
$\overset{b}{CH_3}\overset{a}{CH_2}COCH_3$	(a) 2.47 (b) 1.05	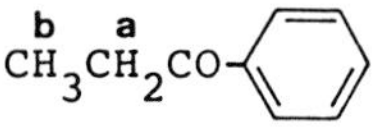 $\overset{b}{CH_3}\overset{a}{CH_2}CO-C_6H_5$	(a) 2.92 (b) 1.18
$\overset{c}{CH_3}\overset{b}{CH_2}\overset{a}{CH_2}COCH_3$	(a) 2.32 (b) 1.56 (c) 0.93	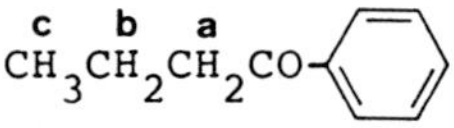 $\overset{c}{CH_3}\overset{b}{CH_2}\overset{a}{CH_2}CO-C_6H_5$	(a) 2.86 (b) 1.72 (c) 1.02
$(\overset{b}{CH_3})_2\overset{a}{CH}COCH_3$	(a) 2.54 (b) 1.08	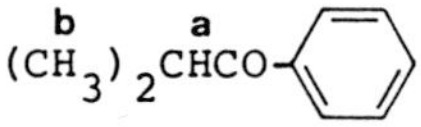 $(\overset{b}{CH_3})_2\overset{a}{CH}CO-C_6H_5$	(a) 3.58 (b) 1.22
$(\overset{a}{CH_3})_3CCOCH_3$	(a) 1.12		
$\overset{a}{CH_3}CO\overset{b}{CH_2}COCH_3$	(a) 2.17 (b) 3.62		

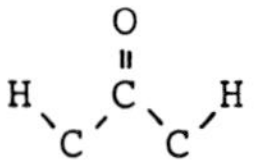

H–C–C(=O)–C–H $\quad J_{HCCOCH} = 0\text{–}0.5$

larger if conformation fixed ("W-effect"):

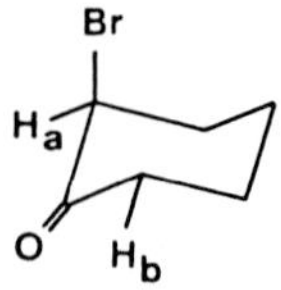

2-bromocyclohexanone (Br, H_a, O, H_b) $\quad J_{ab} = 1$

^{1}H-Chemical Shifts (δ in ppm relative to TMS) and Coupling Constants (J in Hz) in Cyclic Ketones

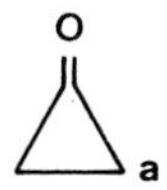

(a) 1.65

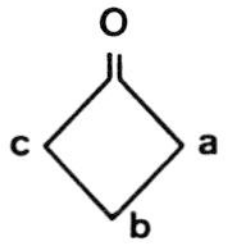

(a,c) 3.03
(b) 1.96
$J_{ac(cis)}$ = 1.9
$J_{ac(trans)}$ = 1.0

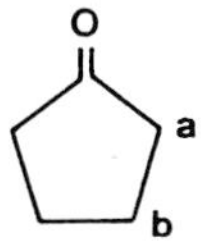

(a) 2.06
(b) 2.02

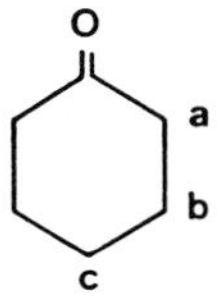

(a) 2.22
(b,c) ca. 1.8

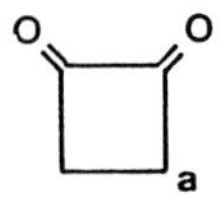

(a) 3.0

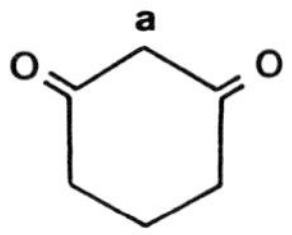

(a) 3.4

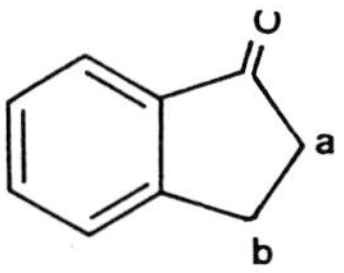

(a) 2.60
(b) 3.08

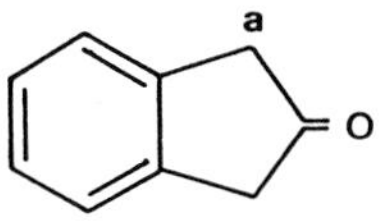

(a) 3.50

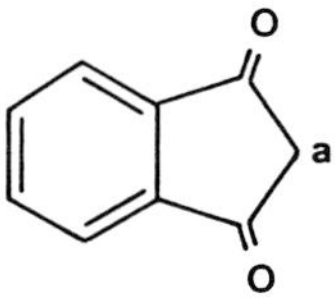

(a) 3.23

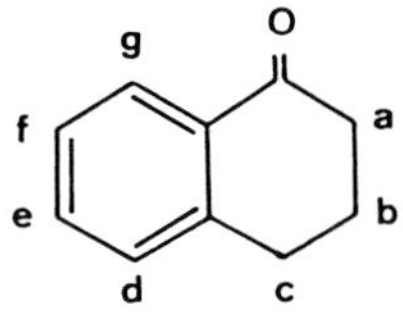

(a) 2.58
(b) 2.18
(c) 2.92
(d,e,f) 7.0-7.4
(g) 8.0

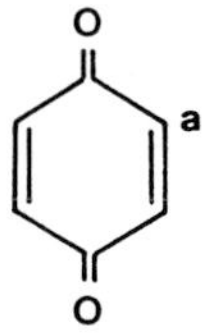

(a) 6.78

4-oxopyran: see p. H70.

4-oxo-1-thiopyran: see p. H105.

^{1}H-Chemical Shifts in Aliphatic Carboxylic Acids

(δ in ppm relative to TMS)

alkyl-COOH (a); C_6H_5-COOH (a)	(a) 10-13

The position of the signals depends on the solvent, the concentration and the presence of other exchangeable protons (e.g. alcohols, water, etc.).

Compound	δ
$\overset{a}{CH_3}COOH$	(a) 2.08
$\overset{b}{CH_3}\overset{a}{CH_2}COOH$	(a) 2.36 (b) 1.16
$\overset{c}{CH_3}\overset{b}{CH_2}\overset{a}{CH_2}COOH$	(a) 2.31 (b) 1.68 (c) 1.00
$(\overset{b}{CH_3})_2\overset{a}{CH}COOH$	(a) 2.56 (b) 1.21
$(\overset{a}{CH_3})_3CCOOH$	(a) 1.23
$HOOC\overset{a}{CH_2}COOH$	(a) 3.4
$HOOCCH_2\overset{a}{CH_2}COOH$	(a) 2.6

^{1}H-Chemical Shifts (δ in ppm relative to TMS) and Coupling Constants (J in Hz) in Aliphatic Esters

Formates

Compound	δ	Coupling constants
$HCOOCH_3$ (H of HCO = a)	(a) 8.03	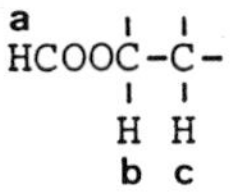
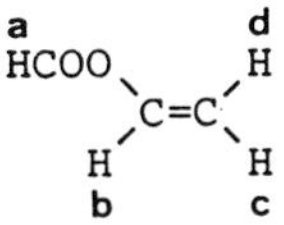 $HCOO(H_b)C=C(H_c)(H_d)$ (H of HCO = a)	(a) 8.1	J_{ab} = -0.7; J_{ac} = 1.6; J_{ad} = 0.8
$HCOOC(H_b)-C(H_c)-$ (H of HCO = a)		J_{ab} = ∿ -1; J_{ac} = ∿0.5

Acetates

Compound	δ
$\overset{a}{CH_3}COO\overset{b}{CH_3}$	(a) 2.01 (b) 3.67
$\overset{a}{CH_3}COOC=C$	(a) 2.1
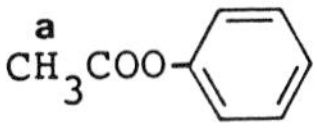 $\overset{a}{CH_3}COO-C_6H_5$	(a) 2.1

Propionates

Compound	δ
$\overset{b}{CH_3}\overset{a}{CH_2}COOCH_3$	(a) 2.28 (b) 1.12

Isobutyrates

Compound	δ
$(\overset{b}{CH_3})_2\overset{a}{CH}COOCH_3$	(a) 2.48 (b) 1.15

Methyl esters

Compound	δ	Compound	δ
$\overset{a}{CH_3}OCOCH_3$	(a) 3.67	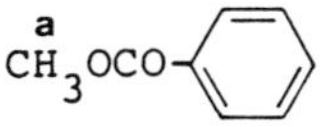 $\overset{a}{CH_3}OCO-C_6H_5$	(a) 3.88
$\overset{a}{CH_3}OCOC=C$	(a) 3.8	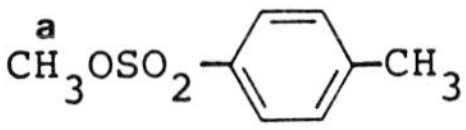 $\overset{a}{CH_3}OSO_2-C_6H_4-CH_3$	(a) 3.70

Methyl ester of:

	δ		δ
boric acid	3.5	perchloric acid	4.3
silicic acid	3.6	nitric acid	4.2
carbonic acid	3.8	sulfuric acid	3.94
phosphoric acid	3.8	sulfurous acid	3.58

¹H-NMR

ALIPHATIC ESTERS, LACTONES

Ethyl esters

$\overset{b}{CH_3}\overset{a}{CH_2}OCOCH_3$	(a) 4.05 (b) 1.21 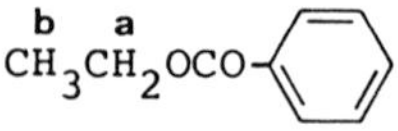
$CH_3\overset{a}{CH_2}OCOCF_3$	(a) 4.3 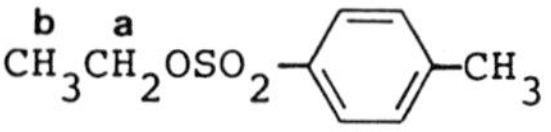
$\overset{b}{CH_3}\overset{a}{CH_2}OCOC{=}C$	(a) 4.3 (b) 1.3
$\overset{b}{CH_3}\overset{a}{CH_2}OCO-C_6H_5$	(a) 4.37 (b) 1.38
$\overset{b}{CH_3}\overset{a}{CH_2}OSO_2-C_6H_4-CH_3$	(a) 4.07 (b) 1.30
$\overset{b}{CH_3}\overset{a}{CH_2}ONO$	(a) 4.78 (b) 1.39

Isopropyl esters

$(\overset{b}{CH_3})_2\overset{a}{CH}OCOCH_3$	(a) 4.94 (b) 1.22 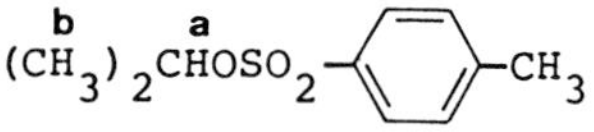
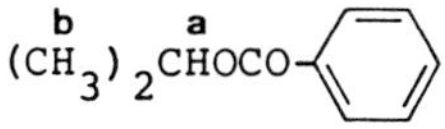 $(\overset{b}{CH_3})_2\overset{a}{CH}OCO-C_6H_5$	(a) 5.22 (b) 1.37
$(\overset{b}{CH_3})_2\overset{a}{CH}OSO_2-C_6H_4-CH_3$	(a) 4.70 (b) 1.25

¹H-Chemical Shifts (δ in ppm relative to TMS) and Coupling Constants (J in Hz) in Lactones

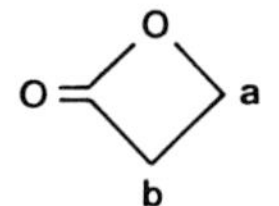

(a) 4.29
(b) 3.56

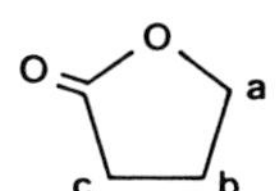

(a) 4.3
(b) 2.1
(c) 2.3

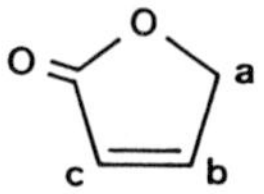

(a) 4.92
(b) 7.63
(c) 6.15

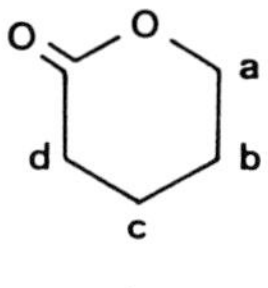

(a) 4.09
(b), (c) ∿1.6
(d) 3.31

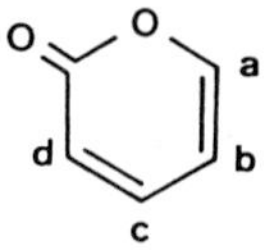

(a) 7.77
(b) 6.43
(c) 7.56
(d) 6.38

J_{ab} = 5.0, J_{bc} = 6.3, J_{cd} = 9.4
J_{ac} = 2.4, J_{bd} = 1.5, J_{ad} = 1.3

^{1}H-Chemical Shifts (δ in ppm relative to TMS) and Coupling Constants (J in Hz) in Aliphatic Amides

Chemical Shifts of the NH-Protons

alkyl-CONH$_2$ (a)	(a) 5-7
C_6H_5-CONH$_2$ (a)	
alkyl-CONH-alkyl (a)	(a) 6-8.5
C_6H_5-CONH-alkyl (a)	
alkyl-CONH-C_6H_5 (a)	(a) 7.5-9.5
C_6H_5-CONH-C_6H_5 (a)	

The signals for the NH-protons are often broad because the ^{14}N-^{1}H coupling is only partly eliminated by the quadrupole relaxation.

Vicinal Coupling H-C-N-H

The HC-NH-CO coupling depends on the conformation analogous to the HC-CH coupling (see p. H20). For free rotation:

$$J_{\underline{H}C-N\underline{H}-CO} = \sim 7$$

This coupling which causes splitting of the C-H signal can be clearly observed even in those cases where the signal for the NH-proton is broad and featureless.

Slow Rotation

The rotation around the CO-N bond is usually so slow that two separate signals are observed for the two conformers. In general the following holds:

for NC$\underline{H}_3$, NCH$_2$C$\underline{H}_3$ and NCH(C$\underline{H}_3$)$_2$ $\quad \delta_{cis\ to\ O} \lessdot \delta_{trans\ to\ O}$

for NC$\underline{H}$(CH$_3$)$_2$ and NC(C$\underline{H}_3$)$_3$ $\quad \delta_{trans\ to\ O} \lessdot \delta_{cis\ to\ O}$

for N-C$\underline{H}_2$- $\quad \delta_{cis\ to\ O} \sim \delta_{trans\ to\ O}$

ALIPHATIC AMIDES

Formamides

In the more stable conformer of monosubstituted formamides the substituent occupies the cis-position relative to the carbonyl oxygen:

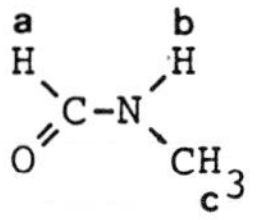

H(a)-C(=O)-N(H(b))(CH$_3$(c))   I (∿90%)

H(a)-C(=O)-N(CH$_3$(c))(H(b))   II (∿10%)

I: (a) 8.1
(b) 7.9
(c) 2.74

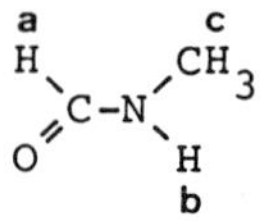

II: (a) 8.1
(b) 7.9
(c) 2.88

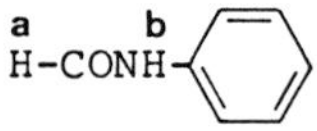

H(a)-CONH(b)-C$_6$H$_5$

(a) 8.2-8.7
(b) 7.5-9.5

H(a)-C(=O)-N(CH$_3$(b))(CH$_3$(c))

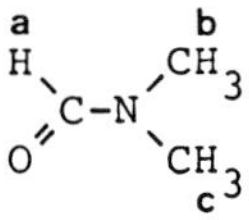

(a) 8.02
(b) 2.97
(c) 2.88

$|J|_{ab} = \sim 0.3$
$|J|_{ac} = \sim 0.7$

In the more stable conformer of disubstituted formamides the larger substituent occupies the trans-position relative to the carbonyl oxygen:

H-C(=O)-N(CH(b)(CH$_3$(c))$_2$)(CH$_3$(a))   I (∿70%)   ⇄   H-C(=O)-N(CH$_3$(a))(CH(b)(CH$_3$(c))$_2$)   II (∿30%)

I: (a) 2.71
(b) 4.12
(c) 1.19

II: (a) 2.83
(b) 4.78
(c) 1.10

Acetamides

In mono-substituted acetamides the only observable conformer has the substituent cis to the carbonyl oxygen:

CH_3(a)NH(b)COCH$_3$(c)

(a) 2.71
(b) 8.1
(c) ∿2.0

CH_3(a)CH_2(b)NH(c)COCH$_3$(d)

(a) 1.12
(b) 3.21
(c) 8.2
(d) ∿2.0

^{1}H-Chemical Shifts in Acid Halides and Anhydrides

(δ in ppm relative to TMS)

$\overset{a}{CH_3}COCl$	(a) 2.8	$\overset{a}{CH_3}COOCOCH_3$	(a) 2.2
$\overset{a}{CH_3}COBr$	(a) 2.7		
succinic anhydride (a = ring CH_2)	(a) 3.01	maleic anhydride (a = ring CH)	(a) 7.1

^{1}H-Chemical Shifts in Carbonic Acid Derivatives

(δ in ppm relative to TMS)

$O{=}C(O\overset{a}{CH_2}\overset{b}{CH_2}\overset{c}{CH_2}\overset{d}{CH_3})(OCH_2CH_2CH_2CH_3)$

(a) 4.13
(b,c) 1.2-1.7
(d) 0.93

ethylene carbonate (a = ring CH_2)	(a) 4.20	ethylene trithiocarbonate (a = ring CH_2)	(a) 3.94
$\overset{a}{CH_3}\overset{b}{NH}CONHCH_3$	(a) 2.7 (b) ~6	$\overset{a}{CH_3}\overset{b}{CH_2}NHCONHCH_2CH_3$	(a) 1.26 (b) 3.27
$\overset{a}{CH_3}\overset{b}{CH_2}\overset{c}{NH}CO\overset{d}{NH_2}$	(a) 1.5 (b) 3.0 (c) 5.5 (d) 5		
$\overset{a}{CH_3}\overset{b}{NH}COO\overset{c}{CH_2}\overset{d}{CH_3}$	(a) 2.78 (b) 5.16 (c) 4.14 (d) 1.23		

^{1}H-Chemical Shifts in Oximes, Imines, Hydrazones and Azines

(δ in ppm relative to TMS)

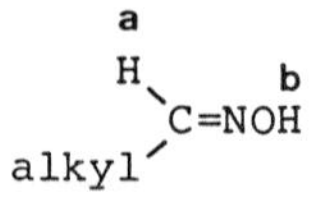

H^a(alkyl)C=NOH^b	(a) 6.8-7.9 (b) 7-10	$\delta_{a\ syn} > \delta_{a\ anti}$
H^a(aryl)C=NOH^b	(a) 7.2-8.6 (b) 7-10	$\delta_{a\ syn} > \delta_{a\ anti}$
H^a(alkyl)C=N-NH-aryl	(a) 6.1-7.7	$\delta_{a\ syn} > \delta_{a\ anti}$

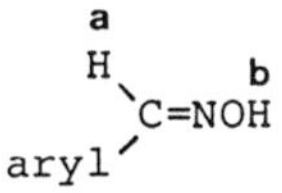

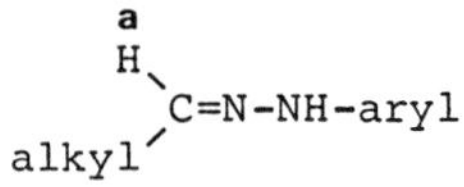

H^a(CH_3^b)C=N-OH^c (OH anti to H)	(a) 6.8 (b) 1.9 (c) 9	H^a(CH_3^b)C=N-OH^c (OH syn to H)	(a) 7.4 (b) 1.9 (c) 9

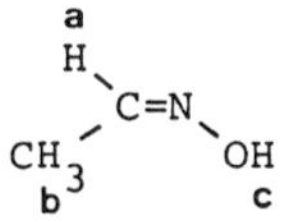

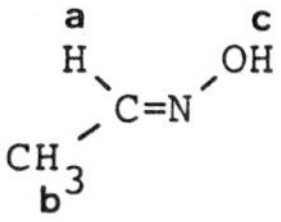

In aldoximes and ketoximes $\Delta\delta = \delta_{syn} - \delta_{anti}$ depends on the dihedral angle Θ (H-C-C=N):

Θ	$\Delta\delta$ (ppm)
0°	1
60°	0
115°	-0.3

H
θ
N

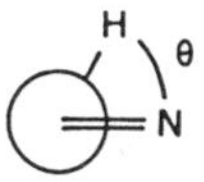

$\overset{a}{CH_3}$-N=$\overset{b}{CH}$-C_6H_5	(a) 3.4 (b) 8.40
$\overset{a}{CH_3}$-C($\overset{b}{H}$)=N-N=C(H)-CH_3	(a) 2.03 (b) 7.89
$\overset{a}{CH_3}$-C($\overset{b}{CH_3}$)=N-N=C(CH_3)-CH_3	(a) 2.00 (b) 1.83

^{1}H-Chemical Shifts (δ in ppm relative to TMS) and Coupling Constants (J in Hz) in Nitriles, Isonitriles, Cyanates and Isocyanates

For thiocyanates and isothiocyanates see p. H115.

$\overset{a}{CH_3}CN$	(a) 1.98	$(\overset{b}{CH_3})_2\overset{a}{CH}CN$	(a) 2.67 (b) 1.35
$\overset{b}{CH_3}\overset{a}{CH_2}CN$	(a) 2.35 (b) 1.31	$(\overset{a}{CH_3})_3CCN$	(a) 1.37
$\overset{a}{CH_3}NC$	(a) 2.85	$(\overset{b}{CH_3})_2\overset{a}{CH}NC$	(a) 4.83 (b) 1.45

In isonitriles the quadrupole relaxation of the nitrogen nucleus is so small that the ^{14}N-^1H couplings become observable:

$-\overset{\beta}{CH_2}-\overset{\alpha}{CH_2}-{}^{14}NC$ $\quad |J|_{H_\alpha N} = 1.8\text{-}2.8$, $\quad |J|_{H_\beta N} = 2.5\text{-}3.5$

$\overset{b}{CH_3}\overset{a}{CH_2}OCN$	(a) 4.54 (b) 1.45
$\overset{b}{CH_3}\overset{a}{CH_2}NCO$	(a) 3.37 (b) 1.20

^{1}H-Chemical Shifts (δ in ppm relative to TMS) and Coupling Constants (J in Hz) in Saturated Alicyclics

For unsaturated alicyclics see p. H230 - H240.

cyclopropane	0.22	geminal coupling:	J_{gem} = -3...-9
		vicinal couplings:	J_{cis} = 7...13
			J_{trans} = 4...9.5
		always $J_{cis} > J_{trans}$	
cyclobutane	1.96	geminal coupling:	J_{gem} = -11...-17
		vicinal couplings:	J_{cis} = 4...12
			J_{trans} = 2...10
		"long-range" coupling:	J = 0.5...2.0
cyclopentane	1.51	geminal coupling:	J_{gem} = -8...-18
		vicinal couplings:	J_{cis} = 5...10
			J_{trans} = 5...10
cyclohexane	1.44 (for -100°C: H_{ax}: 1.1, H_{eq}: 1.6)	geminal coupling:	J_{gem} = -11...-14
		vicinal couplings:	$J_{ax,ax}$ = 8...13
			$J_{aq,ax}$ = 2...6
			$J_{eq,eq}$ = 2...5
		generally $J_{eq,ax} \sim J_{eq,eq} + 1$	
adamantane (H_a, H_b)	(a) 1.8 (b) 1.9		

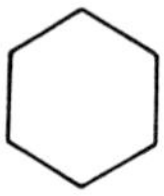

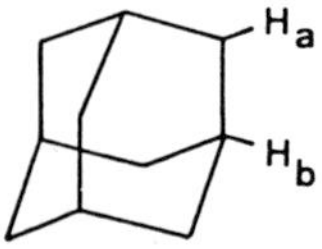

Couplings in Condensed Alicycles: see A.P. Marchand, Stereochemical Applications of NMR Studies in Rigid Bicyclic Systems, Verlag Chemie International, Deerfield Beach, Florida, 1982.

In condensed alicyclics "long-range" couplings (coupling over 4 or more bonds) are often observed. They are particularly large if the bonds between the two protons are w-shaped:

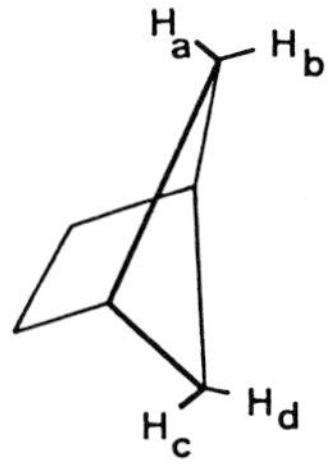

$J_{ac} = \sim 7$

$J_{ad} = J_{bd} = \sim 0$

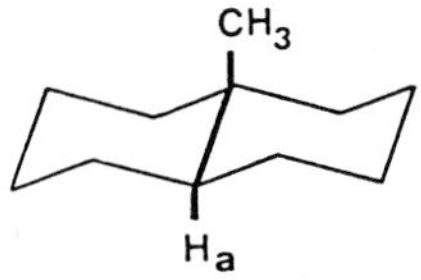

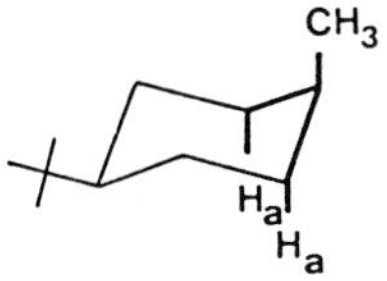

the signal of the methyl group is significantly broadened due to the "long-range" coupling with H_a

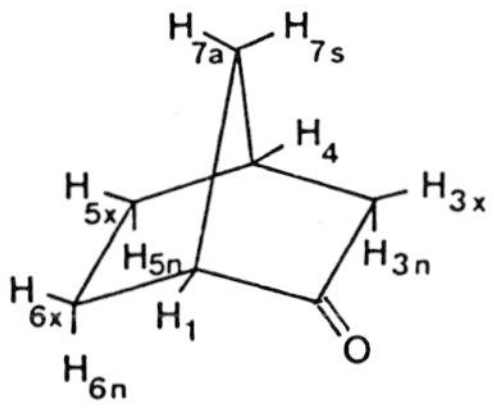

	δ	H	J	H	J
H_1	2.41	1,4	1.2	5n,5x	-12.8
H_{3n}	1.73	1,5n	-0.3	5n,6n	9.1
H_{3x}	1.95	1,5x	0.2	5n,6x	4.7
H_4	2.61	1,6n	0.1	5n,7a	-0.1
H_{5n}	1.41	1,6x	4.7	5n,7s	2.1
H_{5x}	1.76	1,7a	1.2	5x,6n	4.6
H_{6n}	1.44	1,7s	1.6	5x,6x	12.1
H_{6x}	1.76				
H_{7a}	1.51	3n,3x	-17.6	6n,6x	-12.3
H_{7s}	1.69	3n,7a	4.2	6n,7a	-0.1
		3x,4	4.8	6n,7s	2.3
		3x,5x	2.3		
				7a,7s	-10.2
		4,5n	0.1		
		4,5x	4.3		
		4,6n	-0.5		
		4,6x	0.7		
		4,7a	2.1		
		4,7s	1.6		

^{1}H-Chemical Shifts of Axial and Equatorial Protons in Substituted Cyclohexanes[1)] (δ in ppm relative to TMS)

Substituent X	$\delta_{H_{ax}}$	$\delta_{H_{eq}}$
-D	1.12	1.60
$-CH_3$	1.27	1.93
-phenyl	2.47	2.98
-Cl	3.63	4.34
-Br	3.81	4.62
-I	3.98	4.72
-OH	3.38	3.89
$-OCOCH_3$	4.46	4.98
$-NH_2$	2.52	3.15
$-NHCH_3$	2.08	2.70
$-NO_2$	4.23	4.43
-SH	2.57	3.43

[1)] These values were determined by measurements at low temperatures or on 4-tert.-butyl-substituted cycloalkanes.

Effect of the Substituent on the Chemical Shift of Protons in Cyclohexane (in ppm, a negative sign indicates shielding by the substituent)

Substituent X,	position	H_1 ax or eq	H_2 ax	H_2 eq	H_3 ax	H_3 eq	H_4 ax	H_4 eq
$-CH_3$	eq	0.15	-0.31	-0.03	0.03	0	-0.06	-0.02
	ax	0.33	0.25	-0.20	0.27	-0.26	-0.06	-0.02
-Cl	eq	2.51						
	ax	2.74			0.6			
-OH	eq	2.26	-0.03	0.18	0.07	0.01	0.01	0.06
	ax	2.29	0.23	-0.02	0.46	-0.27	0.02	-0.08
$-OCOCH_3$	eq	3.34						
	ax	3.38	0.35	∿0.7				
$-NO_2$	eq	3.11	1.1	0.3				
	ax	2.83	0.5	1.0				
-SH	eq	1.45	-0.4	-0.3				
	ax	1.83		-0.1	0.8			

^{1}H-NMR

ALKENES

^{1}H-Chemical Shifts (δ in ppm relative to TMS) and Coupling Constants (J in Hz) in Alkenes

Ethylene

H₂C=CH₂ (structure) 5.28

geminal coupling: $J_{gem} = 2.5$

vicinal couplings: $J_{cis} = 11.6$

$J_{trans} = 19.1$

The coupling constants depend strongly on the electronegativity of the substituents. They decrease with increasing electronegativity of the substituents and their number (the contribution of the geminal coupling constant may, however, increase because J_{gem} is often negative in substituted ethylenes). J_{trans} is always > J_{cis}.

(H_b)(H_c)C=C(H_a)(R)

R	J_{ab}	J_{ac}	J_{bc}
$-CH_3$	10.0	16.8	2.1
-F	4.7	12.7	-3.2
$-OCH_3$	7.0	14.1	-2.0
$-SCH_3$	10.3	16.4	-0.3
$-COCH_3$	10.7	18.7	1.3
-Li	19.3	23.9	7.1

Coupling Over More than Three Bonds ("Long-Range" Coupling)

If there is a double bond between two coupling nuclei, coupling across four (allylic coupling) or five (homoallylic coupling) bonds may be observed.

Allylic Couplings

H_b, H_c, H_a, ϕ (structure)

cisoid: $J_{ab} = -3.....+2$

transoid: $J_{ac} = -3.5...+2.5$

the magnitude of the coupling constant depends on the conformation:

ϕ	J_{ab}	J_{ac}
0°	-3.0	-3.5
90°	+1.8	+2.2
180°	-3.0	-3.5
270°	0	0.8

The frequently used rough rule that $|J|_{cisoid} > |J|_{transoid}$ does not hold generally. It most often holds in acyclic systems.

Homoallylic coupling

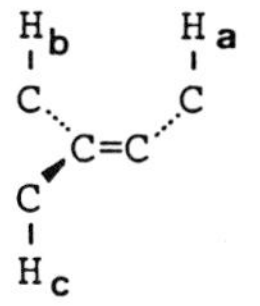

cisoid: $|J|_{ab} = 0...3$

transoid: $|J|_{ac} = 0...3$

generally: $J_{H-C=C-CH_3} \sim -J_{CH_3-C=C-CH_3}$

In acyclic systems the relationship $|J|_{cisoid} < |J|_{transoid}$ generally holds.

Large homoallylic couplings are generally observed in cyclic systems:

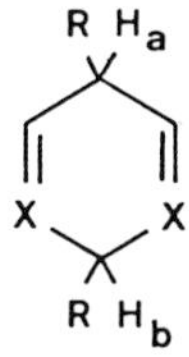

$J_{ab} = 5...11$

$J_{ab} = 0...7$

X : CH, N R : any substituent X : O, NH

Butadiene

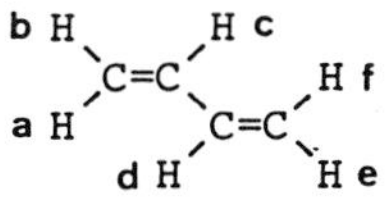

(a) 5.16

(b) 5.06

(c) 6.27

J_{ab} = 1.8	J_{bc} = 10.2
J_{ac} = 17.1	J_{bd} = -0.9
J_{ad} = -0.8	J_{be} = 1.3
J_{ae} = 0.6	J_{cd} = 10.4
J_{af} = 0.7	

Allene

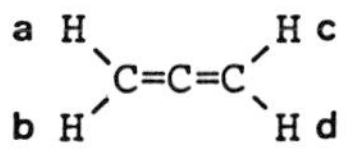

4.67

$J_{ab} = -9$

$J_{ac} = -6$

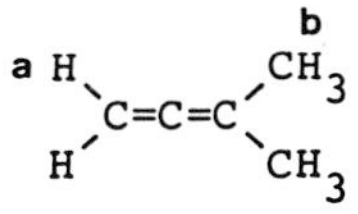

$|J|_{ab} = 3.0$

^{1}H-NMR

ALKENES, ADDITIVITY RULE

The Chemical Shift of Protons at a Double Bond

(δ in ppm relative to TMS)

$$\delta_{C=CH} = 5.25 + Z_{gem} + Z_{cis} + Z_{trans}$$

R_{cis} and R_{trans} on one carbon, H and R_{gem} on the other carbon of C=C

	Substituent R	Z_{gem}	Z_{cis}	Z_{trans}
C	-H	0	0	0
	-alkyl	0.45	-0.22	-0.28
	-alkyl ring[1]	0.69	-0.25	-0.28
	$-CH_2$-aromatic	1.05	-0.29	-0.32
	$-CH_2X$, X: F, Cl, Br	0.70	0.11	-0.04
	$-CHF_2$	0.66	0.32	0.21
	$-CF_3$	0.66	0.61	0.32
	$-CH_2O$	0.64	-0.01	-0.02
	$-CH_2N$	0.58	-0.10	-0.08
	$-CH_2S$	0.71	-0.13	-0.22
	$-CH_2CO$, CH_2CN	0.69	-0.08	-0.06
	-C=C isolated	1.00	-0.09	-0.23
	-C=C conjugated[2]	1.24	0.02	-0.05
	-C≡C	0.47	0.38	0.12
	-aromatic free rotation	1.38	0.36	-0.07
	-aromatic fixed[3]	1.60	-	-0.05
	-aromatic o-substituted	1.65	0.19	0.09
HAL	-F	1.54	-0.40	-1.02
	-Cl	1.08	0.18	0.13
	-Br	1.07	0.45	0.55
	-I	1.14	0.81	0.88

1) The increment for "alkyl ring" is to be used if the substituent and the double bond are part of a cyclic structure.

2) The increment "C=C conjugated" is to be used if either the double bond or the C=C substituent is conjugated to other substituents.

3) The increment "aromatic, fixed" is to be used if the double bond conjugated to an aromatic ring is part of a fused ring (such as in 1,2-dihydronaphthalene).

	Substituent R	z_{gem}	z_{cis}	z_{trans}
O	-OR, R aliphatic	1.22	-1.07	-1.21
	-OR, R unsaturated	1.21	-0.60	-1.00
	-OCOR	2.11	-0.35	-0.64
N	$-NH_2$	0.80	-1.26	-1.21
	-NHR, R aliphatic	0.80	-1.26	-1.21
	$-NR_2$, R aliphatic	0.80	-1.26	-1.21
	-NHR, R unsaturated	1.17	-0.53	-0.99
	-NRR', R unsaturated, R' any substituent	1.17	-0.53	-0.99
	-NCOR	2.08	-0.57	-0.72
	-N=N-phenyl	2.39	1.11	0.67
	$-NO_2$	1.87	1.30	0.62
S	-SR	1.11	-0.29	-0.13
	-SOR	1.27	0.67	0.41
	$-SO_2R$	1.55	1.16	0.93
	-SCOR	1.41	0.06	0.02
	-SCN	0.94	0.45	0.41
	$-SF_5$	1.68	0.61	0.49
O=C<	-CHO	1.02	0.95	1.17
	-CO isolated	1.10	1.12	0.87
	-CO conjugated [1)]	1.06	0.91	0.74
	-COOH isolated	0.97	1.41	0.71
	-COOH conjugated [1)]	0.80	0.98	0.32
	-COOR isolated	0.80	1.18	0.55
	-COOR conjugated [1)]	0.78	1.01	0.46
	$-CONR_2$	1.37	0.98	0.46
	-COCl	1.11	1.46	1.01
	-CN	0.27	0.75	0.55
	$-PO(OCH_2CH_3)_2$	0.66	0.88	0.67
	$-OPO(OCH_2CH_3)_2$	1.33	-0.34	-0.66

1) The increment "conjugated" is to be used if either the double bond or the substituent is conjugated to additional substituents.

^{1}H-Chemical Shifts in Substituted Isobutenes

(δ in ppm relative to TMS)

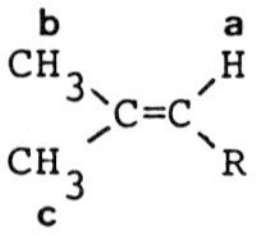

$$\overset{b}{CH_3}\diagdown C{=}C \diagup \overset{a}{H}$$
$$\underset{c}{CH_3}\diagup \qquad \diagdown R$$

R	δ_a	δ_b	δ_c
-H		1.70	1.70
$-C(CH_3)_3$	5.13	1.68	1.62
-C≡CH	5.17	1.80	1.88
-Br	5.78	1.75	1.75
$-OCOCH_3$	6.79	1.65	1.65
-CHO	5.63	1.91	2.11
$-COCH_3$	5.97	1.86	2.06
$-COOCH_3$	5.62	1.84	2.12
-COCl	6.01	1.97	2.12

^{1}H-Chemical Shifts (δ in ppm relative to TMS) and Coupling Constants (J in Hz) in Acetylene Derivatives

Compound	δ
H-C≡C-H	1.80
H-C≡C-alkyl	1.7-1.9
H-C≡C-C=C	2.6-3.1
H-C≡C-C≡C	1.7-2.4
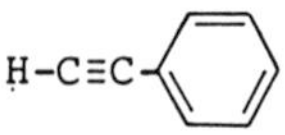 H-C≡C-C$_6$H$_5$	2.7-3.4
H-C≡C-O-alkyl	1.3
H-C≡C-CO	2.1-3.3

Compound	J
CH_3-C≡C-H	|J| = 2.9
CH_3-C≡C-CH_3	|J| = 2.7
H-C≡C-C≡C-H	|J| = 2.2

Compound	δ
$\overset{a}{CH_3}$-C≡C-$\overset{b}{H}$	(a) 1.80 (b) 1.80
$\overset{b}{CH_3}\overset{a}{CH_2}$-C≡CH	(a) 2.16 (b) 1.15
$\overset{b}{(CH_3)_2}\overset{a}{CH}$-C≡CH	(a) 2.59 (b) 1.15
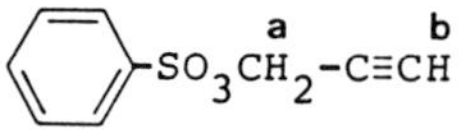 C$_6$H$_5$-$SO_3\overset{a}{CH_2}$-C≡$\overset{b}{CH}$	(a) ∿4.7 (b) 2.55
$CH_3CONH\overset{a}{CH_2}$-C≡$\overset{b}{CH}$	(a) 4.06 (b) 2.25

^{1}H-Chemical Shifts (δ in ppm relative to TMS) and Coupling Constants (J in Hz) in Unsaturated Alicyclics

(a,b) 7.01
(c) 0.92

in derivatives: J_{ab} = 0.5-1.5
J_{bc} = 1.8

(a,b) 5.95
(c,d) 2.57

geminal coupling: $J_{cc'}$ = -12.0
vicinal couplings: J_{ab} = 2.7 (in derivatives: 2.5-4.0)
J_{bc} = -0.8
$J_{cd(cis)}$ = 4.4
$J_{cd(trans)}$ = 1.7
"long-range" coupling: J_{ac} = 1.6

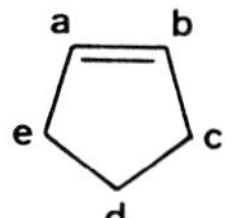

(a,b) 5.60
(c,e) 2.28
(d) 1.90

in derivatives: J_{ab} = 5.0-7.0
J_{bc} = 0.5

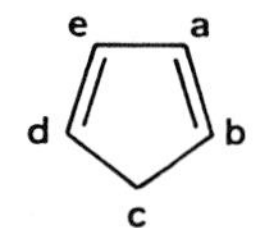

(a,e) 6.5
(b,d) 6.4
(c) 2.90

vicinal couplings: J_{ab} = 5.1
J_{bc} = 1.2
J_{ae} = 1.9
"long-range" couplings: J_{ac} = -1.3
J_{ad} = 1.1
J_{bd} = 1.9

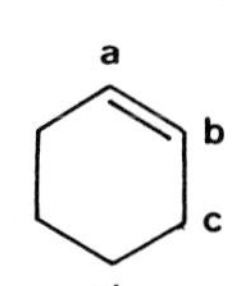

(a,b) 5.59
(c) 1.96
(d) 1.65

in derivatives: J_{ab} = 8.5-11.0
J_{bc} = 1.5

(a,d)	5.8	vicinal couplings:	$J_{ab} = 9.4$
(b,c)	5.9		$J_{bc} = 5.1$
(e,f)	2.15	"long-range" couplings:	$J_{ac} = 1.1$
			$J_{ad} = 0.9$

(a,b)	5.71	in derivatives:	$J_{ab} = 9.0\text{-}12.5$
(c)	2.11		$J_{bc} = 3.7$

(a,f)	5.26	geminal coupling:	$J_{gg'} = -13.0$
(b,e)	6.09	vicinal couplings:	$J_{ab} = 8.9$
(c,d)	6.50		$J_{bc} = 5.5$
(g)	2.22		$J_{cd} = 11.2$
			$J_{ag} = 6.7$
		"long-range" couplings:	$J_{ac} + J_{ad} = 1.5$
			$J_{ae} = J_{af} = 0$
			$J_{bd} = 0.8$
			$J_{be} = -0.6$
			$J_{bf} = 0$

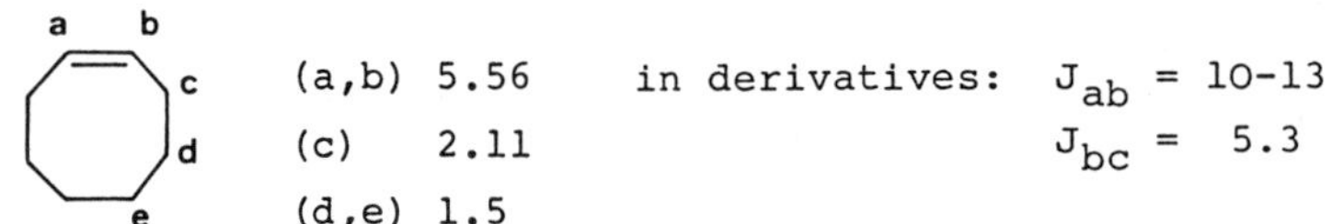

(a,b)	5.56	in derivatives:	$J_{ab} = 10\text{-}13$
(c)	2.11		$J_{bc} = 5.3$
(d,e)	1.5		

^{1}H-Chemical Shifts (δ in ppm relative to TMS) and Coupling Constants (J in Hz) in Alicyclic Rings Condensed with Aromatic Systems

(See also: cyclic ketones, p. H130.)

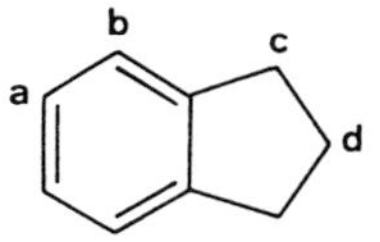

(a,b) ∿7.2
(c) 2.91
(d) 2.04

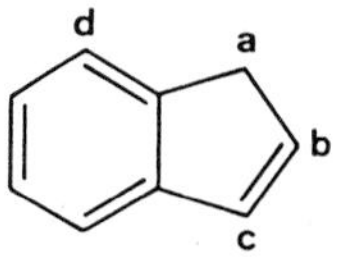

(a) 3.33	J_{ab} = 2.0	J_{bc} = 5.8
(b) 6.50	J_{ac} = 2.0	J_{cd} = 0.7
(c) 6.82		

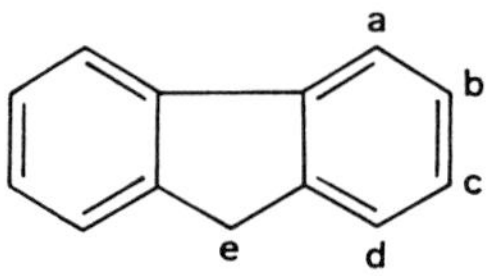

(a) 7.84	J_{ab} = 7.5	J_{bc} = 6.5
(b) 7.38	J_{ac} = 1.6	J_{bd} = 1.6
(c) 7.28	J_{ad} = 0.2	J_{cd} = 7.2
(d) 7.55		
(e) 3.87		

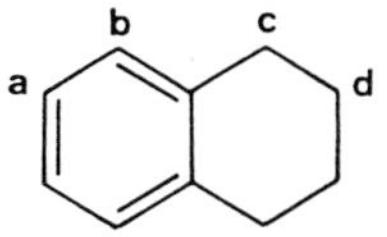

(a) 6.93
(b) 7.01
(c) 2.85
(d) 1.60

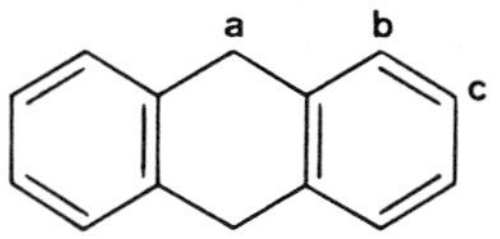

(a) 3.91
(b) 7.31
(c) 7.19

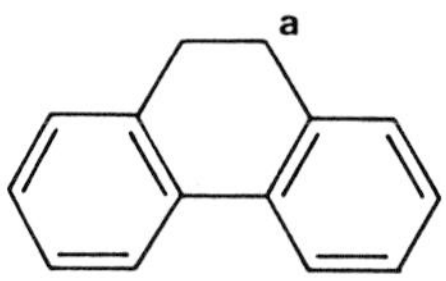

(a) 2.86

For acenaphthene and acenaphthylene see p. H250.

^{1}H-Chemical Shifts (δ in ppm relative to TMS) and Coupling Constants (J in Hz) in Aromatic Hydrocarbons

(Substituted benzenes: see p. H255, H260.)

Benzene: 7.26 — in derivatives: $J_{ortho} = 6.0-9.0$, $J_{meta} = 1.0-3.0$, $J_{para} = 0-1.0$

"Long-range" couplings with the protons of substituents:

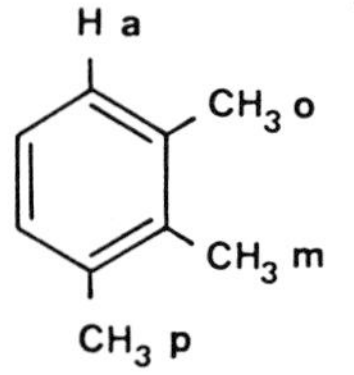

H a; CH_3 o; CH_3 m; CH_3 p

$J_{ao} = -0.6...-0.9$
$J_{am} = 0...+0.3$
$J_{ap} = \sim-0.6$

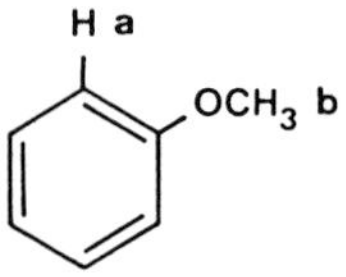

H a; OCH_3 b

$J_{ab} = 0-0.8$

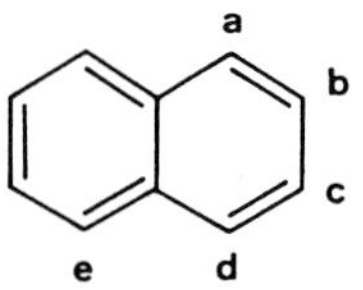

Naphthalene (positions a, b, c, d, e):

(a) 7.66
(b) 7.30

$J_{ab} = 8.5$ (in derivatives: 8-9)
$J_{bc} = 7.5$ (in derivatives: 5-7)
$J_{ac} = 1.4$ (in derivatives: 1-2)
$J_{ad} = 0.7$ (in derivatives: $\sim$1)
J_{ae} in derivatives: $\sim$1

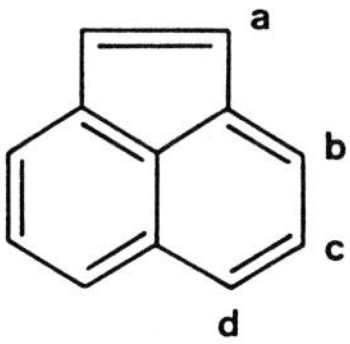

(a) 7.15	J_{ab} = 0	J_{bc} = 7
(b) 7.90	J_{ac} = 0	J_{bd} = 0.6
(c) 7.58	J_{ad} = 0	J_{cd} = 8
(d) 7.79		

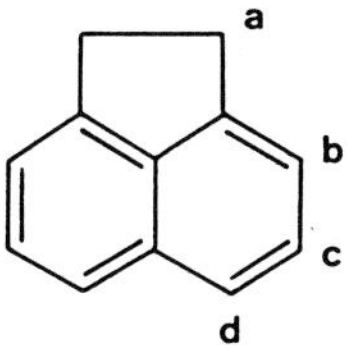

(a) 3.34	J_{bc} = 6.7	J_{ab} = 1.5
(b) 7.11	J_{cd} = 8.1	J_{ad} = 0.5
(c) 7.31	J_{bd} = 1.2	
(d) 7.46		

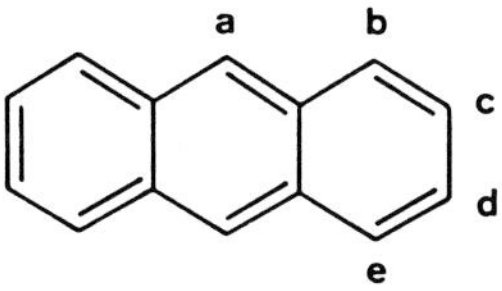

(a) 8.31	J_{bc} = 8.4	J_{be} = 0.5
(b) 7.91	J_{bd} = 1.5	J_{cd} = 6.0
(c) 7.39		

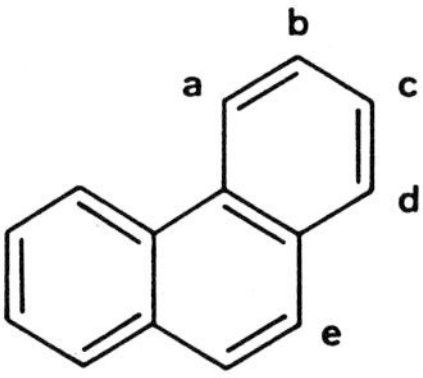

(a) 8.93	J_{ab} = 8.4	J_{bc} = 7.2
(b) 7.88	J_{ac} = 1.2	J_{bd} = 1.3
(c) 7.82	J_{ad} = 0.7	J_{cd} = 8.1
(d) 8.12		
(e) 7.71		

Effect of a Substituent on the Chemical Shift of the Ring-Protons in Benzene (δ in ppm relative to TMS)

(Benzene ring with substituent X; ring positions labelled 2, 3, 4)

$$\delta_{H_i} = 7.26 + Z_i$$

	Substituent X	Z_2	Z_3	Z_4
	-H	0	0	0
	$-CH_3$	-0.20	-0.12	-0.22
	$-CH_2CH_3$	-0.14	-0.06	-0.17
	$-CH(CH_3)_2$	-0.13	-0.08	-0.18
	$-C(CH_3)_3$	0.02	-0.08	-0.21
	$-CH_2Cl$	0.00	0.00	0.00
	$-CF_3$	0.32	0.14	0.20
C	$-CCl_3$	0.64	0.13	0.10
	$-CH_2OH$	-0.07	-0.07	-0.07
	$-CH{=}CH_2$	0.06	-0.03	-0.10
	-CH=CH-phenyl	0.15	-0.01	-0.16
	-C≡CH	0.15	-0.02	-0.01
	-C≡C-phenyl	0.19	0.02	0.00
	-phenyl	0.37	0.20	0.10
H	-F	-0.26	0.00	-0.20
A	-Cl	0.03	-0.02	-0.09
L	-Br	0.18	-0.08	-0.04
	-I	0.39	-0.21	0.00
	-OH	-0.56	-0.12	-0.45
	$-OCH_3$	-0.48	-0.09	-0.44
	$-OCH_2CH_3$	-0.46	-0.10	-0.43
O	-O-phenyl	-0.29	-0.05	-0.23
	$-OCOCH_3$	-0.25	0.03	-0.13
	-OCO-phenyl	-0.09	0.09	-0.08
	$-OSO_2CH_3$	-0.05	0.07	-0.01

	Substituent X	Z_2	Z_3	Z_4
N	$-NH_2$	-0.75	-0.25	-0.65
	$-NHCH_3$	-0.80	-0.22	-0.68
	$-N(CH_3)_2$	-0.66	-0.18	-0.67
	$-N^+(CH_3)_3\ I^-$	0.69	0.36	0.31
	$-NHCOCH_3$	0.12	-0.07	-0.28
	$-N(CH_3)COCH_3$	-0.16	0.05	-0.02
	$-NHNH_2$	-0.60	-0.08	-0.55
	-N=N-phenyl	0.67	0.20	0.20
	-NO	0.58	0.31	0.37
	$-NO_2$	0.95	0.26	0.38
S	-SH	-0.08	-0.16	-0.22
	$-SCH_3$	-0.08	-0.10	-0.24
	-S-phenyl	0.06	-0.09	-0.15
	$-SO_3CH_3$	0.60	0.26	0.33
	$-SO_2Cl$	0.76	0.35	0.45
C=O	-CHO	0.56	0.22	0.29
	$-COCH_3$	0.62	0.14	0.21
	$-COCH_2CH_3$	0.63	0.13	0.20
	$-COC(CH_3)_3$	0.44	0.05	0.05
	-CO-phenyl	0.47	0.13	0.22
	-COOH	0.85	0.18	0.27
	$-COOCH_3$	0.71	0.11	0.21
	$-COOCH(CH_3)_2$	0.70	0.09	0.19
	-COO-phenyl	0.90	0.17	0.27
	$-CONH_2$	0.61	0.10	0.17
	-COCl	0.84	0.22	0.36
	-COBr	0.80	0.21	0.37
	-CH=N-phenyl	~0.6	~0.2	~0.2
	-CN	0.36	0.18	0.28
	$-Si(CH_3)_3$	0.22	-0.02	-0.02
	$-PO(OCH_3)_2$	0.48	0.16	0.24

^{1}H-Chemical Shifts (δ in ppm relative to TMS) and Coupling Constants (J in Hz) in Non-Condensed Heteroaromatics

For non-aromatic heterocycles see: cyclic ethers, amines and thioethers.

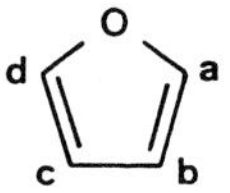

Furan (O; ring positions a, b, c, d)

(a,d)	7.38	J_{ab}	= 1.8	J_{ad}	= 1.5
(b,c)	6.30	J_{ac}	= 0.9	J_{bc}	= 3.4

Substituted furans: see p. H285, H290.

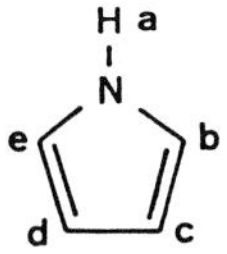

Pyrrole (N–H a; ring positions b, c, d, e)

(a)	7-12 (very solvent-dependent, broad)				
(b,e)	6.62	J_{ab}	= 2.6	J_{bd}	= 1.3
(c,d)	6.05	J_{ac}	= 2.3	J_{be}	= 2.1
		J_{bc}	= 2.6	J_{cd}	= 3.5

Substituted pyrroles: see p. H295, H300.

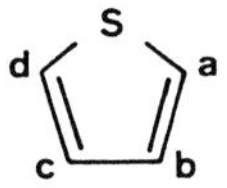

Thiophene (S; ring positions a, b, c, d)

(a,d)	7.20	J_{ab}	= 4.8	J_{ad}	= 2.8
(b,c)	6.96	J_{ac}	= 1.0	J_{bc}	= 3.5

Substituted thiophenes: see p. H305, H310.

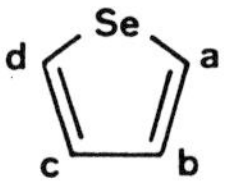

Selenophene (Se; ring positions a, b, c, d)

(a,d)	7.70	J_{ab}	= 5.4	J_{ad}	= 2.5
(b,c)	7.12	J_{ac}	= 1.1	J_{bc}	= 3.6

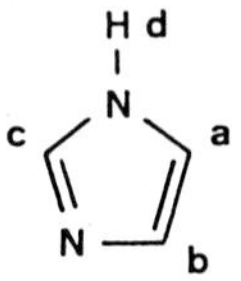

Imidazole (N–H d; a, b, c)

	δ	
(a,b)	7.13	in derivatives: J_{ab} = 1-2
(c)	7.70	J_{ac} = 1-2
(d)	13.4	J_{bc} = 0.5-1.5

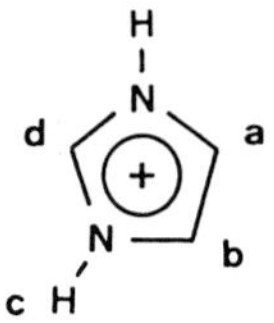

Imidazolium (H, a, b, c H, d)

	δ	
(a,b)	7.6	$J_{ac} + J_{bc}$ = 4.4
(c)	∿12	J_{cd} = 2.4
(d)	8.6	J_{ad} = 1.4

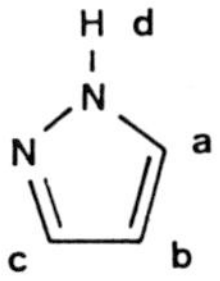

Pyrazole (N–H d; a, b, c)

	δ		
(a,c)	7.55	J_{ab} = 2.1	in derivatives: J_{ab} = 2-3
(b)	6.25		J_{bc} = 1-2
(d)	13.7		J_{ac} = 0.5-1

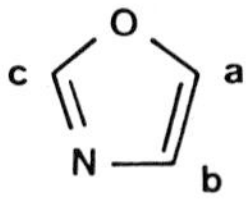

Oxazole

	δ	
(a)	7.69	J_{ab} = 0.8
(b)	7.09	J_{ac} = 0.5
(c)	7.95	

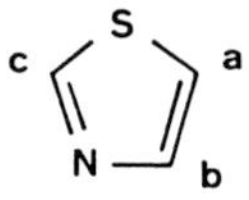

Thiazole

	δ	
(a)	7.41	J_{ab} = 3.2
(b)	7.98	J_{ac} = 1.9
(c)	8.88	J_{bc} = 0

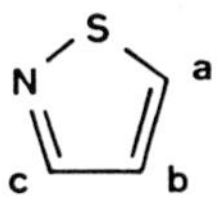

Isothiazole

	δ	
(a)	8.72	J_{ab} = 4.7
(b)	7.26	J_{ac} < 0.4
(c)	8.56	J_{bc} = 1.7

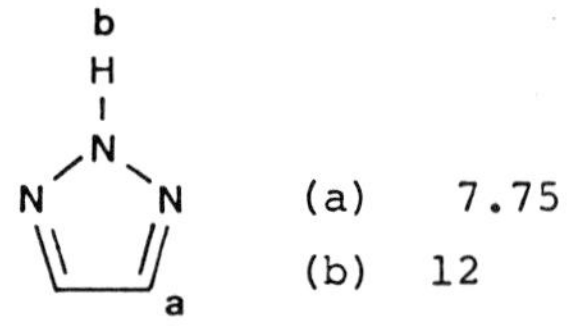

(a) 7.75
(b) 12

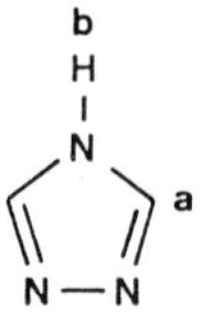

(a) 8.27
(b) 13.5

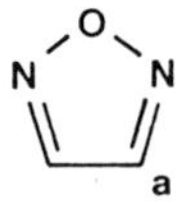

(a) 8.19

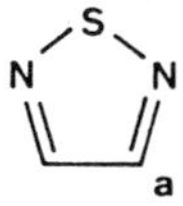

(a) 8.58

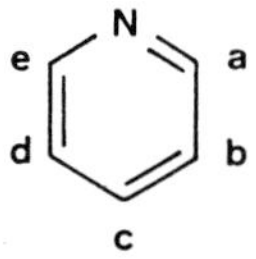

(a,e) 8.60	J_{ab} = 5.5	in derivatives:	J_{ab} = 4-6
(b,d) 7.25	J_{ac} = 1.9		J_{ac} = 0-2.5
(c) 7.64	J_{ad} = 0.9		J_{ad} = 0-2.5
	J_{ae} = 0.4		J_{ae} = 0-0.6
	J_{bc} = 7.6		J_{bc} = 7-9
	J_{bd} = 1.6		J_{bd} = 0.5-2

Substituted pyridines: see p. H315, H320.

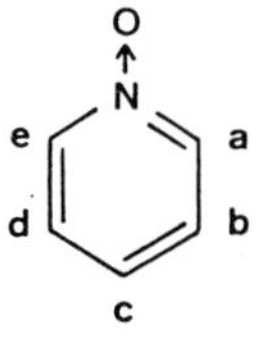

(a,e) 8.10
(b,d) 7.28
(c) 7.08

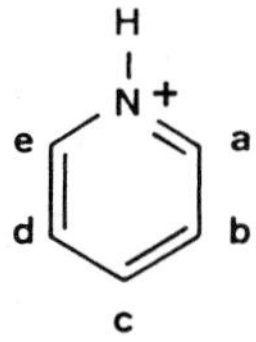

(a,e) 9.23	J_{ab} = 6.0	J_{ae} = 1.0
(b,d) 8.50	J_{ac} = 1.5	J_{bc} = 8.0
(c) 9.04	J_{ad} = 0.8	J_{bd} = 1.4

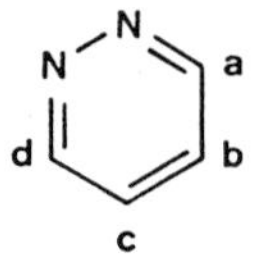

(a,d)	9.24	J_{ab} = 4.9
(b,c)	7.55	J_{ac} = 2.0
		J_{ad} = 3.5
		J_{bc} = 8.4

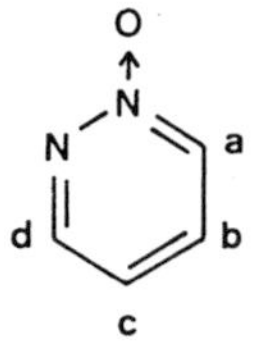

(a)	8.26	J_{ab} = 6.5
(b)	7.83	J_{ac} = 1.0
(c)	7.22	J_{ad} = 1.0
(d)	8.54	J_{bc} = 8.0
		J_{bd} = 2.5

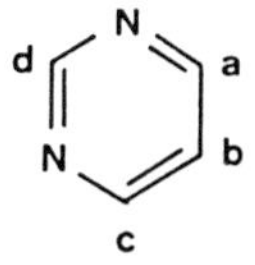

(a,c)	8.78	J_{ab} = 5.0
(b)	7.36	J_{ac} = 2.5
(d)	9.26	J_{ad} = 1.5
		J_{bd} = 0

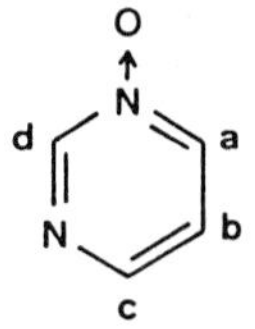

(a)	8.43	J_{ab} = 6.8	J_{bc} = 4.9
(b)	7.34	J_{ac} = 1.6	J_{bd} = 1.0
(c)	8.24	J_{ad} = 2.0	J_{cd} = 0
(d)	8.98		

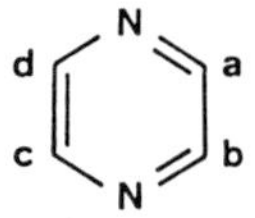

(a)	8.63	J_{ab} = 1.8
		J_{ac} = 1.8
		J_{ad} = 0.5

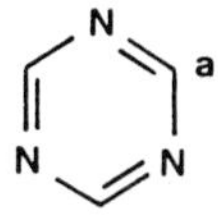

(a) 9.25

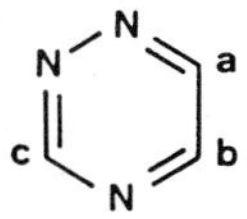

(a)	9.48
(b)	8.84
(c)	9.88

Effect of a Substituent on the Chemical Shift of the Ring-Protons in Monosubstituted Furan (δ in ppm relative to TMS)

(For couplings see p. H265.)

$$\delta_{H_2} = 7.38 + Z_{i2}$$
$$\delta_{H_3} = 6.30 + Z_{i3}$$
$$\delta_{H_4} = 6.30 + Z_{i4}$$
$$\delta_{H_5} = 7.38 + Z_{i5}$$

2- or 5-Substituent (i = 2 or 5)		$Z_{23} = Z_{54}$	$Z_{24} = Z_{53}$	$Z_{25} = Z_{52}$
	-H	0	0	0
C	$-CH_3$	-0.42	-0.12	-0.17
C	$-CH_2OH$	-0.11	-0.05	-0.08
C	$-CH_2NH_2$	-0.24	-0.06	-0.10
C	-CH=CHCHO	0.70	0.35	0.42
HAL	-Br	-0.02	0.03	-0.01
HAL	-I	0.12	-0.13	-0.01
O	$-OCH_3$	-1.34	-0.23	-0.68
N	$-NO_2$	1.21	0.55	0.51
S	$-SCH_3$	-0.12	-0.06	-0.09
S	-SCN	0.40	0.06	0.10
C=O	-CHO	0.93	0.31	0.34
C=O	$-COCH_3$	0.81	0.23	0.19
C=O	$-COCF_3$	1.34	0.50	0.64
C=O	-COOH	0.94	0.33	0.41
C=O	$-COOCH_3$	0.85	0.22	0.25
C=O	-COCl	1.20	0.39	0.48
C=O	-CN	0.85	0.32	0.28

3- or 4-Substituent (i = 3 or 4)	$Z_{32} = Z_{45}$	$Z_{34} = Z_{43}$	$Z_{35} = Z_{42}$
-H	0	0	0
$-CH_3$	-0.27	-0.17	-0.15
-I	-0.13	0.04	-0.22
$-OCH_3$	-0.46	-0.28	-0.37
$-SCH_3$	-0.18	-0.05	-0.15
-SCN	0.19	0.19	0.03
-CHO	0.48	0.37	-0.07
$-COCH_3$	0.46	0.36	-0.12
-COOH	0.89	0.54	0.36
$-COOCH_3$	0.45	0.33	-0.14
-CN	0.45	0.22	-0.02

Effect of a Substituent on the Chemical Shift of the Ring-Protons in Monosubstituted Pyrrole (δ in ppm relative to TMS)

(For couplings see p. H265.)

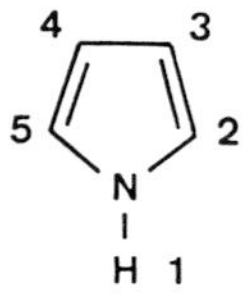

δ_{H_1} = 7-12 (strongly solvent-dependent, generally broad)

$\delta_{H_2} = 6.62 + Z_{i2}$

$\delta_{H_3} = 6.05 + Z_{i3}$

$\delta_{H_4} = 6.05 + Z_{i4}$

$\delta_{H_5} = 6.62 + Z_{i5}$

1-Substituent (i = 1)	$Z_{12} = Z_{15}$	$Z_{13} = Z_{14}$
-H	0	0
$-CH_3$	-0.25	-0.13
$-CH_2CH_3$	-0.16	-0.12
$-CH_2$-phenyl	-0.12	-0.04
-phenyl	0.33	0.14
$-COCH_3$	0.56	0.12

2- or 5-Substituent (i = 2 or 5)	$Z_{23} = Z_{54}$	$Z_{24} = Z_{53}$	$Z_{25} = Z_{52}$
-H	0	0	0
$-CH_3$	-0.33	-0.16	-0.26
$-NO_2$	1.06	0.24	0.43
$-SCH_3$	0.18	0.05	0.10
-SCN	0.48	0.10	0.28
-CHO	0.93	0.27	0.61
$-COCH_3$	0.78	0.10	0.44
$-COOCH_3$	0.79	0.13	0.29
-CN	0.83	0.23	0.51

3- or 4-Substituent (i = 3 or 4)	$Z_{32} = Z_{45}$	$Z_{34} = Z_{43}$	$Z_{35} = Z_{42}$
-H	0	0	0
$-CH_3$	-0.34	-0.20	-0.20
$-NO_2$	1.04	0.70	0.13
$-COCH_3$	0.79	0.63	0.15
$-COOCH_3$	0.90	0.73	0.16

THIOPHENE, SUBSTITUENT EFFECTS

Effect of a Substituent on the Chemical Shift of the Ring-Protons in Monosubstituted Thiophene (δ in ppm relative to TMS)

(For couplings see p. H265.)

(Thiophene ring numbering: S; positions 2, 3, 4, 5)

$$\delta_{H_2} = 7.20 + Z_{i2}$$
$$\delta_{H_3} = 6.96 + Z_{i3}$$
$$\delta_{H_4} = 6.96 + Z_{i4}$$
$$\delta_{H_5} = 7.20 + Z_{i5}$$

	2- or 5-Substituent (i = 2 or 5)	$Z_{23} = Z_{54}$	$Z_{24} = Z_{53}$	$Z_{25} = Z_{52}$
	-H	0	0	0
C	$-CH_3$	-0.36	-0.24	-0.29
	-C≡CH	0.15	-0.16	-0.12
HAL	-Cl	-0.25	-0.22	-0.22
	-Br	-0.05	-0.27	-0.11
	-I	0.13	-0.33	0.01
O	-OH [1)]	-0.72	0.59	-3.10
	$-OCH_3$	-0.94	-0.43	-0.82
N	$-NH_2$	-0.95	-0.45	-0.85
	$-NO_2$	0.82	-0.03	0.30
S	-SH	0.00	-0.20	-0.07
	$-SCH_3$	-0.03	-0.18	-0.05
	-SCN	0.30	-0.05	0.28
	$-SO_2CH_3$	1.03	0.20	0.79
	$-SO_2Cl$	0.73	0.06	0.45
C=O	-CHO	0.65	0.10	0.45
	$-COCH_3$	0.57	0.00	0.28
	-COOH	0.80	0.08	0.40
	$-COOCH_3$	0.70	-0.05	0.20
	-COCl	0.88	0.06	0.44
	-CN	0.47	0.00	0.28

1) present as the keto-form

	3- or 4-Substituent (i = 3 or 4)	$Z_{32} = Z_{45}$	$Z_{34} = Z_{43}$	$Z_{35} = Z_{42}$
	-H	0	0	0
C	$-CH_3$	-0.45	-0.22	-0.14
HAL	-Cl	-0.22	-0.11	-0.03
	-Br	-0.12	-0.08	-0.10
	-I	0.06	0.00	-0.19
O	$-OCH_3$	-1.10	-0.38	-0.20
N	$-NH_2$	-1.25	-0.53	-0.25
	$-NO_2$	0.95	0.60	0.03
S	-SH	-0.22	-0.20	-0.10
	$-SCH_3$	-0.33	-0.10	-0.03
	-SCN	0.25	0.05	0.05
	$-SO_2CH_3$	0.96	0.48	0.46
O=C<	-CHO	0.79	0.45	0.03
	$-COCH_3$	0.68	0.47	-0.02
	-COOH	0.99	0.48	0.24
	$-COOCH_3$	0.78	0.47	-0.05
	-COCl	1.05	0.50	0.03
	-CN	0.63	0.20	0.15

Effect of a Substituent on the Chemical Shift of the Ring-Protons in Monosubstituted Pyridines

(δ in ppm relative to TMS, solvent: dimethyl sulfoxide)

(For couplings and chemical shifts in $CDCl_3$ as the solvent see p. H275.)

$$\delta_{H_2} = 8.59 + Z_{i2}$$
$$\delta_{H_3} = 7.38 + Z_{i3}$$
$$\delta_{H_4} = 7.75 + Z_{i4}$$
$$\delta_{H_5} = 7.38 + Z_{i5}$$
$$\delta_{H_6} = 8.59 + Z_{i6}$$

	2- or 6-Substituent (i = 2 or 6)	$Z_{23} = Z_{65}$	$Z_{24} = Z_{64}$	$Z_{25} = Z_{63}$	$Z_{26} = Z_{62}$
	-H	0	0	0	0
C	$-CH_3$	-0.11	-0.01	-0.16	0.08
C	$-CH_2CH_3$	-0.09	-0.08	-0.15	0.03
C	$-CH_2$-phenyl	0.12	-0.08	-0.20	0.02
C	$-CH_2OH$	0.37	0.30	0.02	0.06
C	$-CH_2NH_2$	0.20	0.07	-0.09	0.05
C	$-CH_2SC_3H_7$	0.04	-0.08	-0.26	-0.06
C	$-CH_2SO_2$-Phenyl	∿0	∿-0.3	∿0	-0.2
C	$-CH=CH_2$	0.11	-0.14	-0.11	0.04
C	-phenyl	0.16	-0.28	-0.40	-0.03
C	-2-pyridyl	1.12	-0.09	-0.26	0.00
HAL	-F	-0.10	0.40	0.12	-0.13
HAL	-Cl	0.32	0.29	0.29	0.20
HAL	-Br	0.41	0.17	0.19	0.02
	-OH	-0.7	0.0	-1.0	-0.9
	$-O-n-C_4H_9$	-0.53	-0.03	-0.49	-0.32
N	$-NH_2$	-0.68	-0.31	-0.78	-0.48
N	$-NHCOCH_3$	0.94	0.16	-0.20	-0.10
N	$-NHCOOCH_2CH_3$	0.59	0.07	-0.24	-0.21
N	$-NHNO_2$	0.34	0.31	-0.03	-0.41
N	$-NO_2$	1.09	0.67	0.74	0.26
O=C<	-CHO	0.93	0.42	0.50	0.44
O=C<	$-COCH_3$	0.82	0.37	0.39	0.28
O=C<	-CO-phenyl	0.62	0.55	0.32	0.28
O=C<	-COOH	0.97	0.43	0.48	0.42
O=C<	$-COO-n-C_4H_9$	0.86	0.39	0.35	0.35
O=C<	$-CONH_2$	1.05	0.47	0.43	0.30
O=C<	$-CSNH_2$	1.41	0.37	0.33	0.25
	-CH=NOH	0.40	0.28	0.01	0.16
	-CN	0.88	0.38	0.55	0.39

	3- or 5-Substituent (i = 3 or 5)	$Z_{32} = Z_{56}$	$Z_{34} = Z_{54}$	$Z_{35} = Z_{53}$	$Z_{36} = Z_{52}$
	-H	0	0	0	0
	$-CH_3$	-0.02	-0.06	-0.09	-0.02
	$-CH_2OH$	0.11	0.15	0.04	-0.04
C	$-CH_2NH_2$	0.16	0.13	0.04	0.00
	$-CH_2SO_2$-phenyl	-0.24	-0.15	-0.22	0.01
	-CH=CHCOOH	0.45	0.52	0.34	0.17
HAL	-Cl	0.20	0.24	0.19	0.09
	-Br	0.20	0.43	0.34	0.18
	-OH	-0.03	-0.37	0.15	-0.24
N	$-NH_2$	-0.06	-0.49	0.02	-0.36
	$-NHCOCH_3$	0.37	0.50	0.06	-0.16
	$-SO_3H$	0.70	1.14	0.81	0.70
O=C	-CHO	0.45	0.42	0.12	0.20
	$-COCH_3$	0.72	0.68	0.30	0.37
	-CO-phenyl	0.47	0.54	0.37	0.34
	$-COOCH_3$	0.62	0.60	0.23	0.34
	$-CSNH_2$	0.68	0.67	0.24	0.26
	-CH=NOH	0.39	0.43	0.19	0.15
	-CN	0.63	0.72	0.43	0.50

	4-Substituent (i = 4)	$Z_{42} = Z_{46}$	$Z_{43} = Z_{45}$
	-H	0	0
	$-CH_3$	0.01	-0.10
	$-CH_2$-phenyl	0.00	-0.15
	$-CH_2OH$	0.07	0.14
C	$-CH_2NH_2$	0.01	0.03
	$-CH_2S-n-C_3H_7$	-0.06	-0.13
	$-CH_2SO_2$-phenyl	-0.09	-0.18
	$-CH=CH_2$	0.12	0.13
	-Cl	0.00	0.05
	-Br	0.09	0.35
	$-OCH_3$	0.02	-0.29
N	$-NH_2$	-0.15	-0.74
	$-NHCOCH_3$	-0.05	0.31
S	$-SCH_2$-phenyl	-0.02	0.04
	-S-phenyl	0.05	-0.16
O=C	-CHO	0.47	0.58
	$-COCH_3$	0.40	0.58
	-CO-phenyl	0.36	0.40
	$-COO-n-C_4H_9$	0.34	0.54
	$-CSNH_2$	0.35	0.68
	-CH=NOH	0.24	0.37
	-CN	0.46	0.62

^{1}H-Chemical Shifts (δ in ppm relative to TMS) and Coupling Constants (J in Hz) in Condensed Heterocycles

Proton	δ	J	J
(a)	6.66	J_{ab} = 2.5	J_{ce} = 0.9
(b)	7.52	J_{ac} = 0.9	J_{cf} = 0.8
(c)	7.42	$J_{ad} = J_{ae} = J_{af}$ = 0	J_{de} = 7.3
(d)	7.19	$J_{bc} = J_{bd} = J_{be} = J_{bf}$ = 0	J_{df} = 1.2
(e)	7.13	J_{cd} = 8.4	J_{ef} = 7.9
(f)	7.49		

Proton	δ	J	J
(a)	6.45	J_{ab} = 3.1	J_{de} = 8.1
(b)	7.26	J_{ac} = 2.0	J_{df} = 1.3
(c)	10.1	J_{ad} = 0.7	J_{dg} = 0.9
(d)	7.40	$J_{ae} = J_{af} = J_{ag}$ = 0	J_{ef} = 7.1
(e)	7.09	J_{bc} = 2.5	J_{eg} = 1.2
(f)	6.99	$J_{bd} = J_{be} = J_{bf} = J_{bg}$ = 0	J_{fg} = 7.8
(g)	7.55		

Proton	δ	J	J
(a)	7.29	J_{ab} = 6	J_{ce} = 2.5
(b)	7.40	J_{ac} = ∿0.8	J_{cf} = 0.7
(c)	7.86	$J_{ad} = J_{ae} = J_{af}$ = 0	J_{de} = 7.2
(d)	7.31	$J_{bc} = J_{bd} = J_{be} = J_{bf}$ = 0	J_{df} = 2.5
(e)	7.33	J_{cd} = 9	J_{ef} = 9
(f)	7.78		

Proton	δ	J	J
(a)	8.08	$J_{bc} = J_{de}$ = 8.2	J_{be} = 0.7
(b,e)	7.70	$J_{bd} = J_{ce}$ = 1.4	J_{cd} = 7.1
(c,d)	7.26		

Proton	δ	J	J
(a)	8.20	J_{ac} = 0.8	J_{cf} = 1.0
(b)	12.4	$J_{ad} = J_{ae} = J_{af}$ = 0	J_{de} = 7.0
(c)	7.60	J_{cd} = 7.9	J_{df} = 1.2
(d)	7.34	J_{ce} = 1.2	J_{ef} = 7.8
(e)	7.20		
(f)	7.85		

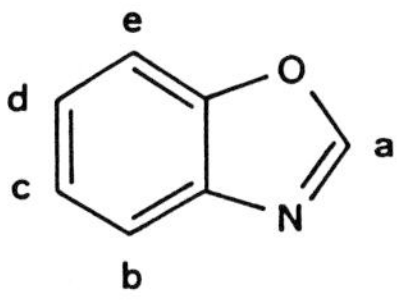

Benzoxazole (a–e)

(a)	8.1
(b), (c), (d), (e)	7.2-7.9

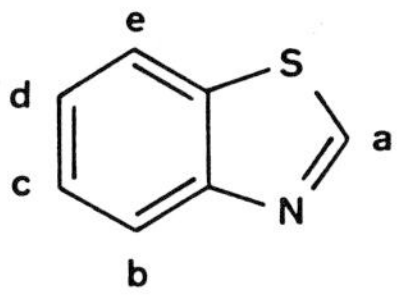

Benzothiazole (a–e)

(a)	9.23	in derivatives: J_{bc} = 8
(b)	8.23	J_{bd} = 1.5
(c)	7.55	J_{be} = 0.5
(d)	7.55	J_{cd} = 7.5
(e)	8.12	J_{ce} = 1.5
		J_{de} = 8

Benzotriazole (1H; a–d)

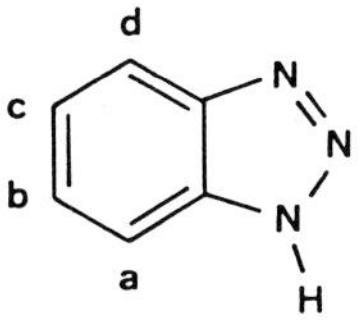

(a,d)	7.98	J_{ab} = 8.6
(b,c)	7.45	J_{ac} = 0.8
		J_{ad} = 1.0

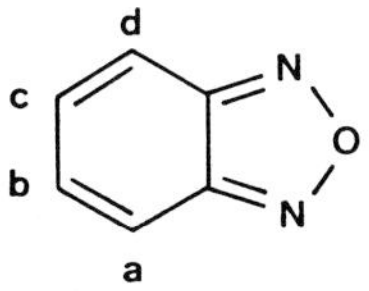

Benzofurazan (a–d)

(a,d)	7.64	J_{ab} = 9.3
(b,c)	7.24	J_{ac} = 1.2
		J_{ad} = 0.9
		J_{bc} = 6.4

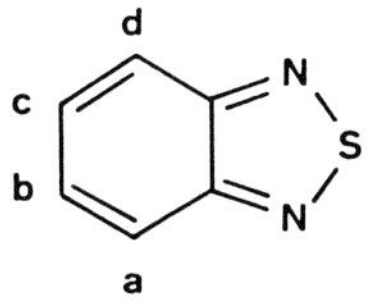

2,1,3-Benzothiadiazole (a–d)

(a,d)	7.96	J_{ab} = 9.2
(b,c)	7.52	J_{ac} = 1.0
		J_{ad} = 0.8
		J_{bc} = ∿6

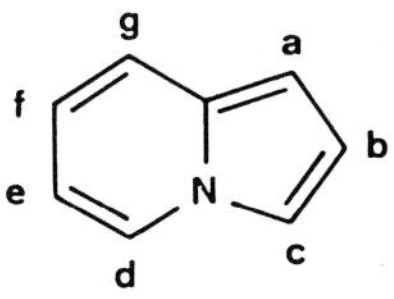

Indolizine (a–g)

(a)	6.28	J_{ab} = 3.9	J_{de} = 6.8
(b)	6.64	J_{ac} = 1.2	J_{df} = 1.0
(c)	7.14	J_{ad} = 1.0	J_{dg} = 1.2
(d)	7.76	J_{bc} = 2.7	J_{ef} = 6.4
(e)	6.31	J_{be} = 0.5	J_{eg} = 1.0
(f)	6.50	J_{cg} = 0.5	J_{fg} = 9.0
(g)	7.25		

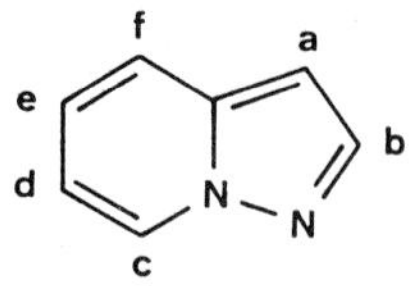

Proton	δ		J		J
(a)	6.38		J_{ab} = 2.2		J_{cf} = 1.0
(b)	7.80		J_{ac} = 0.9		J_{de} = 7.0
(c)	8.39		J_{be} = 0.5		J_{df} = 1.2
(d)	6.62		J_{cd} = 6.9		J_{ef} = 8.9
(e)	6.97		J_{ce} = 1.0		
(f)	7.44				

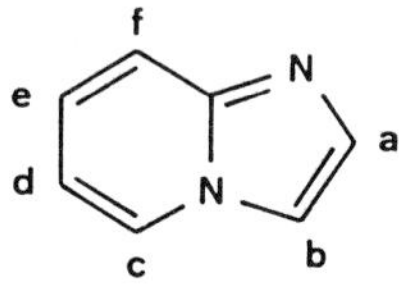

Proton	δ		J		J
(a)	7.48		J_{ab} = 1.0		J_{cf} = 1.0
(b)	7.48		J_{bc} = 0.7		J_{de} = 6.8
(c)	8.09		J_{cd} = 6.9		J_{df} = 1.0
(d)	6.65		J_{ce} = 1.2		J_{ef} = 9.3
(e)	7.03				
(f)	7.51				

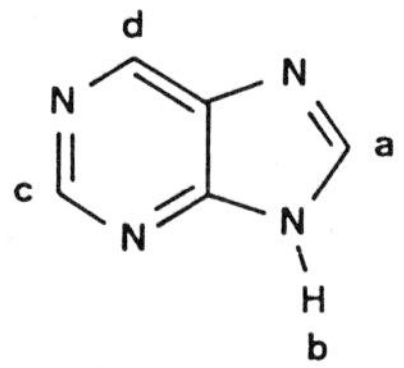

(a)	8.68
(b)	11
(c)	8.99
(d)	9.19

Proton	δ		J		J
(a)	8.00		J_{ab} = 8.3		J_{df} = 1.1
(b)	7.26		J_{ac} = 1.8		J_{dg} = 0.5
(c)	8.81		J_{ad} = 0.8		J_{ef} = 6.8
(d)	8.05		J_{bc} = 4.3		J_{eg} = 1.6
(e)	7.61		J_{de} = 8.2		J_{fg} = 8.2
(f)	7.43				
(g)	7.68				

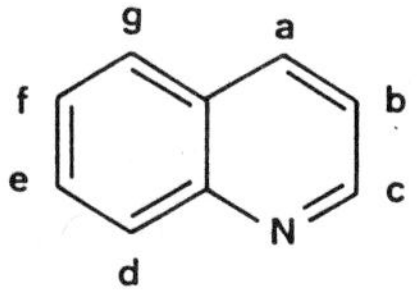

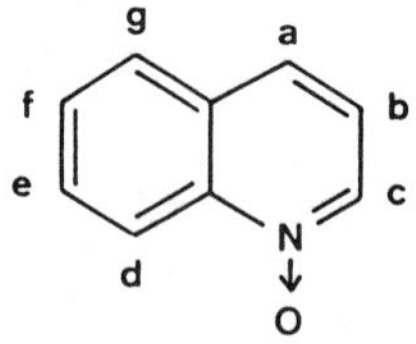

(a)	7.74		J_{ab} = 8.5
(b)	7.27		J_{ac} = 1.1
(c)	8.57		J_{bc} = 6.0
(d)	8.75		

Proton	δ		
(a)	7.50	J_{ab} = 6.0	J_{df} = 1.3
(b)	8.45	J_{ad} = 0.8	J_{dg} = 0.9
(c)	9.15	J_{bc} = 0.8	J_{ef} = 7.0
(d)	7.87	J_{cg} < 0.5	J_{eg} = 1.1
(e)	7.50	J_{de} = 8.2	J_{fg} = 8.7
(f)	7.57		
(g)	7.71		

Proton	δ	
(b)	8.14	J_{ab} = 7.0
(c)	8.77	J_{bc} = 1.7

Proton	δ		
(a)	7.73	J_{ab} = 5.7	J_{cf} = 0.8
(b)	9.10	J_{ac} = 0.8	J_{de} = 6.9
(c)	8.30	J_{cd} = 8.6	J_{df} = 1.5
(d)	7.57	J_{ce} = 1.3	J_{ef} = 7.8
(e)	7.57		
(f)	7.57		

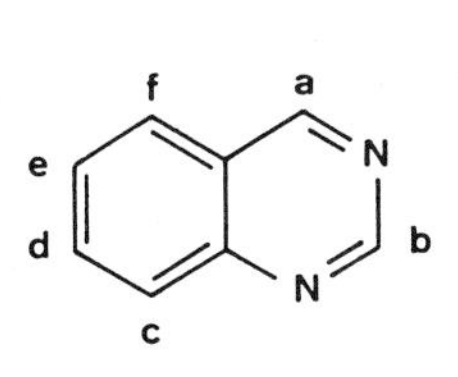

Proton	δ		
(a)	9.29	J_{ab} = 0	J_{cf} = 0.8
(b)	9.23	J_{ac} = 0.5	J_{de} = 6.9
(c)	8.01	J_{cd} = 8.5	J_{df} = 1.2
(d)	7.83	J_{ce} = 1.2	J_{ef} = 7.9
(e)	7.58		
(f)	7.84		

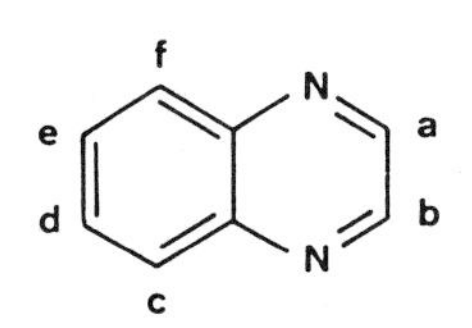

Proton	δ		
(a,b)	8.74	J_{ab} = 1.8	J_{de} = 6.9
(c,f)	8.07	J_{cd} = 8.4	J_{df} = 1.6
(d,e)	7.68	J_{ce} = 1.6	J_{ef} = 8.4
		J_{cf} = 0.6	

^{1}H-NMR

CONDENSED HETEROAROMATICS

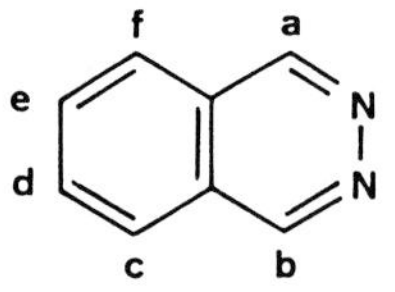

Shift		Coupling	Coupling
(a,b)	9.44	J_{ac} = 0.4	J_{cf} = 0.6
(c,f)	7.93	J_{cd} = 8.2	J_{de} = 6.8
(d,e)	7.85	J_{ce} = 1.2	

Shift		Coupling	Coupling
(a)	7.80	J_{ab} = 9.8	J_{de} = 8.6
(b)	6.45	J_{cd} = 8.5	J_{df} = 2.0
(c)	7.20	J_{ce} = 1.8	J_{ef} = 8.5
(d)	7.45	J_{cf} = 0.0	
(e)	7.22		
(f)	7.63		

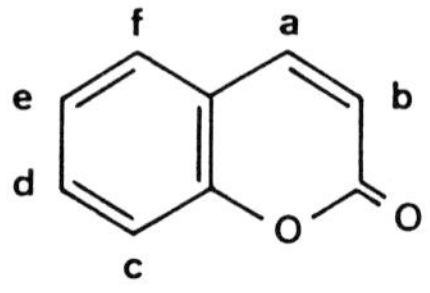

Shift		Coupling	Coupling
(a)	6.34	J_{ab} = 6.1	J_{de} = 7.0
(b)	7.88	J_{cd} = 8.4	J_{df} = 1.8
(c)	7.47	J_{ce} = 1.1	J_{ef} = 8.0
(d)	7.68	J_{cf} = 0.5	
(e)	7.43		
(f)	8.21		

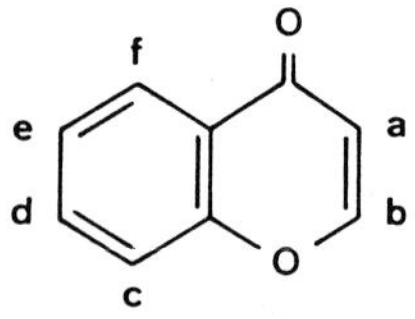

(a,b)	5.77	J_{cd} = 7.9
(c,f)	6.52	J_{ce} = 1.5
(d,e)	6.71	J_{cf} = 0.4
		J_{de} = 7.9

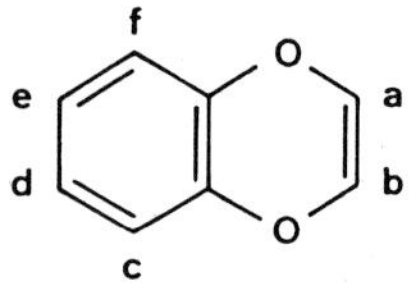

(a,b)	6.42	J_{cd} = 7.8
(c,f)	7.19	J_{ce} = 1.3
(d,e)	7.12	J_{cf} = 1.1
		J_{de} = 7.1

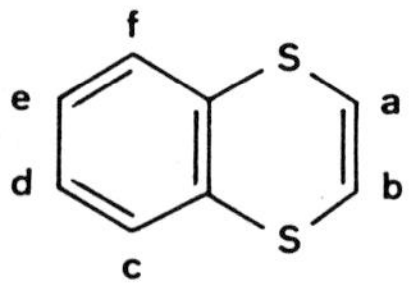

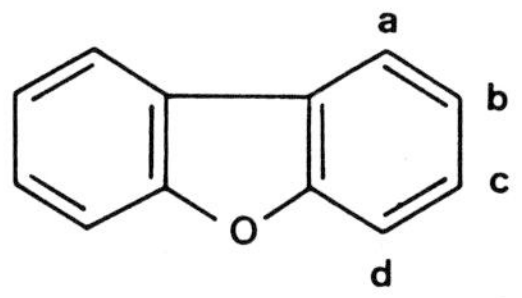

	δ		
(a)	7.84	J_{ab} = 7.6	J_{bc} = 7.3
(b)	7.23	J_{ac} = 1.3	J_{bd} = 0.9
(c)	7.35	J_{ad} = 0.6	J_{cd} = 8.5
(d)	7.48		

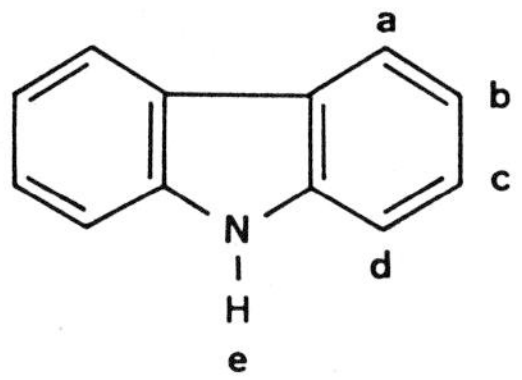

	δ		
(a)	8.08	J_{ab} = 7.8	J_{bc} = 7.2
(b)	7.16	J_{ac} = 1.2	J_{bd} = 0.9
(c)	7.36	J_{ad} = 0.7	J_{cd} = 8.2
(d)	7.49	J_{ae} = 0.7	
(e)	10.3		

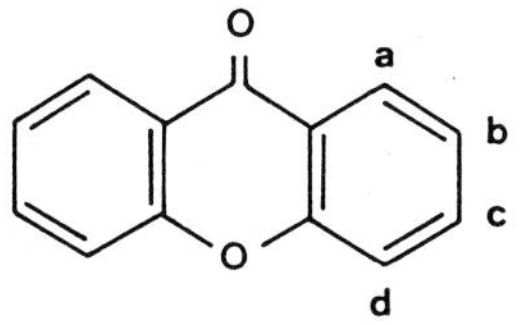

	δ		
(a)	8.36	J_{ab} = 8.0	J_{bc} = 7.1
(b)	7.38	J_{ac} = 1.8	J_{bd} = 1.1
(c)	7.73	J_{ad} = 0.5	J_{cd} = 8.4
(d)	7.50		

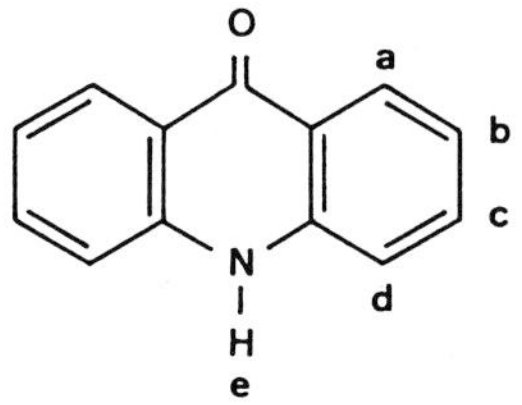

	δ		
(a)	8.27	J_{ab} = 8.2	J_{bc} = 7.0
(b)	7.27	J_{ac} = 1.4	J_{bd} = 1.0
(c)	7.74	J_{ad} = 0.4	J_{cd} = 8.6
(d)	7.57	J_{ae} = 0.4	
(e)	11.7		

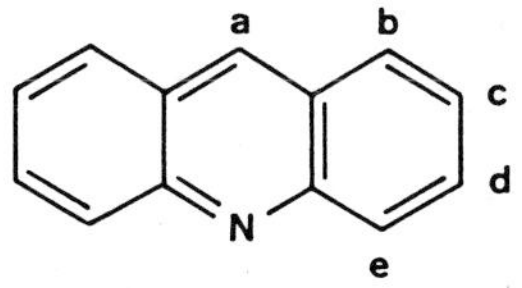

	δ		
(a)	9.09	J_{ab} = 0.4	J_{be} = 0.6
(b)	8.19	J_{ae} = 0.9	J_{cd} = 6.6
(c)	7.64	J_{bc} = 8.2	J_{ce} = 1.2
(d)	7.89	J_{bd} = 1.4	J_{de} = 9.0
(e)	8.22		

^{1}H-NMR

AMINO ACIDS

^{1}H-Chemical Shifts (δ in ppm relative to TMS) and Coupling Constants (|J| in Hz) in Amino Acids (Solvent: CF_3COOH or CF_3COOD, Frequency: 220 MHz)

see: B. Bak, C. Dambmann, F. Nicolaisen, E.J. Pedersen, N.S. Bhacca, J. Mol. Spectr. 26, 78 (1968)

H_3N^+-CH_2COOH (b: H_3; a: CH_2)

δ		J		in D_2O	
(a)	4.28	(a,b)	5.7	(a)	3.58
(b)	7.47				

H_3N^+—CH(CH$_3$)—COOH (a: CH; b: CH_3; c: H_3)

δ		J		in D_2O	
(a)	4.49	(a,b)	7.3	(a)	3.79
(b)	1.86	(a,c)	5.8	(b)	1.49
(c)	7.41				

H_3N^+—CH[CH(CH$_3$)$_2$]—COOH (a: α-CH; b: CH; c: $(CH_3)_2$; d: H_3)

δ		J	
(a)	4.32	(a,b)	4.2
(b)	2.60	(a,c)	5.7
(c)	1.25	(b,c)	6.9
(d)	7.33		

H_3N^+—CH[CH$_2$CH(CH$_3$)$_2$]—COOH (a: α-CH; b,c: CH_2; d: CH; e,f: $(CH_3)_2$; g: H_3)

δ		J	
(a)	4.48	(a,b)	∿6.7*
(b)	∿2	(a,c)	∿6.7*
(c)	∿2	(a,g)	5.5
(d)	∿2	(d,e)	6.1
(e)	1.11	(d,f)	5.7
(f)	1.10		
(g)	7.38		

*average value

H_3N^+—CH[CH(CH$_3$)CH$_2$CH$_3$]—COOH (a: α-CH; b: CH; c: CH_3; d,e: CH_2; f: CH_3; g: H_3)

δ		J	
(a)	4.41	(a,b)	3.6
(b)	2.28	(a,g)	5.5
(c)	1.21	(b,c)	7.0
(d)	1.55	(b,d)	6.1
(e)	1.70	(b,e)	8.4
(f)	1.10	(d,e)	13.6
(g)	7.35	(d,f)	7.0
		(e,f)	7.0

$H_3\overset{+}{N}$(d)–CH(a)(CH$_2$(b,c)OH)–COOH

	δ		J
(a)	4.65	(a,b)	4.0
(b)	4.51	(a,c)	4.0
(c)	4.56	(a,d)	6
(d)	7.70	(b,c)	13.5

$H_3\overset{+}{N}$(d)–CH(a)(CH(b)(OH)–CH$_3$(c))–COOH

	δ		J
(a)	4.44	(a,b)	4.5
(b)	4.82	(a,d)	5.5
(c)	1.67	(b,c)	6.5
(d)	7.63		

$H_3\overset{+}{N}$(e)–CH(a)(CH$_2$(b,c)–SH(d))–COOH

	δ		J
(a)	4.68	(a,b) (a,c)	5.0*
(b)	3.36		
(c)	3.41	(a,e)	5.3
(d)	1.84	(b,c)	15.5
(e)	7.58	(b,d) (c,d)	9.1*

$H_3\overset{+}{N}$(f)–CH(a)(CH$_2$(b,c)–CH$_2$(d)–S–CH$_3$(e))–COOH

	δ		J
(a)	4.67	(a,b)	7.7
(b)	2.50	(a,c)	4.4
(c)	2.65	(a,f)	5.5
(d)	2.96	(b,c)	15.7
(e)	2.27	(b,d) (c,d)	6.5*
(f)	7.73		

*average value

$H_3\overset{+}{N}$(g)–CH(a)(CH$_2$(b,c)–C$_6$H$_5$ (d, e, f))–COOH

	δ		J
(a)	4.68	(a,b)	4.5
(b)	3.64	(a,c)	8.5
(c)	3.37	(b,c)	14.5
(d)	7.3		
(e) (f)	7.4-7.5		
(g)	7.33		

$H_3\overset{+}{N}$(f)–CH(a)(CH$_2$(b,c)–C$_6$H$_4$(d, e)–OH)–COOH

	δ		J
(a)	4.64	(a,b)	4.5
(b)	3.60	(a,c)	8.5
(c)	3.34	(b,c)	15.0
(d)	7.27	(d,e)	8.5
(e)	7.03		
(f)	7.4		

^{1}H-NMR

AMINO ACIDS

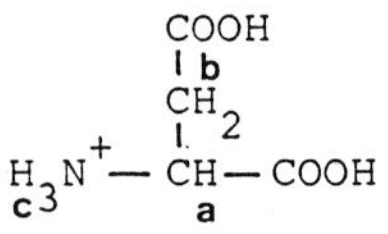

```
        COOH
         | b
        CH2
         |
H3N+ — CH — COOH
c        a
```

Proton	δ	Coupling	J
(a)	4.76	(a,b)	4.7
(b)	3.55	(a,c)	5.1
(c)	7.73		

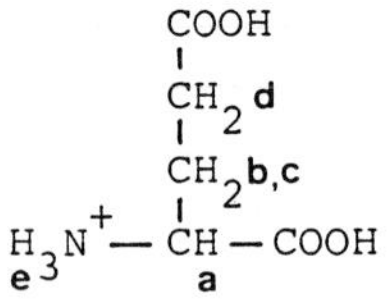

```
        COOH
         |
        CH2 d
         |
        CH2 b,c
         |
H3N+ — CH — COOH
e        a
```

Proton	δ	Coupling	J
(a)	4.60	(a,b) (a,c)	5.6*
(b)	2.63	(a,e)	5.5
(c)	2.55	(b,c)	15.5
(d)	3.01	(b,d) (c,d)	6.2*
(e)	7.71		

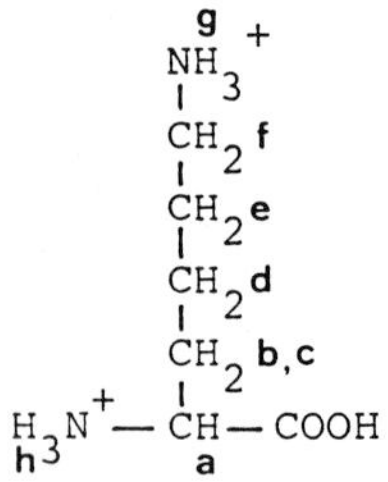

```
        g  +
        NH3
         |
        CH2 f
         |
        CH2 e
         |
        CH2 d
         |
        CH2 b,c
         |
H3N+ — CH — COOH
h        a
```

Proton	δ	Coupling	J
(a)	4.52	(a,b) (a,c)	5.6*
(b)	2.35	(a,h)	5.8
(c)	2.26	(b,c)	∿15
(d)	∿1.8	(b,d) (c,d)	∿6.0*
(e)	2.00	(e,f)	∿6.0
(f)	3.38	(f,g)	∿6.0
(g)	6.97		
(h)	7.60		

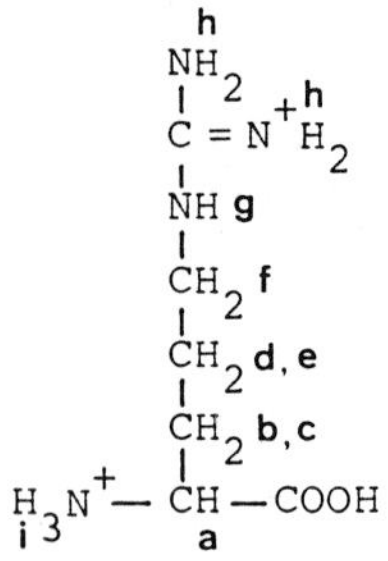

```
        h
        NH2
         |      h
        C = N+H2
         |
        NH g
         |
        CH2 f
         |
        CH2 d,e
         |
        CH2 b,c
         |
H3N+ — CH — COOH
i        a
```

Proton	δ	Coupling	J
(a)	4.46	(a,b) (a,c)	5.3*
(b)	2.34	(a,i)	5.5
(c)	2.25	(b,c)	∿15
(d)	∿2.08	(b,d) (b,e) (c,d) (c,e)	∿6*
(e)	∿2.00	(d,f) (e,f)	6.5*
(f)	3.43	(f,g)	5.3
(g)	6.50		
(h)	6.19		
(i)	7.60		

*average value

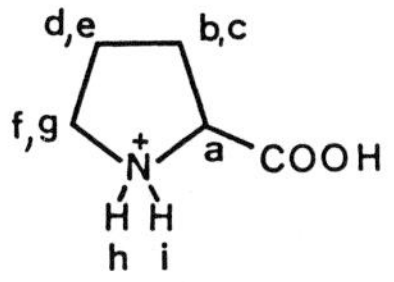

Shift		Coupling	
(a)	4.78	(a,b)	9.0
(b)	2.74	(a,c)	7.0
(c)	2.50	(a,h)	6.6
(d)	2.35	(a,i)	4.4
(e)	2.29	(b,c)	14.0
(f)	3.82	(b,d)	∿6.8*
(g)	3.76	(b,e)	∿6.8*
(h)	8.08	(c,d)	∿6.8*
(i)	7.71	(c,e)	∿6.8*

in D_2O relative to $(CH_3)_3SiCH_2CH_2CH_2SO_3Na$ (see L. Pogliani, M. Ellenberger, J. Valat, Org. Magn. Res. 7, 61 (1975)):

pH = 2.0				pH = 7.0				pH = 13.0			
(a)	4.33	(a,b)	8.5	(a)	3.74	(a,b)	8.4	(a)	2.81	(a,b)	8.6
(b)	2.42	(a,c)	6.5	(b)	1.96	(a,c)	6.2	(b)	1.45	(a,c)	6.6
(c)	2.14	(b,c)	-13.5	(c)	1.68	(b,c)	-13.5	(c)	1.04	(b,c)	-12.0
(d)	2.06	(b,d)	7.5	(d)	1.63	(b,d)	7.6	(d)	1.07	(b,d)	8.1
(e)	2.04	(b,e)	5.5	(e)	1.60	(b,e)	5.4	(e)	1.05	(b,e)	5.9
(f)	3.46	(b,f)	-0.4	(f)	3.04	(b,f)	-0.4	(f)	2.36	(b,f)	-0.6
(g)	3.39	(b,g)	0.0	(g)	2.95	(b,g)	0.0	(g)	2.08	(b,g)	0.0
		(c,d)	5.5			(c,d)	5.6			(c,d)	6.7
		(c,e)	7.5			(c,e)	7.8			(c,e)	8.5
		(d,e)	-13.0			(d,e)	-13.0			(d,e)	-11.0
		(d,f)	5.5			(d,f)	5.7			(d,f)	5.5
		(d,g)	7.5			(d,g)	7.9			(d,g)	8.1
		(e,f)	7.5			(e,f)	7.9			(e,f)	7.7
		(e,g)	5.5			(e,g)	5.7			(e,g)	5.7
		(f,g)	-11.0			(f,g)	-11.0			(f,g)	-10.5

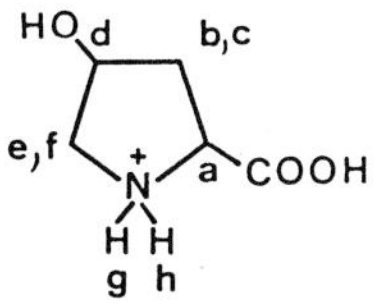

Shift		Coupling	
(a)	∿5	(a,b)	8.2
(b)	2.95	(a,c)	10.4
(c)	2.56	(b,c)	15.0
(d)	∿5	(b,d)	<2
(e)	3.9*	(c,d)	4.2
(f)	3.9*		
(g)	8.60		
(h)	8.00		

*average value

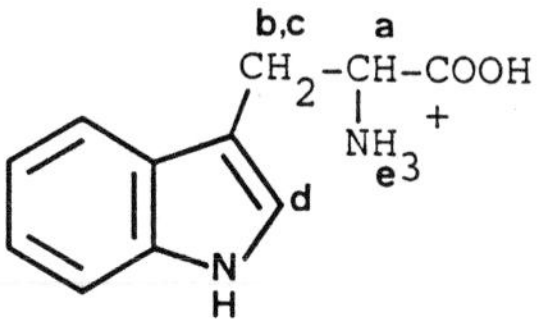

^{1}H-NMR

AMINO ACIDS

(Tryptophan: indole-3-CH_2(b,c)-CH(a)-COOH, NH_3^+ (e); indole H-2 (d))

	δ		J
(a)	4.71	(a,b)	4.0
(b)	3.79	(a,c)	8.0
(c)	3.57	(b,c)	15.5
(d)	7.25		
(e)	7.05		

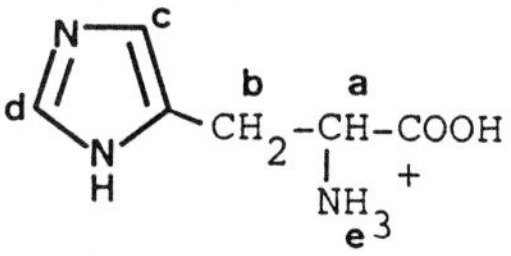

(Histidine: imidazol-4-yl (c, d)-CH_2(b)-CH(a)-COOH, NH_3^+ (e))

	δ		J
(a)	4.91	(a,b)	6.4
(b)	3.87	(c,d)	1.4
(c)	7.66		
(d)	8.73		
(e)	7.82		

^{1}H-NMR

^{19}F-^{1}H COUPLING CONSTANTS

Coupling Constants (J in Hz) between ^{1}H and ^{19}F (Spin Quantum Number I = 1/2, Natural Abundance 100 %)

Compound	Coupling	Remarks
c b a CH-CH-CH-F	$\|J\|_{aF}$ = 45-80	(generally 45-50)
	$\|J\|_{bF}$ = 0-30	conformation-dependent: synclinal: 0-5 antiperiplanar: 10-25
	$\|J\|_{cF}$ = 0-4	
CH_3F	$\|J\|$ = 81	
b a CH_3CH_2F	$\|J\|_{aF}$ = 46.7	$\|J\|_{bF}$ = 25.2
H_b, H_c C=C H_a, F	$\|J\|_{aF}$ = 85	(in derivatives: 70-90)
	$\|J\|_{bF}$ = 52	(in derivatives: 10-50)
	$\|J\|_{cF}$ = 20	(in derivatives: -3...+20)
H-C≡C-F	$\|J\|$ = 21	
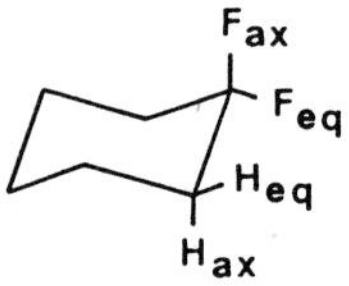 cyclohexane with F_{ax}, F_{eq}, H_{eq}, H_{ax}	$\|J\|_{F_{ax}H_{ax}}$ = 34	$\|J\|_{F_{eq}H_{eq}} < 8$
	$\|J\|_{F_{ax}H_{eq}}$ = 12	$\|J\|_{F_{eq}H_{ax}} < 8$
fluorobenzene (F; o, m, p)	$\|J\|_{oF}$ = 9.0	(in derivatives: 6-10)
	$\|J\|_{mF}$ = 5.7	(in derivatives: 4-8)
	$\|J\|_{pF}$ = 0.2	(in derivatives: 0-3)
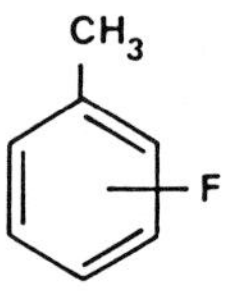 CH_3-benzene with F	$\|J\|_{ortho}$ = 2.5	
	$\|J\|_{meta}$ = 0.0	
	$\|J\|_{para}$ = 1.5	

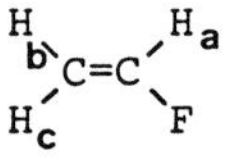

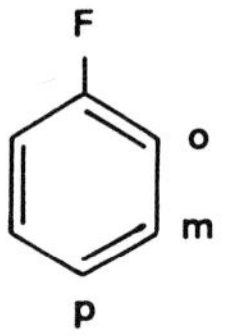

Coupling Constants (J in Hz) Between ^{1}H and ^{31}P (Spin Quantum Number I = 1/2, Natural Abundance 100 %)

Compound	Coupling	Compound	Coupling
>P-H	$\|J\|$ = 180-200	$(CH_3CH_2O)_2P(=O)H^a$	$\|J\|_{aP}$ = 630
$(CH_3^aCH_2^b)_3P$	$\|J\|_{aP}$ = 13.7 $\|J\|_{bP}$ = 0.5	$[(CH_3)_2N]_3P$	$\|J\|$ = 8.8
$(CH_3^aCH_2^b)_4P^+$	$\|J\|_{aP}$ = 18.0 $\|J\|_{bP}$ = 13.0	$[(CH_3)_2N]_3PO$	$\|J\|$ = 9.5
$(CH_3^aCH_2^b)_3P{=}O$	$\|J\|_{aP}$ = 16.3 $\|J\|_{bP}$ = 11.9	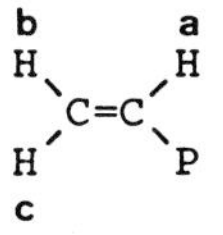 $H^b(H^c)C{=}C(H^a)P$	$\|J\|_{aP}$ = 10-40 $\|J\|_{bP}$ = 30-60 $\|J\|_{cP}$ = 10-30
$(CH_3^aCH_2^bO)_3P{=}O$	$\|J\|_{aP}$ = 0.8 $\|J\|_{bP}$ = 8.4	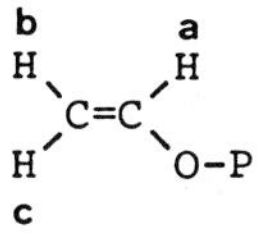 $H^b(H^c)C{=}C(H^a)O\text{-}P$	$\|J\|_{aP}$ = ∿7 $\|J\|_{bP}$ = ∿3 $\|J\|_{cP}$ = ∿1
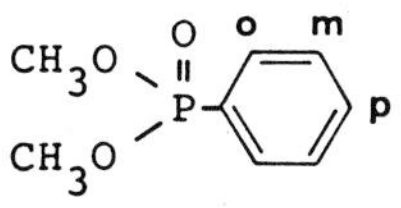 $(CH_3O)_2P(=O)C_6H_5$ (o, m, p)	$\|J\|_{oP}$ = 13.3 $\|J\|_{mP}$ = 4.1 $\|J\|_{pP}$ = 1.2		

^{1}H-NMR Spectra of the Most Common Solvents

(δ in ppm relative to TMS)

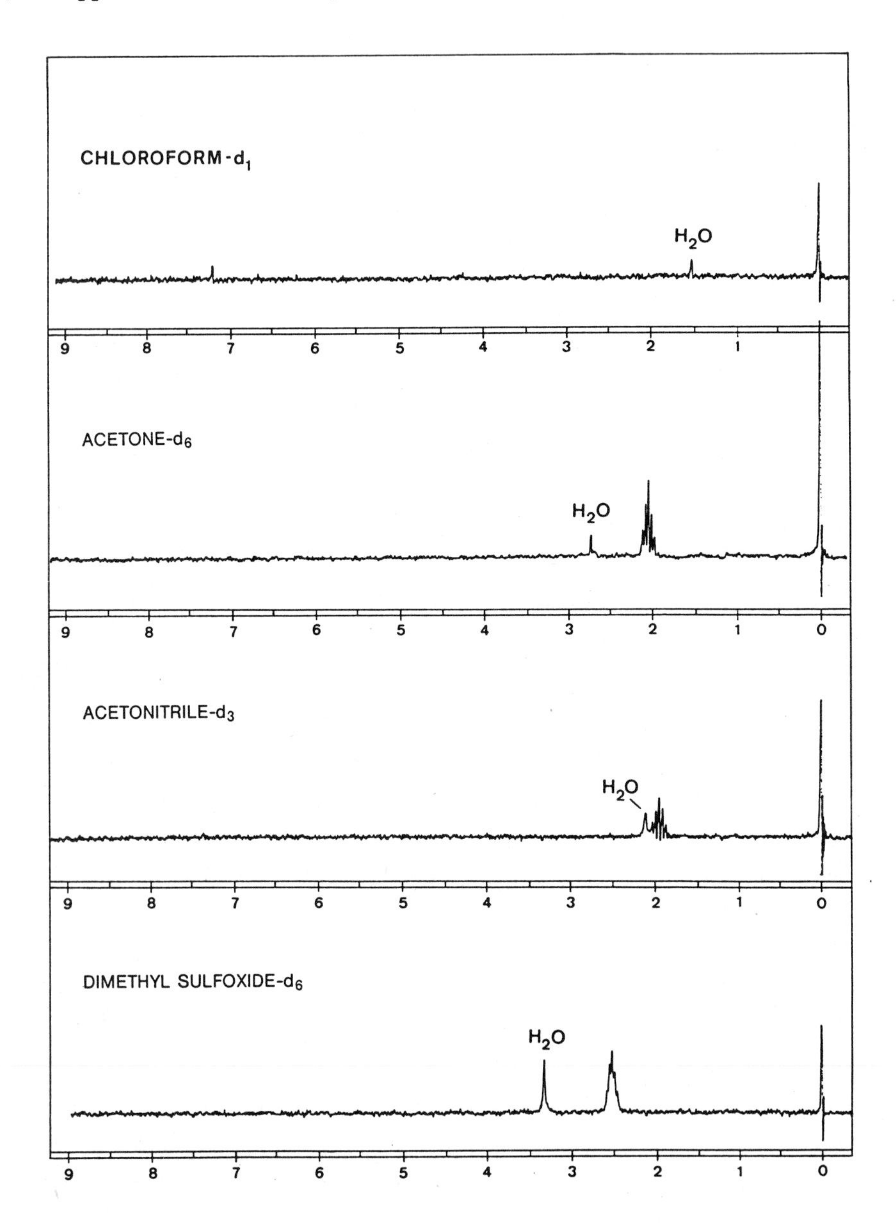

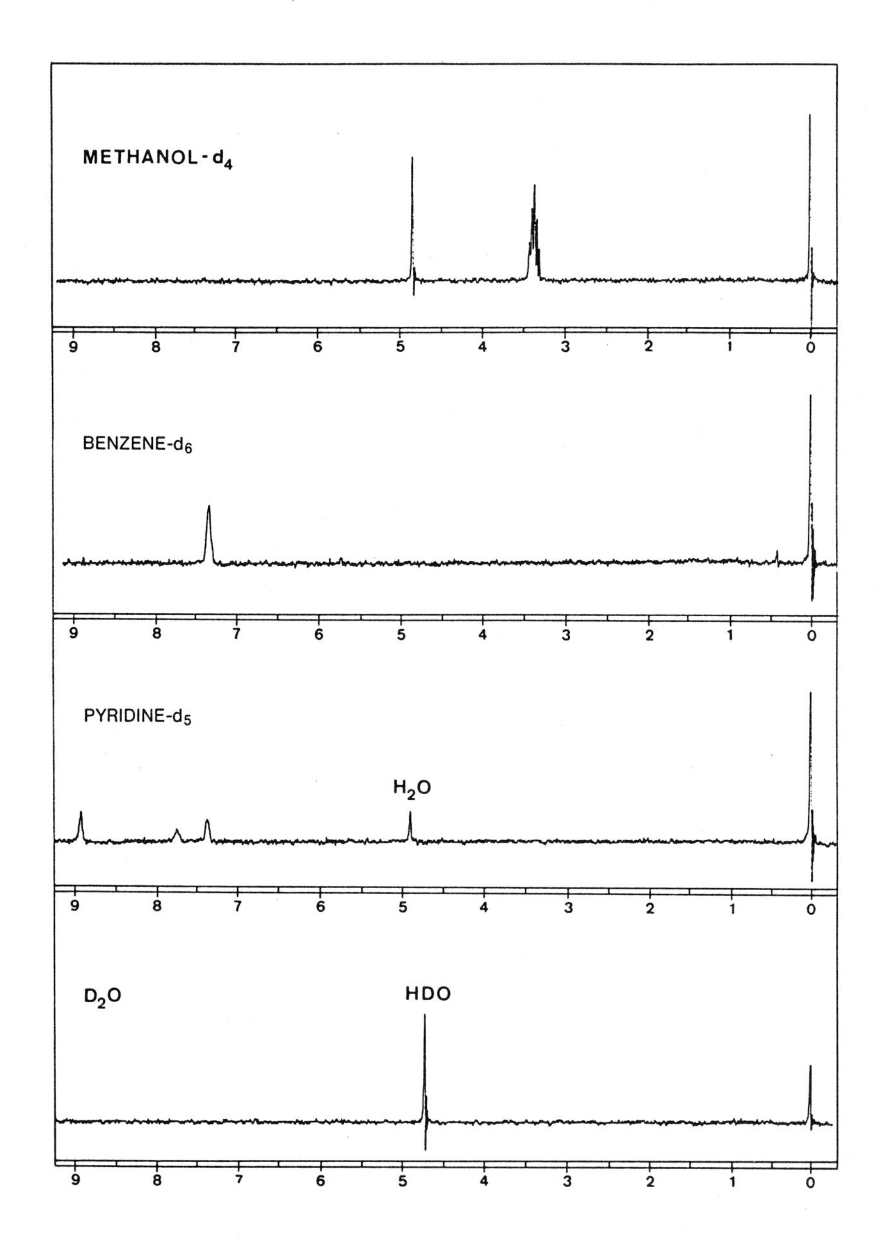
METHANOL-d_4
9 8 7 6 5 4 3 2 1 0
BENZENE-d_6
9 8 7 6 5 4 3 2 1 0
PYRIDINE-d_5
H_2O
9 8 7 6 5 4 3 2 1 0
D_2O
HDO
9 8 7 6 5 4 3 2 1 0

Characteristic IR-Absorption Bands of Alkyl Groups (in cm^{-1})

Assignment	Range	Comments
C-H st	3000 - 2840	intensity variable, frequently multiplet beyond the normal range: $O-CH_3$ (methyl ethers): 2850-2815 $O-CH_2$ (ethers): 2880-2835 $O-CH_2-O$ (methylenedioxy: 2880-2835, 2780-2750 O-CH-O (acetals): ∿2820, weak O◁ , N◁ : 3050-3000 CHO (aldehydes): 2900-2800, 2780-2680 (Fermi resonance) for amines: $N-CH_3$, often also for $N-CH_2-$: 2820-2780; in aliphatic dimethylamines in addition 2780-2760 △ : 3100-3050, 3035-2995 ⬡ (cyclohexanes): in addition ∿2700, weak (comb) hal-C-H st: 3080-2900
$-CH_3$ δ as	1470 - 1430	medium; coincides with CH_2 δ beyond the normal range: CH_3CO (methyl ketones, acetals): } 1440-1400 $CH_3-C=C$: } 1440-1400

$-CH_2$ δ	1475 - 1450	medium, coincides with CH_3 δ as beyond the normal range: CH_2-C=C, CH_2-C≡C: ∿1440 CH_2-C=O, CH_2-C≡N, CH_2-X (X: NO_2, halogen, S, P): ∿1425
$-CH_3$ δ sy	1395 - 1365	medium doublet in compounds with geminal methyl groups: $-CH(CH_3)_2$: ∿1385, ∿1370, equal intensity (γ: 1175-1140, doublet) $>C(CH_3)_2$: ∿1385, ∿1365, 1385 weaker than 1365 (γ: 1220-1190, frequently doublet) $-C(CH_3)_3$: ∿1390, ∿1365, equal intensity, sometimes triplet (γ: 1250-1200, doublet, often incompletely resolved) $-N(CH_3)_2$: no doublet solid-state spectra sometimes exhibit a doublet in the absence of geminal methyl groups beyond the normal range: O_2S-CH_3: 1325-1310 $-S-CH_3$ (sulfides): 1330-1290 $P-CH_3$: 1310-1280 $Si-CH_3$: 1275-1260, strong, sharp band

Characteristic IR-Absorption Bands of Alkyl Groups (in cm^{-1})
(continued)

Assignment	Range	Comments
$-CH_3$ γ	1250 - 800	intensity variable; no practical significance strong band in compounds with geminal methyl groups: $-CH(CH_3)_2$: 1175-1140, doublet $>C(CH_3)_2$: 1220-1190, generally doublet $-C(CH_3)_3$: 1250-1200, doublet, often not resolved beyond the normal range: $\equiv SiCH_3$: ∿765; $>Si(CH_3)_2$: ∿855, ∿800; $-Si(CH_3)_3$: ∿840, ∿765
$-CH_2$ γ	770 - 720	medium, sometimes doublet $C-(CH_2)_n-C$: for $n \geq 4$ at ∿720 for $n < 4$ at higher wave numbers, weaker in cyclohexanes at ∿890 beyond the normal range: cycloalkanes: 1060-800, a number of bands; unreliable
-C-D st	2200 - 2080	In general: substitution of L by the isotope L'. $\nu_{X-L'} = \nu_{X-L} \cdot \sqrt{\frac{1/m_X + 1/m_{L'}}{1/m_X + 1/m_L}}$

Characteristic IR-Absorption Bands of Alkenyl Groups (in cm^{-1})

Assignment	Range	Comments
$=CH_2$ st	3095 - 3075	medium, often a number of bands
>CH st	3040 - 3010	CH st in aromatics and three-membered rings fall in the same range; in cyclic compounds: cyclopropene ∿3075, cyclobutene ∿3060, cyclopentene ∿3045, cyclohexene ∿3020
=CH δ ip	1420 - 1290	no practical significance

Characteristic IR-Absorption Bands of Alkenyl Groups (in cm^{-1})

(continued)

Assignment	Range	Comments
=CH δ oop	1005 - 675	a number of bands sub-ranges:

	C=C	C=C-C=O	C=C-OR	C=C-O-C=O
$-CH{=}CH_2$:	1005-985	∿980	∿960	∿950
	920-900	∿960	∿815	∿870
	(also overtone at 1850-1800)	∿810		
${>}C{=}CH_2$:	900-880	∿940	∿795	
	(also overtone at 1850-1780)	∿810		
$_{H}{>}C{=}C{<}^{H}$:	990-960	∿975	∿960	∿950
$^{H}{>}C{=}C{<}^{H}$:	725-675	∿820		
${>}C{=}C{<}^{H}$:	840-800	∿820		

in the same range also arC-H δ oop, C-O-C γ and C-N-C γ in saturated heterocycles, OH δ oop in carboxylic acids, NH γ, N-O st, S-O st, CH_2 γ, C-F δ (?), C-Cl st

C=C st	1690 - 1635	variable intensity; weak if compound highly symmetric, strong for N-C=C and O-C=C sub-ranges: $-CH=CH_2$: 1650-1635 $>C=CH_2$: 1660-1640 HC=CH (trans) : 1690-1665, weak HC=CH (cis) : 1665-1635 $>C=CH-$: 1690-1660 } weak, often absent $>C=C<$: 1690-1650 } weak, often absent beyond the normal range: C=C-X, X = O, N, S: at lower frequency (to ∿1590), increased intensity; in vinyl ethers often doublet due to rotational isomers at lower frequency if conjugated with:

C=C	C≡C	C≡N	phenyl	C=O	cyclopropyl	oxiranyl
∿1650 ∿1600	∿1600	∿1620	∿1630	∿1630	∿1640	∿1640

Characteristic IR-Absorption Bands of Alkenyl Groups: Examples (in cm^{-1})

in and on small rings (C=C st):

		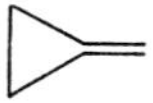			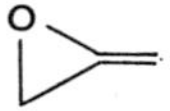
∿1640		∿1780			∿1650
			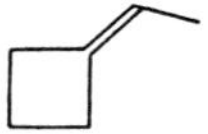		
∿1570	∿1640	∿1680	∿1690		
	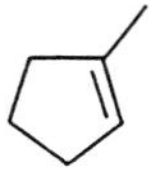	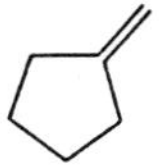	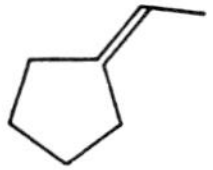	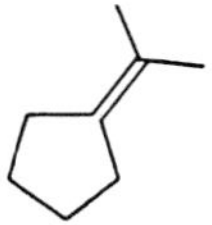	
∿1610	∿1660	∿1660	∿1670	∿1690	∿1570
	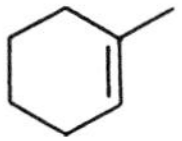	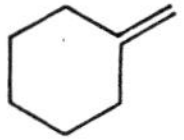	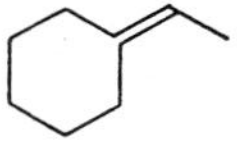	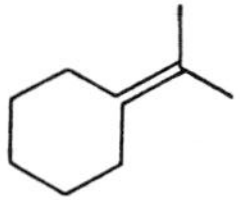	
∿1650	∿1675	∿1650	∿1665	∿1670	∿1615

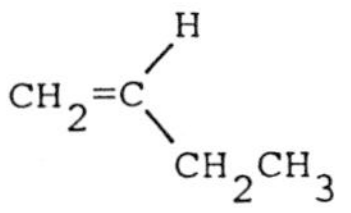

$CH_2=C(H)CH_2CH_3$ — 1645; 994, 912

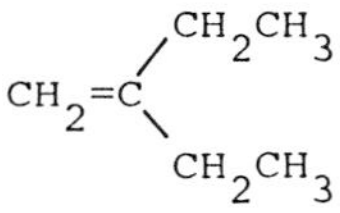

$CH_2=C(CH_2CH_3)_2$ — 1647; 889, 669

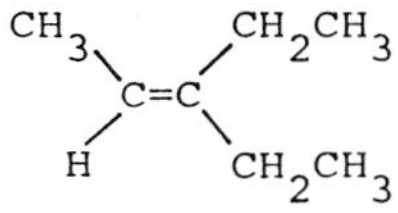

$CH_3(H)C=C(CH_2CH_3)_2$ — 1667; 825

trans-$CH_3CH_2CH_2(H)C=C(H)CH_2CH_2CH_3$ — 1670; 968

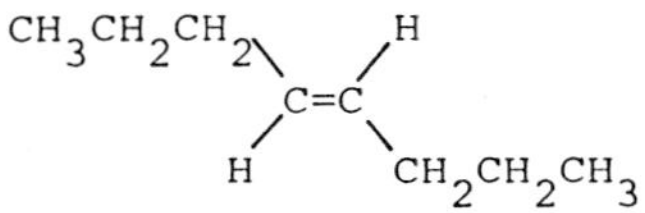

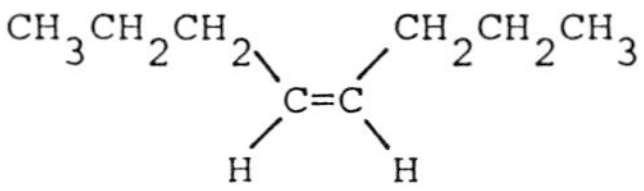

cis-$CH_3CH_2CH_2(H)C=C(H)CH_2CH_2CH_3$ — 1650; 709

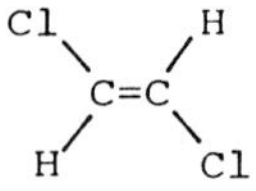

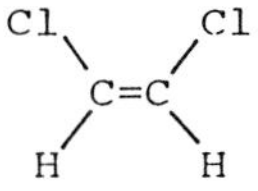

trans-$Cl(H)C=C(H)Cl$ — 1575; 895

$Cl_2C=CH_2$ — 1595; 845, 695

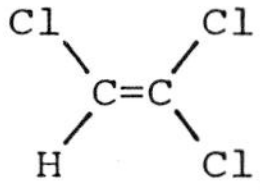

$Cl(H)C=CCl_2$ — 1587; 929, 840

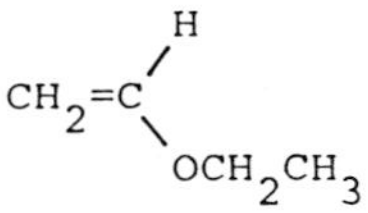

	neat liquid	CCl_4
$CH_2=C(H)OCH_2CH_3$	1610	1634
		1608
	987	964
	810	943

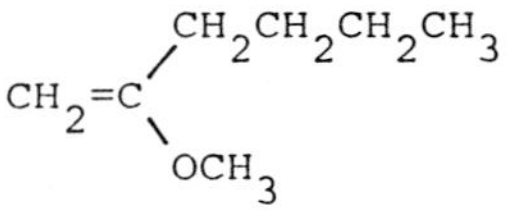

$CH_2=C(CH_2CH_2CH_2CH_3)OCH_3$ — 1655, 1592; 958, 793

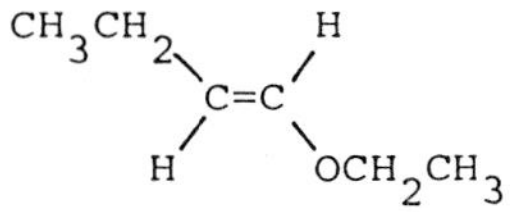

trans-$CH_3CH_2(H)C=C(H)OCH_2CH_3$ — 1670, 1652; 937, 925

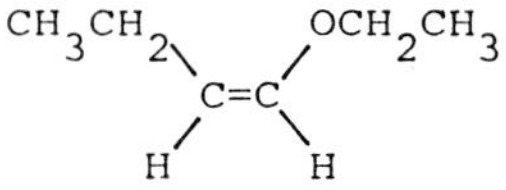

cis-$CH_3CH_2(H)C=C(H)OCH_2CH_3$ — 1663

$H(CH_3)C=C(CH(CH_3)_2)OCH_2CH_3$ — 1660

$CH_3(H)C=C(CH(CH_3)_2)OCH_2CH_3$ — 1673

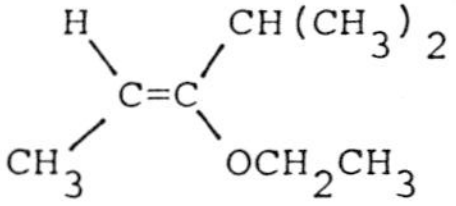

$(CH_3CH_2)_2C=C(CH_2CH_3)OCH_2CH_3$ — 1663

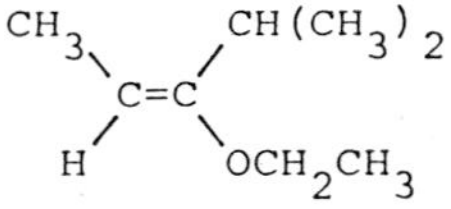

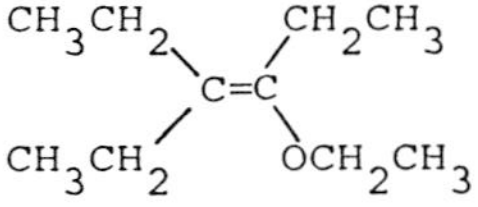

Characteristic IR-Absorption Bands of Alkenyl Groups (in cm^{-1}) (continued)

$CH_2=CH-N(CH_2CH_3)(CH_2CH_2CH_2CH_3)$	1628
$CH_3CH_2CH_2CH_2-CH=CH-N(CH_3)_2$	1650
$CH_3(H)C=C(CH_2CH_3)N(CH_3)_2$	1640
$(CH_3)_2C=CH-N(CH_2CH_3)_2$	1662
$CH_2=C(CH_3)-CH=CH-CH_3$	1652 1612
$CH_2=CH-CH=CH-CH=CH_2$ (H, H cis)	1830 1621 987 818
$CH_2=CH-CH=CH-CH=CH_2$ (H, H trans)	1800 1621 941 899
$CH_3-C\equiv C-CH=CH-CH_3$	1607 (2270)
$CH_2=CH-C_6H_5$	1636
$CH_2=CHCN$	1645 1612
$CH_2=CHCHO$	1618 (1704)
$CH_2=CHCOCH_3$	1618 (1684)
$CH_2=CHCOOH$	1635 1615 (1730) (1706)
$CH_2=CHCOOCH_3$	1637 (1735)

Characteristic IR-Absorption Bands of Alkynyl Groups (in cm^{-1})

Assignment	Range	Comments
≡C-H st	3340 - 3250	strong, sharp in the same range also OH st and NH st
C≡C st	2260 - 2100	weak, sharp R-C≡C-H: at the lower end of the range cited R-C≡C-R: usually 2 bands (Fermi resonance); often missing if substitution symmetrical sub-ranges: C-C≡C-H ∿2120; C-C≡C-C ∿2220; C-C≡C-C≡N ∿2240; C-C≡C-COOH ∿2240; C-C≡C-$COOCH_3$ ∿2240,2140 in the same range also C≡Z st, X=Y=Z st, Si-H st
≡C-H δ	700 - 600	weak, no practical significance

Characteristic IR-Absorption Bands of Aromatic Compounds (in cm^{-1})

Assignment	Range	Comments
arC-H st	3080 - 3030	often a number of bands of lower intensity in the same range also CH st of alkenes and small rings
arC-C	1625 - 1575	medium, often doublet; generally weak in benzene derivatives with a center of symmetry in the ring in the same range also C=C st, C=N st, C=O st, N=O st, C-C in heterocycles, NH δ
	1525 - 1475	medium, often doublet; weak for
		in the same range also N=O st, C-C in heterocycles, C=O st, B-N st, CH_3 δ, CH_2 δ, NH δ
comb	2000 - 1650	very weak; useful for the determination of substitution pattern in 6-membered aromatic rings, see p. I55 in the same range also H_2O δ, C=O st, B-H···B st, $\overset{+}{N}$-H st
arC-H δ ip	1250 - 950	a number of bands of variable intensity, no practical significance

arC-H δ oop	900 - 650	one or more strong bands; in 6-membered aromatic rings useful for the determination of substitution pattern, see p. I55 in the same range also =C-H δ oop, C-O-C γ and C-N-C γ in saturated heterocycles, OH δ oop in carboxylic acids, NH δ, N-O st, S-O st, CH_2 γ, C-F δ (?), C-Cl st

5-ring-heteroaromatics:

	pyrrole (N-H)	furan (O)	thiophene (S)
NH st free:	3500-3400	–	–
hydrogen bonded:	3400-2800		
CH st	~3100	~3100	~3100
γ	1590-1560	1610-1560	1535-1515
intensity variable, generally multiplets	1540-1500	1510-1475	1455-1410
CH δ oop generally strong	770- 710	990- 725	935- 700

Characteristic IR-Absorption Bands of Aromatic Compounds (in cm^{-1})
(continued)

Determination of substitution pattern in 6-membered aromatic rings:

not to be used for ring systems with strongly conjugated substituents such as C=O, NO_2, C≡N
position and shape of bands related to the number of adjacent H-atoms

number of adjacent H	type of substitution in benzene derivatives	band positions comb	δ oop, γ
		2000 1600	
5	mono-		∿900, 770-730, 710-690
4	o-di-	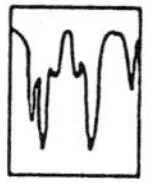	770-735
(1 + 3)	m-di-		900-860, 865-810*, 810-750, 725-680

		2000 1600	
3	vic-tri-		800-770, 720-685, 780-760*
(1 + 2)	asym-tri-		900-860, 860-800, 730-690
2	p-di- 1,2,3,4-tetra-		860-780
1	1,3,5-tri- 1,2,3,5-tetra- 1,2,4,5-tetra- penta-		900-840; in 1,3,5-trisubstituted benzenes also: 850-800, 730-675*
0	hexa-		

*: band sometimes missing

Characteristic IR-Absorption Bands of Compounds of the Type X≡Y (in cm^{-1})
(Nitriles, Cyanates, Thiocyanates, Isonitriles, Diazo Compounds)

Assignment	Range	Comments
-C≡N st	2260 - 2240	intensity medium to strong, sharp. For $O-CH_2-C≡N$ and $N-CH_2-C≡N$ lower intensity beyond the normal range: C=C-C≡N, C_6H_5-C≡N: 2240-2215 XC-C≡N, X : Cl, Br, I: 2240-2230; $-CF_2-C≡N$: ∿2275 >N-C≡N ⟷ $>\overset{+}{N}=C=\overset{-}{N}$: 2225-2175 >N-C=C-C≡N: 2210-2185 C≡N→O: 2305-2280; N=O st: 1395-1365 $(C≡N)^-$: 2200-2070 $(OC≡N)^-$: 2220-2130; $(OC≡N)^-$ st sy: 1335-1290 -S-C≡N: 2170-2135; C-S st: 725-550, often doublet $(SC≡N)^-$: 2090-2020; C-S st: 750
$-\overset{+}{N}≡\overset{-}{C}$	2150 - 2110	strong
$-\overset{+}{N}≡N$	2310 - 2130	medium; frequency depends on anion
		in the region for X≡Y st also C≡C st, X=Y=Z st as, $\overset{+}{N}H$ st, PH st, POH st, SiH st, BH st

Characteristic IR-Absorption Bands of Compounds of the Type X=Y=Z (in cm^{-1})

(Allenes, Ketenes, Ketenimines, Diazo Compounds, Isocyanates, Isothiocyanates, Carbodiimides, Azides)

Assignment	Range	Comments
(C=C)=C-H st	3050 - 2950	in this range also CH st of other compounds
C=C=C st as	1950 - 1930	strong, doublet in $X-C=C=CH_2$ if X not alkyl
		ring strain increases frequency: (cyclopropylidene)$=C=CH_2$: ∿2020
		for $C=C=CH_2$: overtone at ∿1700, weak
		C=C=C st sy: 1075-1060, weak, absent if high symmetry
$(C=C)=CH_2$ δ oop	∿850	strong
C=C=O st as	2155 - 2130	very strong, $CH_2=C=O$: ∿2155; $(C_6H_5)_2-C=C=O$: ∿2130
C=C=N st as	2050 - 2000	very strong, sometimes doublet

Characteristic IR-Absorption Bands of Compounds of the Type X=Y=Z (in cm^{-1})
(Allenes, Ketenes, Ketenimines, Diazo Compounds, Isocyanates, Isothiocyanates, Carbodiimides, Azides)
(continued)

Assignment	Range	Comments
$C=\overset{+}{N}=\overset{-}{N}$ st as	2050 - 2010	very strong sub-ranges: $R-CH=\overset{+}{N}=\overset{-}{N}$: 2050-2035 } R = ar or al $R_2-C=\overset{+}{N}=\overset{-}{N}$: 2035-2010 } R = ar or al beyond the normal range: $R-CO-C=\overset{+}{N}=\overset{-}{N}$: 2100-2050 C=O st: R = al: ∿1645; R = ar: ∿1615 $C=\overset{+}{N}=\overset{-}{N}$ st sy: ∿1350, strong O=(cyclohexadienylidene)=$\overset{+}{N}=\overset{-}{N}$: 2180-2010 C=O st: 1645-1560
-N=C=O st as	2275 - 2230	strong, sharp; CH_3NCO: ∿2230; (phenyl)-NCO: ∿2275; $-CF_2NCO$: ∿2300 N=C=O st sy: ∿1390, weak beyond the normal range: $(N=C=O)^-$: 2220-2130

-N=C=S st as	2150 - 2050	very strong, generally doublet (Fermi resonance) R-N=C=S st sy: R = al: ∿950; R = ar: 700-650 beyond the normal range: $(N{=}C{=}S)^{-}$: 2090-2020
-N=C=N- st as	2150 - 2100	very strong, doublet if aromatic substituent -N=C=N st sy: ∿1390, very weak
$-N{=}\overset{+}{N}{=}\overset{-}{N}$ st as	2250 - 2080	very strong, in acyl azides often doublet (C=O st: ∿1700) $-N{=}\overset{+}{N}{=}\overset{-}{N}$ st sy: 1350-1180, strong other related systems: C=S=O st: ∿1100, ∿1050 N=S=O st: ∿1250, ∿1135 in the region for X=Y=Z st as also C≡C st, X≡Y st, $\overset{+}{N}H$ st, PH st, POH st, SiH st, BH st

Characteristic IR-Absorption Bands of Alcohols and Phenols (in cm^{-1})

Assignment	Range	Comments
-OH st	3650 - 3200	intensity variable sub-ranges: -OH free: 3650-3590, sharp -OH hydrogen bonded: 3550-3450, broad -OH polymer: 3500-3200, broad, often a number of bands beyond the normal range: R-O-H ⟷ R=O···H (chelates): 3200-2500, often very broad in the same range also NH st, ≡CH st: ∿3300, sharp, H_2O
-OH δ ip	1450 - 1200	medium, no practical significance
C-O st	1260 - 970	strong, often doublet sub-ranges: $-CH_2$-OH: 1075-1000 >CH-OH : 1125-1000 ≥C-OH : 1210-1100 arC-OH : 1260-1180 in this range also bands for C-F st, C-N st, N-O st, P-O st, C=S st, S=O st, P=O st, Si-O st, Si-H δ
-OH δ oop	<700	medium, no practical significance

Characteristic IR-Absorption Bands of Ethers (in cm^{-1})

Assignment	Range	Comments
C-O-C st as	1310 - 1000	strong, sometimes split (C-O-C st sy weak) sub-ranges for non-cyclic ethers: $-CH_2-O-CH_2-$: 1150-1085 (OCH_3 st: 2850-2815; OCH_2 st: 2880-2835) $>CH-O-CH<$: 1170-1115, often split C=C-O-C-al : 1225-1180 (st sy: 1125-1080, medium) arC-O-C-al : 1275-1200 (st sy: 1075-1020, medium) sub-ranges for cyclic ethers : ∿1270, ∿840 (CH st: 3050-3000) : ∿1030, ∿980 : ∿1070, ∿915; 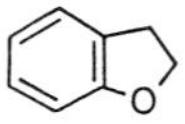 : ∿1235 : ∿1100, ∿815 ketals, acetals: 4 to 5 bands in the range indicated, in cyclic compounds also: 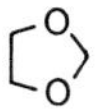 : ∿950; 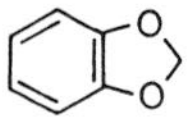 : ∿925 (CH_2 st: ∿2780) 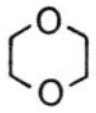 : ∿880; 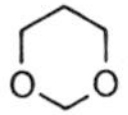 : ∿800 in acetals CH st: ∿2820, weak strong bonds in the same range also for C-O st, C-F st, C-N st, N-O st, P-O st, C=S st, S=O st, P=O st, Si-O st, Si-H δ

Characteristic IR-Absorption Bands of Peroxides and Hydroperoxides (in cm^{-1})

Assignment	Range	Comments
O-O-H st	3450 - 3200	intensity variable sub-ranges: -OOH free: ∿3450 -OOH hydrogen bonded: about 30 cm^{-1} higher than in the corresponding alcohols; in peracids at ∿3300 in the same range also OH st, NH st, ≡CH st, H_2O
C-O-O st	1200 - 1000	strong, about 20 cm^{-1} lower than in the corresponding alcohols strong bands in the same range also for C-O st, C-F st, C-N st, N-O st, P-O st, C=S st, S=O st, P=O st, Si-O st, Si-H δ
O-O st	1000 - 800	medium or weak; assignment uncertain also: C=O st in peracids: 1760-1745 in diacylperoxides: 1820-1770, two bands

Characteristic IR-Absorption Bands of Amines (in cm^{-1})

Assignment	Range	Comments
$-NH_2$ st	3500 - 3300	intensity variable, generally 2 sharp bands, $\Delta\nu$ = 65 - 75 at lower wave numbers (often <3200) if hydrogen bonded, broader hydrogen bonded and free form often simultaneously observed in primary aromatic amines an additional band at ~3200 in the same range also OH st, ≡CH st
>NH st	3450 - 3300	intensity variable, only one band at lower wave numbers (often <3200) if hydrogen bonded, broader hydrogen bonded and free form often simultaneously observed in the same range also OH st, ≡CH st: ~3300, sharp, H_2O
$-\overset{+}{N}H_3$ st $>\overset{+}{N}H_2$ st $\geq\overset{+}{N}H$ st	3000 - 2000	medium, broad, very structured band sub-ranges for major maximum: $\overset{+}{N}H_3$: 3000-2700; also $\overset{+}{N}H_3$ comb: ~2000 $\overset{+}{N}H_2$: 3000-2700 $\overset{+}{N}H$: 2700-2250 in the same range also OH st, NH st, CH st, SH st, PH st, SiH st, BH st, X=Y=Z st, X≡Y st

Characteristic IR-Absorption Bands of Amines (in cm^{-1})

(continued)

Assignment	Range	Comments
$-NH_2$ δ	1650 - 1590	medium or weak
>NH δ	1650 - 1550	weak
$-\overset{+}{N}H_3$ δ, $>\overset{+}{N}H_2$ δ, $\geq\overset{+}{N}H$ δ	1600 - 1460	medium, weak in aliphatic amines, often more than one band
C-N st	1400 - 1000	medium, no practical significance
$-NH_2$ δ, >NH δ	850 - 700	medium or weak, in primary amines 2 bands
		also: P-N-C st: 1110-930, 770-680

Characteristic IR-Absorption Bands of Halogen Compounds (in cm^{-1})

Assignment	Range	Comments
C-F st	1400 - 1000	strong, sharp, often more than one band (rotational isomers), often not resolved
		sub-ranges:
		alCF : 1100-1000 } FC-H st: 3080-2990
		alCF_2: 1150-1000 } FC-H st: 3080-2990
		alCF_3: 1350-1100
		C=CF : 1350-1150; C=CF_2 st: ∿1745
		arCF : 1250-1100
		strong bands in the same range also for C-O st, NO_2 st sy, C=S st, S=O st
CF_2 } CF_3 }	780 - 680	medium or weak, assignment uncertain (C-F δ?)
		also:
		S-F st : 815-755, strong
		P-F st : 1110-760
		Si-F st: 980-820
		B-F st : 1500-800

Characteristic IR-Absorption Bands of Halogen Compounds (in cm^{-1})
(continued)

Assignment	Range	Comments
C-Cl st	830 - <600	strong, often broad (rotational isomers); absent in aryl chlorides; in $-CCl_3$: 830-700
C-Br st	<700	
C-I st	<600	
		in disubstituted halobenzenes characteristic skeletal vibration:

X-C6H4-R

type of substitution	o	m	p
X : Cl	1055-1035	1080-1075	1095-1090
X : Br	1045-1030	1075-1065	1075-1070
X : I			1060-1055

also:

P-Cl st : <600
Si-Cl st: <625
B-Cl st : 1100-650

Characteristic IR-Absorption Bands of Aldehydes (in cm^{-1})

Assignment	Range	Comments
comb	2900 - 2800 2780 - 2680	weak (Fermi resonance with C-H δ at ∿1390. For extreme position of C-H δ only one band) sub-ranges: al: 2830-2810; 2720-2690 ar: 2830-2810; 2750-2720, for o-substitution often higher in the same range also a weak band of cyclohexanes: ∿2700
C=O st	1765 - 1645	strong sub-ranges: al: 1740-1720; α-halogenated: 1765-1730 ar: 1710-1685; αβ-unsaturated: 1695-1660 with intramolecular H-bonds: 1670-1645

Examples

$CH_3(CH_2)_7CHO$ 1729 $ClCH_2CHO$ 1742 CH_3(H)C=C(H)CHO 1685 C_6H_5-CHO 1705

2-HO-C_6H_4-CHO 1670 $(CH_3)_2N$-C_6H_4-CHO 1696 NO_2-C_6H_4-CHO 1717 (CCl_4), 1710 ($CHCl_3$) NaO-C_6H_4-CHO 1639

Characteristic IR-Absorption Bands of Ketones (in cm^{-1})

Assignment	Range	Comments
C=O st	1775 - 1650	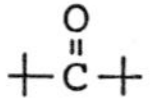 strong sub-ranges: al: ∿1715; branching at α-position shifts to lower frequencies: 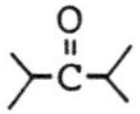 $(CH_3)_2CH-CO-CH(CH_3)_2$: ∿1695; $(CH_3)_3C-CO-C(CH_3)_3$: ∿1685 cyclic ketones: 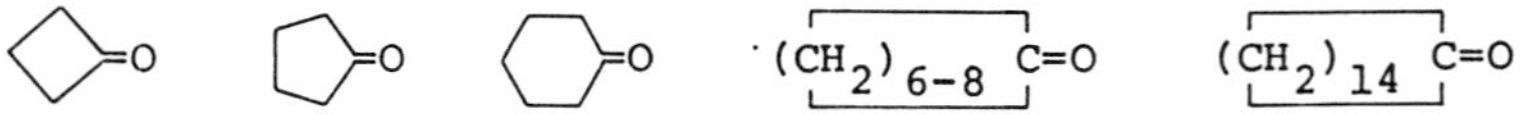 cyclobutanone ∿1775; cyclopentanone ∿1750; cyclohexanone ∿1715; $(CH_2)_{6-8}$ C=O (ring) ∿1705; $(CH_2)_{14}$ C=O (ring) ∿1715 conjugated ketones: αβ-unsaturated: ∿1675, often 2 bands (rotational isomers); C=C st: 1650-1600 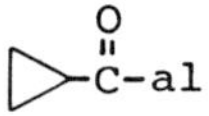 (cyclopropyl-CO-al : ∿1695) αβ,γδ-unsaturated: ∿1665 (cyclopropyl-CO-C=C: ∿1670) αβ,α'β'-unsaturated: ∿1665 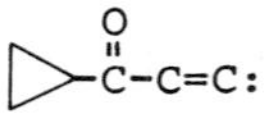 aryl ketones: ∿1690 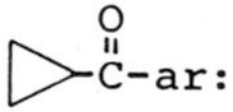 (cyclopropyl-CO-ar: ∿1675) diaryl ketones: ∿1665 (with N or O in p-position to ∿1600)

α-halogenated ketones: shifted towards higher wave numbers, depending on dihedral angle ϕ between C=O and C-hal; largest effect for $\phi = 0^{\circ}$, no effect for $\phi = 90^{\circ}$

maximal shift:

α-chloro:	∿25
α,α-dichloro:	∿45
α,α'-dichloro:	∿45
α-bromo:	∿20
α-iodo:	∿0
α,α-difluoro:	∿60
perfluoro:	∿90

diketones:

α-diketones: al: ∿1720; 5-ring: ∿1775, ∿1760; 6-ring: ∿1760, ∿1730; enolized: ∿1675 (C=C st: ∿1650)
ar: ∿1680; ortho-quinones: ∿1675, with peri-OH: ∿1675, ∿1630

β-diketones: keto form: ∿1720 (sometimes doublet); enol form: ∿1650, with intramolecular H-bonds: ∿1615 (C=C st: ∿1600, strong)

γ-diketones: like normal ketones; para-quinones: ∿1675 (C=C st: ∿1600, with peri-OH: ∿1675, ∿1630

Characteristic IR-Absorption Bands of Ketones: Examples (in cm^{-1})

1691 | 1697 | 1690 ⇌ 1707 | $ClCH_2$-CO-CH_3 1752, 1726 rotamers | 1722

$CH_3C{\equiv}CCOCH_3$ 1678 (2222) | CH_2=CHCOCH=CH_2 1672, 1660 | CCl_3COCCl_3 1780, 1751

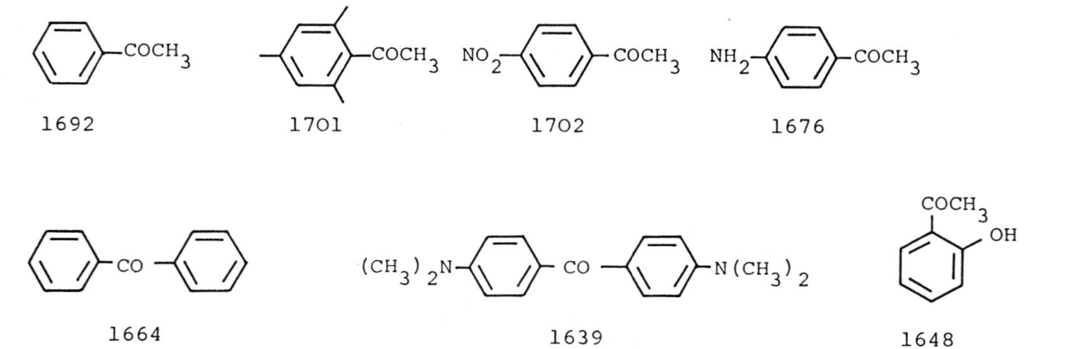

$CH_3(CH_2)_3COCO(CH_2)_3CH_3$

1710

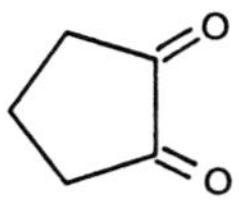

1700
1655

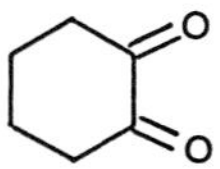

1735

$CH_3COCH_2COCH_3$

1724 (ketoform)
1608 (enolform)

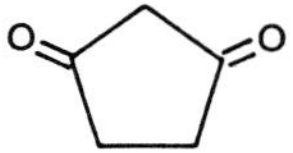

1755
1725
1635
1590

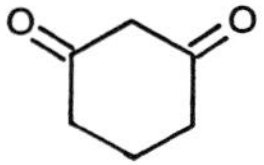

1630
1607

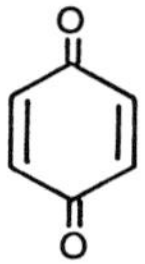

1669

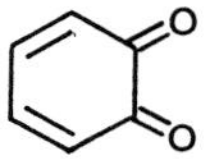

1669

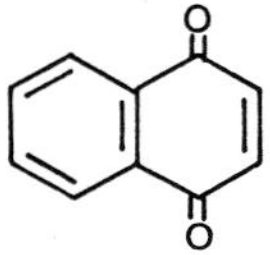

O OH O OH O O O O O

1675

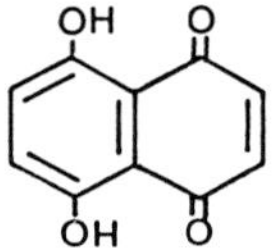

1623

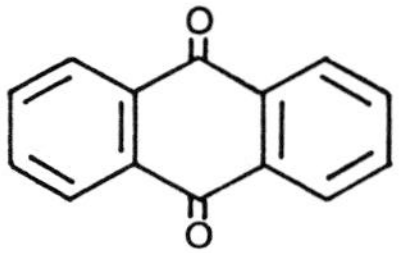

1678

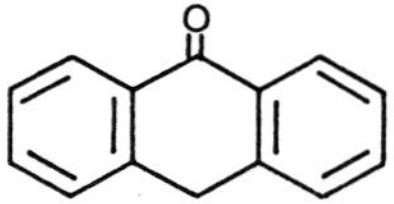

1662

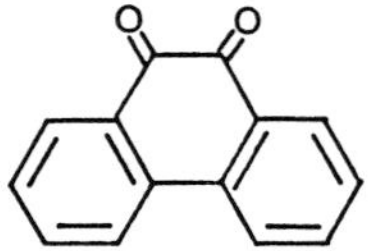

1684

Characteristic IR-Absorption Bands of Esters and Lactones (in cm^{-1})

Assignment	Range	Comments
C=O st	1790 - 1650	strong
		sub-ranges:
		aliphatic esters: 1750-1735
		conjugated esters:
		αβ-unsaturated acid: 1730-1710
		ar acid: 1730-1715
		with intramolecular H-bonds: 1690-1670
		α-halogenated acid: 1790-1740
		vinyl esters: ∿1760 (C=C st: 1690-1650, strong)
		phenol esters: ∿1760
		phenol esters of an ar acid: ∿1735
		diesters: like the corresponding monoesters
		keto esters:
		α-keto: 1755-1725, generally only one band
		β-keto: keto form: ∿1750 (ketone), ∿1735 (ester)
		enol form: ∿1650 (C=C st: ∿1630, strong)
		γ-keto: ∿1740, ∿1715; pseudo esters: ∿1770

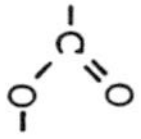

		lactones: 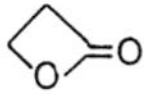 ~1840 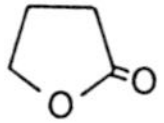 ~1770 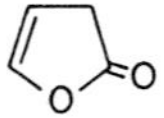 ~1800 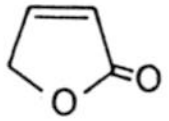 ~1750 (additional band at ~1780 if α-position free) 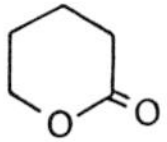 ~1735 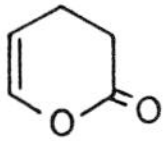 ~1760 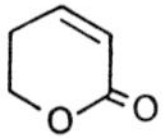 ~1720 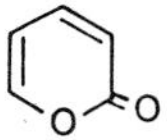 ~1730 (often doublet)
C-O st	1330 - 1050	2 bands: st as: very strong and at higher wave number; st sy: strong, at lower wave number sub-ranges for st as: formates, propionates, higher al esters: ~1185 acetates: ~1240 vinyl esters, phenol esters: ~1210 γ-lactones, δ-lactones: ~1180 methyl esters of al acids: ~1165 in the same range also strong bands for C-F st, C-N st, N-O st, P-O st, C=S st, S=O st, P=O st, Si-O st, Si-H δ

Characteristic IR-Absorption Bands of Esters: Examples (in cm^{-1})

$HCOO(CH_2)_3CH_3$	$CH_3COO(CH_2)_3CH_3$	$CF_3COO(CH_2)_3CH_3$
1730	1743	1787
$CH_3CH_2COOCHBrCH_3$	$CH_3CHBrCOOCH_2CH_3$	
1747	1743	

CH_3COO, H C=C H, CH_3	CH_3COO, CH_3 C=C H, H	CH_3, H C=C H, $COOCH_3$	CH_3, $COOCH_3$ C=C H, H	CH_3, H C=C H, $COOCH=CH_2$
1752 (1675)	1758 (1690)	1726	1724	1730 (1658) (1638)

$CH_3COOSi(CH_3)_3$	$CH_3COCOOCH_3$	$CH_3COCH_2COOCH_3$
1725	1725	1704 (ester) 1690 (ketone) 1645 (enol)

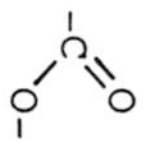

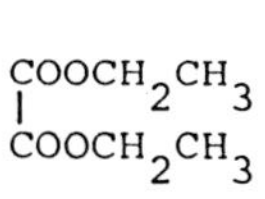

Compound	Wavenumber
$COOCH_2CH_3$–$COOCH_2CH_3$	1774, 1754
$CH_2(COOCH_2CH_3)_2$	1760, 1742
$CH_2COOCH_2CH_3$–$CH_2COOCH_2CH_3$	1740
$CHCOOCH_2CH_3$=$CHCOOCH_2CH_3$	1734
CH_3CH_2OOCCH=$CHCOOCH_2CH_3$	1727
$C_6H_5COOCH_3$	1727
$4\text{-}NO_2C_6H_4COOCH_3$	1737
$4\text{-}(CH_3)_2NC_6H_4COOCH_3$	1715
$2\text{-}HOC_6H_4COOCH_3$	1684
$CH_3COOC_6H_5$	1766
$C_6H_5COOC_6H_5$	1743
$C_6H_4(COO(CH_2)_3CH_3)_2$ (ortho)	1746

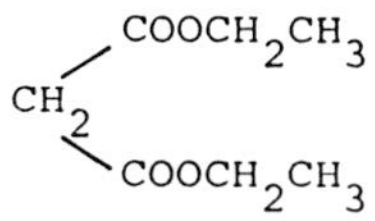

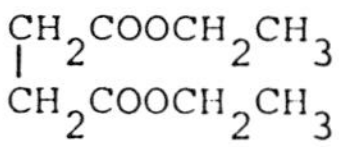

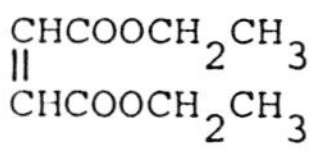

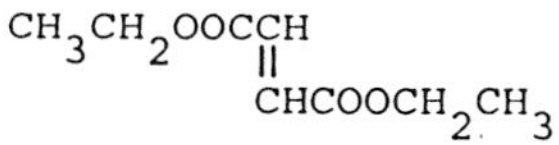

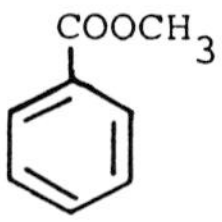

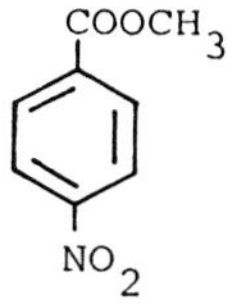

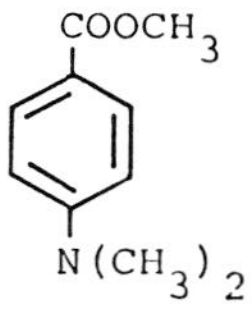

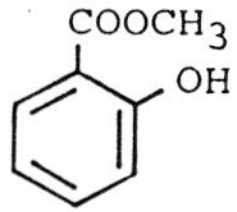

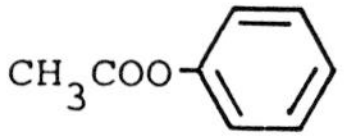

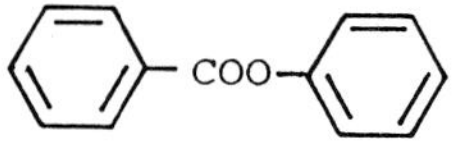

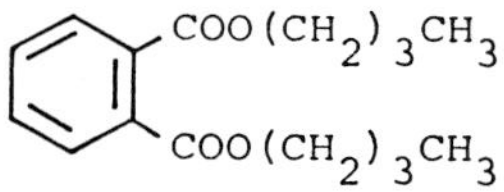

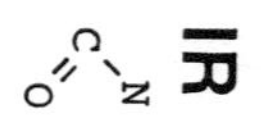

Characteristic IR-Absorption Bands of Amides, Lactams, Imides and Hydrazides (in cm^{-1})

Assignment	Range	Comments
N-H st	3500 - 3100	medium, in primary amides 2 bands, in polypeptides and proteins multiplet sub-ranges: free: 3500-3400 hydrogen bonded: 3350-3100 in primary amides generally 2 bands at ∿3350,3180 in lactams generally 2 bands at ∿3200,3100 in hydrazides: mono-: ∿3200; di-: ∿3100 in imides: ∿3250 in the same range also OH st, ≡CH st: ∿3300, sharp, H_2O

C=O st (amide I)	1740 - 1630	generally strong			
		sub-range:			
		amides:	H_2NCO-	-HNCO-	>NCO-
		free:	∿1690	∿1685	∿1650
		hydrogen bonded:	∿1650	∿1660	∿1650
		lactams: 4-ring: ∿1745			
		5-ring: ∿1700			
		6,7-ring: ∿1650			
		hydrazides: mono-: ∿1670; di-: ∿1600			
		imides: 1740-1670; in 5-ring imides 2 bands at ∿1750,1700			
		polypeptides: 1655-1630			
		isocyanurates: ∿1690, at ∿1770 for aromatic substitution			
		trifluoroacetates: ∿1720, 1755 (shoulder)			
NH δ } N-C=O st sy } (amide II)	1630 - 1510	generally strong			
		sub-range:	H_2NCO-	-HNCO-	lactams
		free:	∿1610	∿1530	} absent
		hydrogen bonded:	∿1630	∿1540	} absent
		polypeptides: 1560-1510			
		trifluoroacetates: ∿1555			
		also:	H_2NCO-	-HNCO-	lactams
		C-N st (?)	∿1400	∿1250	∿1330
		NH δ ip	∿1150		∿1465
		NH δ oop	750-600	∿700	∿800

Characteristic IR-Absorption Bands of Amides: Examples for C=O Stretching Frequencies (in cm^{-1})

NH_2-COH

$CHCl_3$	1709
neat liquid	1672

$CH_3NH-COH$

neat liquid	1672

$(CH_3)_2N-COH$

$CHCl_3$	1673
neat liquid	1670

$NH_2-CO-CH_2CH_2CH_3$

$CHCl_3$	1679
solid	1631

$CH_3CH_2CH_2NH-CO-CH_3$

CCl_4	1690

$(CH_3CH_2CH_2)_2N-CO-CH_3$

CCl_4	1647

$NH_2-CO-CH=CHCH_3$

solid	1677

$(CH_3CH=CH)(CH_3CH_2CH_2)N-CO-CH_3$

CS_2	1675,1650

$NH_2-CO-C_6H_5$

$CHCl_3$	1678
solid	1656

$C_6H_5-NH-CO-CH_3$

$CHCl_3$	1691
solid	1658
CCl_4	1705

$C_6H_5(CH_3CH_2)N-CO-CH_3$

CCl_4	1667

Characteristic IR-Absorption Bands of Vinylogous Amides and Imides: Examples (in cm^{-1})

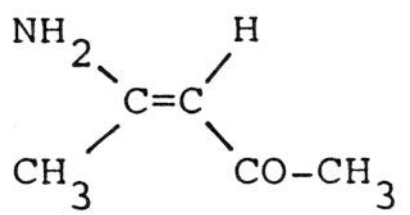

$NH_2(CH_3)C=CH-CO-CH_3$
neat liquid 1700,1625,1540

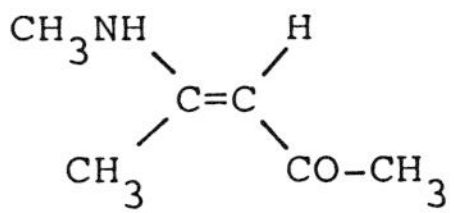

$CH_3NH(CH_3)C=CH-CO-CH_3$
solid 1628,1595

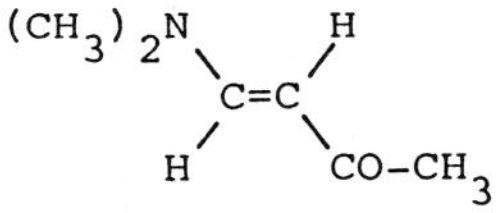

$(CH_3)_2N-CH=CH-CO-CH_3$
solid 1631,1584

$CH_3CONHCOCH_3$
solid 1734,1505

$CH_3CON(CH_3)COCH_3$
solid 1736,1706,1689

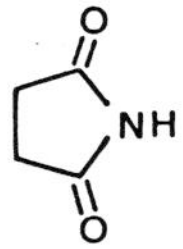

NH (succinimide)
CCl_4 1753,1727
solid 1771,1698

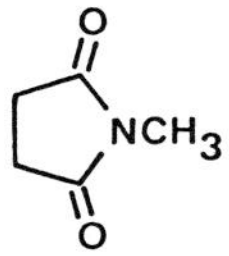

NCH_3 (N-methylsuccinimide)
CCl_4 1721,1705
solid 1760,1690

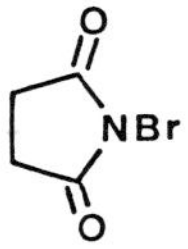

NBr (N-bromosuccinimide)
$CHCl_3$ 1783,1733

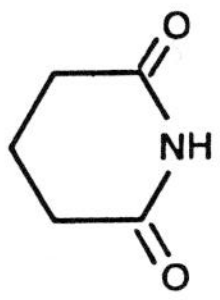

NH (glutarimide)
CCl_4 1742,1730,1718

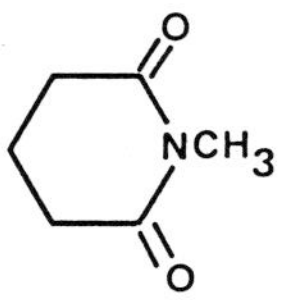

NCH_3 (N-methylglutarimide)
CCl_4 1729,1686
solid 1718,1670

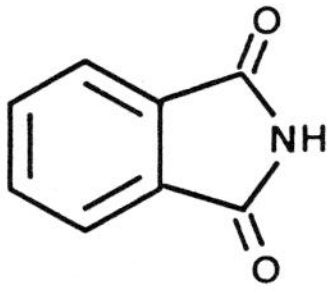

NH (phthalimide)
$CHCl_3$ 1778,1735
solid 1774,1749,1724

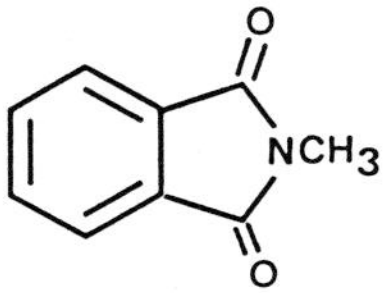

NCH_3 (N-methylphthalimide)
$CHCl_3$ 1772,1712

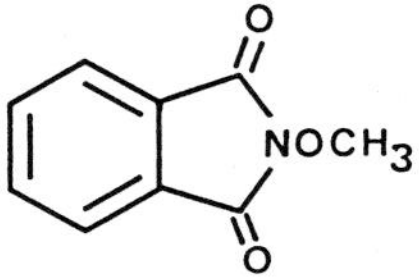

$NOCH_3$ (N-methoxyphthalimide)
solid 1790,1735

Characteristic IR-Absorption Bands of Carbonic Acid Derivatives (in cm^{-1})

Assignment	Range	Comments
C=O st	1820 - 1630	strong sub-ranges: al-O-CO-O-al: ∿1740 (C-O st as: ∿1260) ar-O-CO-O-al: ∿1770 (C-O st as: ∿1230) ar-O-CO-O-ar: ∿1785 (C-O st as: ∿1215) (cyclic -O-C(=O)-O-) : ∿1820 (C-O st as: ∿1200) al-S-CO-O-al: ∿1705 (C-O st as: ∿1150) ar-S-CO-O-al: ∿1725 (C-O st as: ∿1130) al-S-CO-O-ar: ∿1735 (C-O st as: ∿1075) al-S-CO-S-al: ∿1645 (C-S st: ∿880) ar-S-CO-S-al: ∿1650 (C-S st: ∿840) ar-S-CO-S-ar: ∿1715 (C-S st: ∿830) al-O-CO-Cl : ∿1780 (C-O st: ∿1155) ar-O-CO-Cl : ∿1785 al-S-CO-Cl : ∿1770 ar-S-CO-Cl : ∿1775

		H_2N-CO-O- : ∿1700 (NH_2 δ: ∿1620) -NH-CO-O- : ∿1725 (C-N st: ∿1230) H_2N-CO-S- : ∿1700 (NH_2 : ∿1620) -NH-CO-S- : ∿1670 (C-N st: ∿1190) >N-CO-S- : ∿1670 (C-N st: ∿1250) >N-CO-Cl : ∿1740 -NH-CO-NH- : ∿1660 (O=C(CH₂-ring)(NH)₂C=O: 1780-1680, 3 bands)
C=N st	1725 - 1625	strong sub-ranges: -O-C(=NH)-O- : 1690-1645 (C-O st: ∿1325,1100; NH st: 3400-3200) -O-C(=NH_2^+)-O- : 1680-1635 (NH_2^+ δ: 1590-1540) >N-C(=NH)-N< : 1725-1625 (NH st: 3400-3200; NH δ: 1670-1620) -NH-C(=NR)-O-: 1675-1655
C=S st	1225 - 1030	strong sub-ranges: O-CS-O: ∿1225 O-CS-N: ∿1170 N-CS-N: 1340-1130 S-CS-N: ∿1050 S-CS-S: ∿1070

Characteristic IR-Absorption Bands of Carbonic Acid Derivatives: Examples (in cm^{-1})

$NH_2COOCH_2CH_3$	$CH_3NHCOOCH_2CH_3$		
$CHCl_3$ 1725	CCl_4 1727		
$NH_2COOCH_2CH_2OOCNH_2$	$(CH_3)_2NCOOCH_2CH_3$	$(CH_3)_2NCOSCH_3$	
solid 1690	$CHCl_3$ 1684	CCl_4 1662	
$CH_3CH_2OCOOCH_2CH_3$	C_6H_5-OCOO-C_6H_5		
CCl_4 1748	CCl_4 1786		
$CH_3OCOOCH_3$	$CH_3SCOOCH_3$	$CH_3SCOSCH_3$	$CH_3SCSSCH_3$
CCl_4 1758 $CHCl_3$ 1751	CCl_4 1719	CCl_4 1653	neat liquid 1076
CCl_4 1822,1748	CCl_4 1757	CCl_4 1718,1677,1640	CCl_4 1083,1079 solid 1058

NH_2CONH_2
solid 1679,1627

$CH_3NHCONH_2$
solid 1645,1567,1418

$(CH_3)_2NCONH_2$
solid 1656,1610,1511

$CH_3NHCONHCH_3$
solid 1622,1580,1530
$CHCl_3$ 1663,1548

$(CH_3)_2NCON(CH_3)_2$
solid 1645,1560,1497
$CHCl_3$ 1675

$C_6H_5NHCONHC_6H_5$
solid 1650

$CH_3CONHCONH_2$
solid 1667,1634

$(CH_3)_2NC(=NH)N(CH_3)_2$
neat liquid 1600

2-imidazolidinone (HN–C(=O)–NH ring)
CCl_4 1735,1718

2-iminoimidazolidine (HN–C(=NH)–NH ring)
solid 1767,1681,1621

hydantoin (HN–C(=O)–NH ring, C=O)
solid 1776,1697

1-methylhydantoin (CH_3N–C(=O)–NH ring, C=O)
solid 1712,1676

3-methylhydantoin (HN–C(=O)–NCH_3 ring, C=O)
solid 1748,1706

dihydrouracil (HN–C(=O)–NH six-membered ring, C=O)
solid 1767,1695

Characteristic IR-Absorption Bands of Carboxylic Acids (in cm^{-1})

Assignment	Range	Comments
COO-H st	3550 - 2500	intensity variable sub-ranges: COO-H st free: 3550-3500, sharp; only in highly dilute solutions COO-H st hydrogen bonded: 3300-2500, broad, often more than one band in the same range also OH st, NH st, CH st, SiH st, SH st, PH st
C=O st	1800 - 1650	strong free: 1800-1740 } also in dicarboxylic acids hydrogen bonded (dimer): 1740-1650 } also in dicarboxylic acids sub-ranges for hydrogen bonded C=O: al-COOH: 1725-1700 C=C-COOH: 1715-1690 ar-COOH: 1700-1680 hal-C-COOH: 1740-1720 intramolecular H-bonded: 1670-1650

OC-OH st, OC-OH δ	1440 - 1210	no practical significance
OC-OH δ oop	960 - 880	medium, generally broad (only in dimers) in the same range also =CH δ, arCH δ, NH δ
$(COO)^-$ st as	1610 - 1550	very strong; in α-halogen carboxylates near the higher value, with more than one α-halogen beyond the normal range. in polypeptides at ∿1575
$(COO)^-$ st sy	1450 - 1400	strong, no practical significance; in polypeptides at ∿1470
		also: $(COO)^-$ δ: formates: ∿775, weak acetates: ∿925 benzoates: ∿680 CF_3COO^-: ∿600

Characteristic IR-Absorption Bands of Carboxylic Acids: Examples (in cm^{-1})

Compound	Medium	Band
$HCOOH$	neat liquid	1727
	CCl_4	1756,1724
CH_3COOH	neat liquid	1759,1718
	CCl_4	1768,1717
$(CH_3)_3CCOOH$	CCl_4	1704
	solid	1686
$H(CH_3)C=C(H)COOH$ (H and COOH on top, CH_3 and H on bottom)	neat liquid	1700
	CCl_4	1694
$CH_3CH(OH)COOH$	neat liquid	1725
$CH_3CH(Cl)COOH$	neat liquid	1730
$CH_3CH(^{+}NH_3)COO^{-}$	solid	1605
$CH_3CH(^{+}NH_3)COOH$	solid	1740
$CH_3COCOOH$	CCl_4	1788,1725
$HOOC{-}COOH$	solid	1650
$CH_2(COOH)_2$	solid	1735,1703
$HOOC{-}CH_2{-}CH_2{-}COOH$	solid	1724

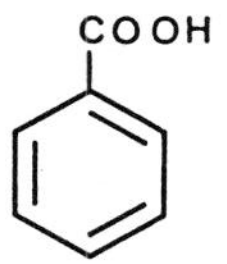

COOH

solid 1690
CCl_4 1730
$CHCl_3$ 1706

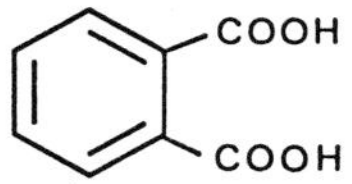

COOH
COOH

solid 1690

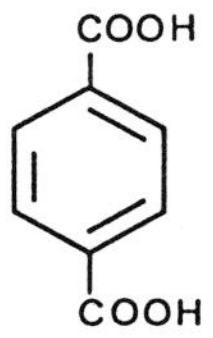

COOH
COOH

solid 1690

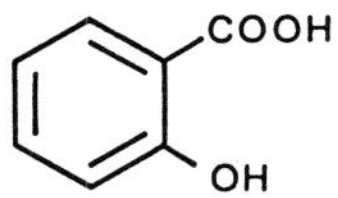

COOH
OH

solid 1667
CCl_4 1696
$CHCl_3$ 1661

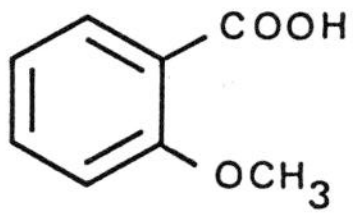

COOH
OCH_3

CCl_4 1750

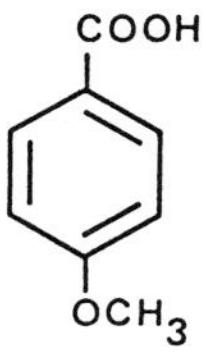

COOH
OCH_3

solid 1693
CCl_4 1693

Characteristic IR-Absorption Bands of Amino Acids (in cm^{-1})

Assignment	Range	Comments
N-H st, O-H st	3400 - 2000	generally strong, broad, very structured sub-ranges: zwitter ion: 3100-2000, distinct side band at 2200-2000 hydrochloride: 3350-2000 Na-salt: 3400-3200
$\overset{+}{N}H_3$ δ as	1660 - 1590	weak; for hydrochloride near the lower value
$\overset{+}{N}H_3$ δ sy	1550 - 1480	medium
COO^- st as	1760 - 1595	strong sub-ranges: zwitter ion: ∿1595 hydrochloride: 1755-1700, in α-amino acids: 1760-1730 Na-salt: ∿1595

Characteristic IR-Absorption Bands of Acid Halides (in cm^{-1})

Assignment	Range	Comments
C=O st	1820 - 1750	strong; the range is for chlorides; fluorides higher (1900-1870), bromides and iodides lower in aromatic acid chlorides and bromides an additional band at ∿1725 (Fermi resonance)
-C(=O)-Hal (?)	1300 - 900	strong; assignment uncertain in aromatic acid chlorides and bromides an additional band at ∿870

Characteristic IR-Absorption Bands of Acid Anhydrides (in cm^{-1})

Assignment	Range	Comments
C=O st	1870 - 1725	strong, 2 bands in this range sub-ranges: linear anhydrides: ∿1820, ∿1760, higher band stronger cyclic anhydrides: 5-ring: ∿1850, ∿1775 } lower band stronger 6-ring: ∿1800, ∿1760 } lower band stronger
C-O-C st as	1050 - 900	strong; linear: ∿1040, cyclic: ∿920

Characteristic IR-Absorption Bands of Compounds with C=N Groups (in cm^{-1})

(Azomethines, Schiff's Bases, Imines, Amidines, Azines, Imino Carbonates)

Assignment	Range	Comments
C=N st	1690 - 1580	generally strong: sub-ranges: R-CH=N-R': R, R' = al: ∿1670; R or R' = conjugated: ∿1645; R, R' = conjugated: ∿1630 CH_3COCH=N-C_6H_5: C=O st: ∿1655; C=N st: ∿1555 R(R")C=N-R': R, R', R" = al: ∿1655; R = conjugated: ∿1645; R, R' = conjugated: ∿1635 R(R')C=NH : R, R' = al: ∿1645; R = conjugated: ∿1625 R-C(NH_2)=N-R : 1685-1580; in R-C(NHR)=N-R additional 1540-1515 CH=N-N=CH: 1670-1600 (RO)$_2$C=NH : 1690-1645 (NH st: ∿3300; C-O st: ∿1325, ∿1100) (RO)$_2$C=NH_2^+ : 1680-1635 (NH_2^+ st: ∿3000; NH_2^+ δ: 1590-1540) S-C=N : 1640-1605; S-S-C=N: 1640-1580 O-C=N : 1690-1645 also: P=N st: 1500-1170

Characteristic IR-Absorption Bands of Oximes (in cm^{-1})

Assignment	Range	Comments
OH st	3600 - 2700	strong sub-ranges: free: ∿3600 hydrogen-bonded: 3300-3100, broad quinone oximes: to ∿2700, more than one band
C=N st	1685 - 1520	generally strong sub-ranges: al: 1685-1650 ar: 1645-1650 quinone oximes: 1580-1520 (C=O st: 1680-1620)
OH δ	1475 - 1315	no practical significance
N-O st	1050 - 400	

Characteristic IR-Absorption Bands of Compounds with N=N Groups (in cm^{-1})

(Azo Compounds, Azoxy Compounds, Azothio Compounds)

Assignment	Range	Comments
N=N st	1500 - 1400	very weak, missing in compounds of high symmetry
$-N=N(\rightarrow O)-$ st as	1480 - 1450	
st sy	1335 - 1315	
$-N=N(\rightarrow S)-$ st as	~1450	
st sy	~1050	
$O\leftarrow N=N\rightarrow O$ st	1410 - 1175	sub-ranges: trans / cis al: 1290-1175 / 1425-1385, 1345-1320 ar: 1300-1250 / ~1410, ~1395 (dimers of C-nitroso compounds; monomers: al: 1585-1540; ar: 1510-1490)

Characteristic IR-Absorption Bands of Nitrites and Nitroso Compounds (in cm^{-1})

Assignment	Range	Comments
N=O st	1680 - 1450	very strong sub-ranges: O-NO (nitrites): trans: 1680-1650; cis: 1625-1610 C-NO (C-nitroso compounds): al: 1585-1540; ar: 1510-1490 (dimers: see p. I200) N-NO (N-nitroso compounds): ∿1450 in nitrites also: comb: 3300-3200, ∿2500, 2300-2250 N-O st: trans: ∿800; cis: very weak ONO δ: trans: ∿600; cis: ∿650 in C-nitroso compounds also: C-N st: al: ∿850; ar: ∿1100 in N-nitroso compounds also: N-N st: ∿1040

Characteristic IR-Absorption Bands of Nitrates, Nitro Compounds and Nitramines (in cm^{-1})

Assignment	Range	Comments
NO_2 st as	1660 - 1490	very strong sub-ranges: $O-NO_2$ (nitrates): 1660-1625 $C-NO_2$ (nitro compounds): al: 1570-1540; ar: 1560-1490 $N-NO_2$ (nitramines): 1630-1530
NO_2 st sy	1390 - 1260	strong sub-ranges: $O-NO_2$ (nitrates): 1285-1270 $C-NO_2$ (nitro compounds): al: 1390-1340 ar: 1360-1310, often 2 bands $N-NO_2$ (nitramines): 1315-1260 in nitrates also: N-O st: ∿870, strong; NO_2 γ: ∿760, NO_2 δ: ∿700

Characteristic IR-Absorption Bands of Mercaptans, Thioethers and Disulfides (in cm^{-1})

Assignment	Range	Comments
S-H st	2600 - 2540	often weak, narrow
S-H δ	915 - 800	weak, no practical significance
C-S st	710 - 570	weak, broad, no practical significance
S-S st	∿500	weak, no practical significance
		also:
		(S-)CH_3 st as: ∿2880; (S-)CH_2 st as: ∿2860
		(S-)CH_3 δ as: ∿1430, δ sy: 1330-1290; (S-)CH_2 δ: ∿1425
		S-F st: 815-755, strong
		S-N st in S-N=O: ∿630
		S-C st in S-C≡N: 725-550, often doublet

Characteristic IR-Absorption Bands of Compounds with C=S Groups (in cm^{-1})

(Thioketones, Thioesters, Dithioacids, Thioamides, Thiolactams)

Assignment	Range	Comments
C=S st	1275 - 1030	strong, narrow sub-ranges: thioketones: 1075-1030 thioesters: 1210-1080 dithioacids: ∿1215 (SH st: ∿2550; SH δ: ∿860) thioacid fluoride: 1125-1075 (perfluorinated: 1130-1105) thioacid chloride: 1100-1065 (perfluorinated: 1100-1075) thioamides: } 1140-1090 (C-N st: 1535-1520; NH δ: 1380-1300) thiolactams: } also: P=S st: 750-580

Characteristic IR-Absorption Bands of Compounds with SO Groups (in cm^{-1})

Assignment	Range	Comments
>S=O st	1225 - 980	strong sub-ranges: R-SO-R : 1060-1015 R-SO-OH : ~1100 (S-O st: 870-810; OH st: free: ~3700; hydrogen bonded: ~2900, ~2500) R-SO-OR : ~1135 (S-O st: 740-720, 710-690) RO-SO-OR: 1225-1195 R-SO-Cl : ~1135 $R-SO_2^-$: ~1030, ~980 R=SO : ~1100, ~1050 (N=SO: ~1250, ~1135)

$>S(=O)_2$: S=O st as, S=O st sy	1420 - 1000	very strong sub-ranges: $R-SO_2-R$: 1370-1290, 1170-1110 $R-SO_2-OR$: 1375-1350, 1185-1165 $R-SO_2-SR$: ∿1340, ∿1150 $RO-SO_2-OR$: 1415-1390, 1200-1185 $R-SO_2-N$: 1365-1315, 1180-1150 (NH st: 3330-3250; NH δ: ∿1750; SN st: 725-650) $R-SO_2$-hal: 1410-1375, 1205-1170 $R-SO_2-OH$: 1355-1340, 1165-1150 (OH st: ∿2900, ∿2400; hydrated: 2800-1650, broad, S=O st like for $R-SO_3^-$) $R-SO_3^-$: 1250-1140, 1070-1030 $RO-SO_3^-$: 1315-1220, 1140-1050
S-O st	870 - 690	variable intensity, in sulfites weak

Characteristic IR-Absorption Bands of Phosphorous Compounds (in cm^{-1})

Assignment	Range	Comments
P-H st	2440 - 2275	weak to medium, generally only 1 band, in $R_3\overset{+}{P}H$ very broad
PO-H st	2700 - 2650	weak, very broad bands
POH comb	2300 - 2250	(weak, very broad bands) in $P{\lt}^{O}_{OH}$ also 1740-1600 (dimer?)
P-O st	1260 - 855	sub-ranges: P-O-C st: al: 1050-970, strong; 830-740, often weak ar: 1260-1160; P^{V}: 995-915; P^{III} : 875-855 P-OH st : 1100-940, broad; for $P(OH)_2$ often 2 bands P-O-P st: 980-900
P=O st	1300 - 960	strong sub-ranges: $R_3P=O$: 1190-1150 (also for R = H) $R_2(R'O)P=O$: 1265-1200 (also for R = H) $R(R'O)_2P=O$: 1280-1240 (also for R = H) $(RO)_3P=O$: 1300-1260 $R(HO)_2P=O$: 1220-1150; $R(HO)PO_2^-$: 1250-990, more than one band; RPO_3^{2-}: 1125-970, 1000-960

$R_2(HO)P{=}O$: 1205-1090; $R_2PO_2^-$: 1200-1090, 1090-995

$RO(HO)_2P{=}O$: ~1250; $RO(HO)PO_2^-$: 1230-1210, 1030-1020;

$ROPO_3^{2-}$: 1140-1050, 1010-970

$(RO)_2(HO)P{=}O$: 1250-1210; $(RO)_2PO_2^-$: 1285-1120, 1120-1050

$R(RO)(HO)P{=}O$: 1220-1170; $R(RO)PO_2^-$: 1245-1150, 1110-1050

$R_2P(=O)-O-P(=O)R_2$: 1240-1205

$(RO)_2P(=O)-O-P(=O)(RO)_2$: 1310-1260

$R(HO)P(=O)-O-P(=O)(HO)R$: ~1195

$RO(R_2N)P(=O)-O-P(=O)(R_2N)RO$: ~1275

$R(RO)P(=O)-O-P(=O)(RO)R$: 1265-1250

$(RO)_2P(=O)-O-P(=O)(R_2N)_2$: ~1300, ~1240

$RO(HO)P(=O)-O-P(=O)(HO)RO$: ~1250

$(R_2N)_2P(=O)-O-P(=O)(R_2N)_2$: ~1235

$R_2(X)P{=}O$	: 1265-1240	for X = F, Cl, Br (for X = F near the upper limit of the range)
$R(X)_2P{=}O$	: 1365-1260	
$(RO)_2(X)P{=}O$	: 1330-1280	
$RO(X)_2P{=}O$	: 1365-1260	

Characteristic IR-Absorption Bands of Phosphorous Compounds (in cm^{-1})
(continued)

Assignment	Range	Comments
P-OH δ	~1280	weak, no practical significance
P-C st	800 - 700	intensity varies widely, no practical significance
P-H δ	1090 - 910	strong, for $(RO)_2HP{=}O$ very strong
		also:
		P-N-C st: 1110-930, 770-680
		P=N-R st: R = al : 1500-1230
		R = ar : 1390-1300
		R = C=O: 1370-1310
		R = PR_2: 1295-1170
		P=S st: 750-580, intensity varies widely (P-S st: <600)
		(P-)CH_3 δ sy: 1310-1280
		P-F st: 905-760 (PF_2: 1110-800, more than one band)
		P-Cl st: <600

Characteristic IR-Absorption Bands of Silicon Compounds (in cm^{-1})

Assignment	Range	Comments
Si-H st	2250 - 2090	medium sub-ranges: R_3SiH: 2160-2090; also for R = H, for SiH_3 2 bands hal-SiH: to ∿2250 (Si-O)SiH: 2220-2120
Si-H δ	1010 - 700	strong, broad, generally 2 bands
(Si-)CH_3 δ sy	1275 - 1260	very strong, sharp, typical for $SiCH_3$; not split for Si$(CH_3)_2$ ($SiCH_3$ δ sy: ∿1410, weak)
(Si-)CH_3 γ	860 - 760	$SiCH_3$: ∿765; Si$(CH_3)_2$: ∿855, ∿800; Si$(CH_3)_3$: ∿840, ∿765
Si-O st	1110 - 1000 900 - <600	Si-O-C : 1110-1000, 850-800 Si-O-Si: 1090-1030, <650 Si-OH : 900- 800 (SiO-H st: 3700-3200; Si-OH δ: ∿1030)
Si-C st	850 - 650	

Characteristic IR-Absorption Bands of Silicon Compounds (in cm^{-1})

(continued)

Assignment	Range	Comments
Si-N st	1250 - 830	sub-ranges: Si-N-Si: 950-830 (Si_2NH st: ∿3400) N-Si-N: 950-830 $Si-NH_2$: 1250-1100 ($SiN-H_2$ st: ∿3570, ∿3390; $Si-NH_2$ δ: ∿1540)
Si-F st	980 - 820	sub-ranges: Si-F : 920-820 $Si-F_2$: 945-870, 2 bands $Si-F_3$: 980-860, 2 bands also: Si-Cl st: <625

Characteristic IR-Absorption Bands of Boron Compounds (in cm^{-1})

Assignment	Range	Comments
B-H st	2640 - 2200	strong; in B-H•••B: 2200-1540, more than one band
B-O st	1380 - 1310	very strong, in haloboroxines to ∿1500 (BO-H st: 3300-3200, very broad)
B-N st	1550 - 1330	very strong
B-C st	1240 - 620	strong, 2 bands if substitution highly asymmetric
		also: B-F st : 1500-800; B-Cl st: 1100-650

INTERFERENCES, OPAQUE REGIONS, SUSPENSION MEDIA

Interferences in Infrared Spectra

Traces of water in carbon tetrachloride or chloroform may give rise to two bands in the vicinity of 3700 and 3600 cm^{-1} as well as one around 1600 cm^{-1}. Water in the vapor phase exhibits many sharp bands between 2000 and 1280 cm^{-1}. If present in high concentration these bands may temporarily block the detector of a double beam instrument. This effect can simulate a shoulder if it occurs while scanning through a steep side of a strong signal. Certain shoulders on carbonyl bands can often be rationalized in this way.

In solution, carbon dioxide shows an absorption band at 2325 cm^{-1}. In solutions which contain amines, CO_2 can form carbonates with traces of water, which leads to the appearance of unexpected bands of protonated nitrogen-containing groups. In improperly balanced double beam instruments gaseous CO_2 can give rise to two signals at approximately 2360 and 2335 cm^{-1} as well as a signal at 667 cm^{-1}.

Commercially available polymers often contain phthalates as plasticizers which can be found in apparently pure samples and give rise to a band at 1725 cm^{-1}. In the course of chemical reactions these phthalates can be transformed to phthalic anhydride which gives a band at 1755 cm^{-1}.

Other frequently encountered contaminants are silicones, which generally exhibit a band at 1265 cm^{-1} together with a broad signal in the region from 1100 to 1000 cm^{-1}.

Opaque Regions and Suspension Media

In the figures shown on pages I270, I275 and I280, the regions shown in black along the wave number scale have very low transparency for that particular solvent which may lead to certain artifacts. These regions should be disregarded in the interpretation of the spectra.

The bands of the suspension matrices are always found superimposed on those of the sample.

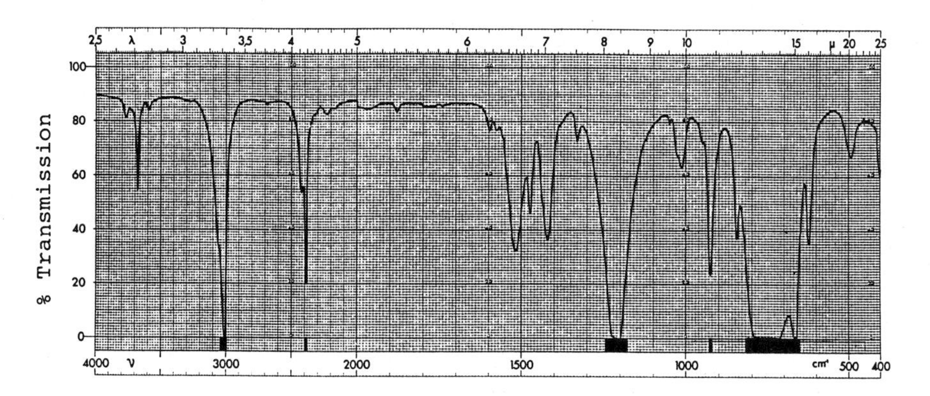

Chloroform: 0.2 mm thickness

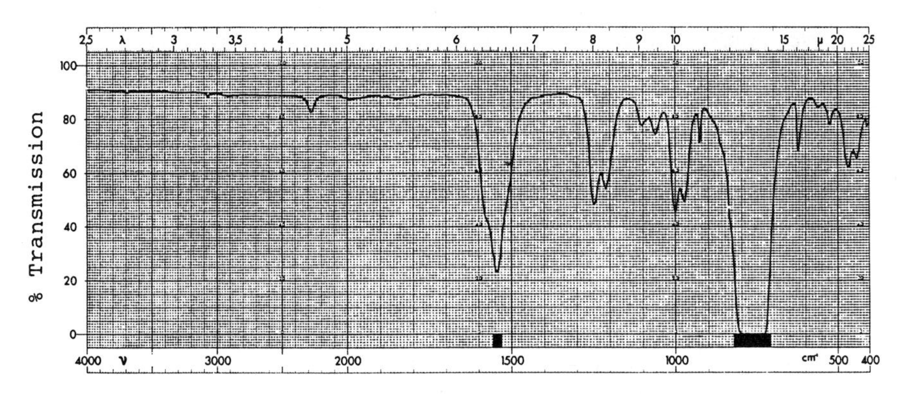

Carbon tetrachloride: 0.2 mm thickness

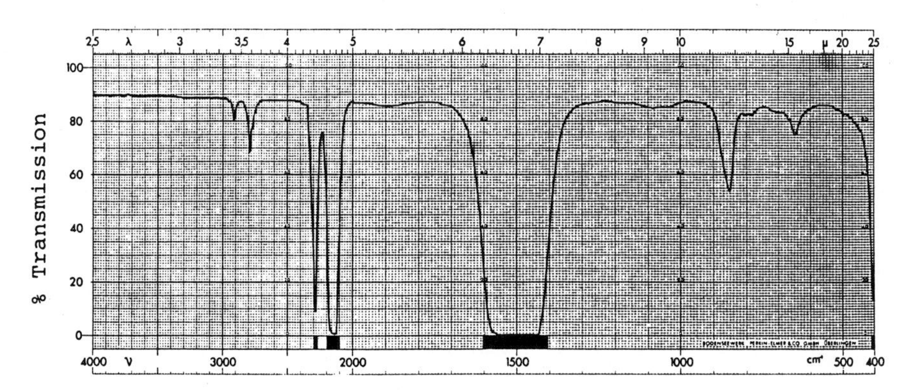

Carbon disulfide: 0.2 mm thickness

IR

OPAQUE REGIONS

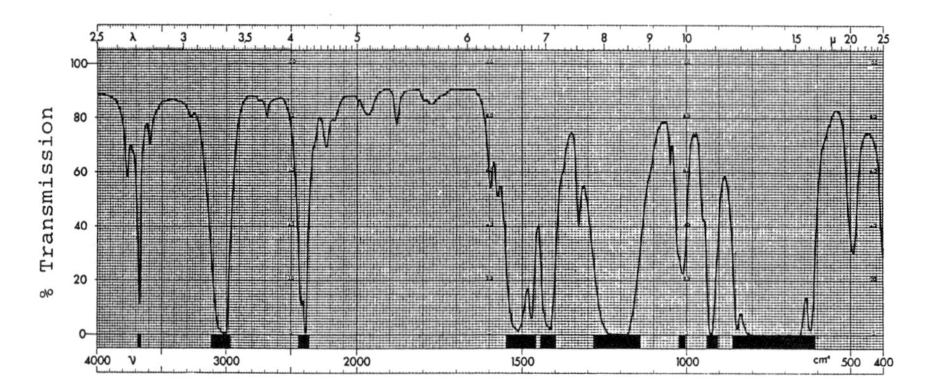

Chloroform: 1.0 mm thickness

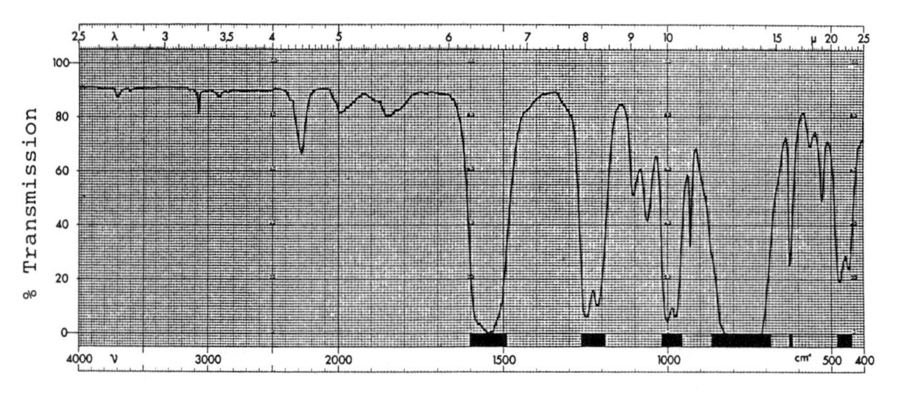

Carbon tetrachloride: 1.0 mm thickness

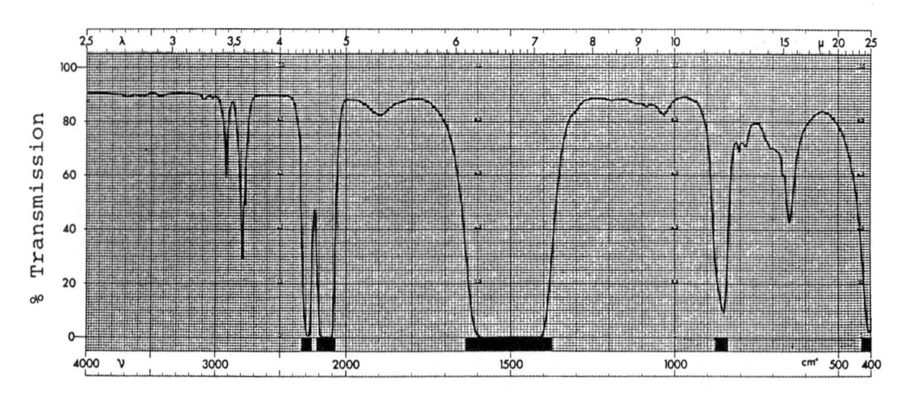

Carbon disulfide: 1.0 mm thickness

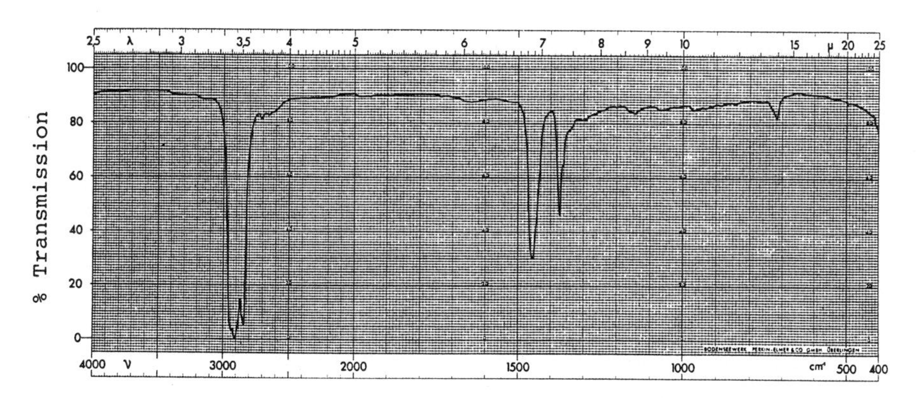

Nujol: ca. 12 μ thickness

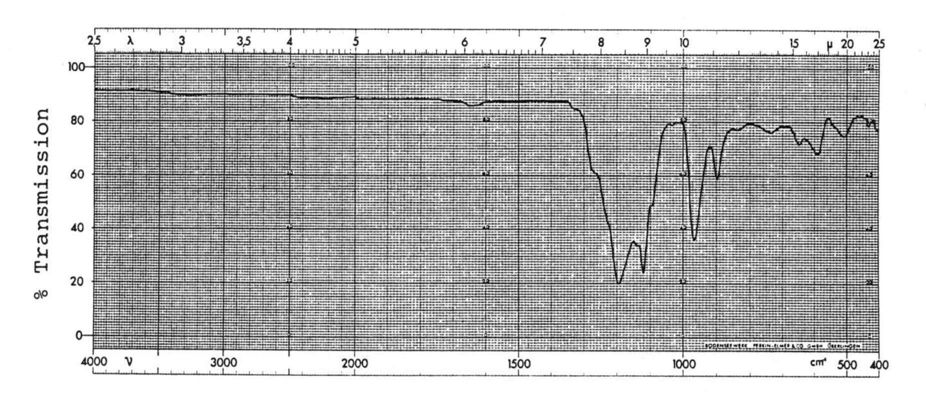

Fluorolube: ca. 12 μ thickness

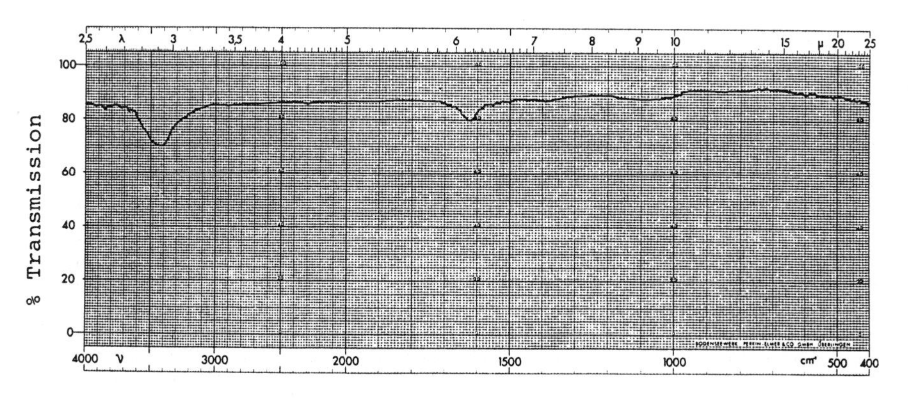

Potassium bromide: pellet

Mass Correlation Table (Note: in the formulae listed in the second column CH_2 can be replaced by N, CH_4 by O, CH_3O by P and O_2 by S as long as that makes chemical sense)

Mass	Ion	Product ion and composition of the neutral particle lost. $M^{+\cdot}$ = molecular ion	Sub-structure or compound type
12	$C^{+\cdot}$		
13	CH^+		
14	$CH_2^{+\cdot}$, N^+, N_2^{++}, CO^{++}		
15	CH_3^+	$M^{+\cdot}-15$ (CH_3)	non-specific; abundant: methyl, N-ethylamines
16	$O^{+\cdot}$, NH_2^+, O_2^{++}	$M^{+\cdot}-16$ (CH_4)	methyl (rare)
		(O)	nitro compounds, sulfones, epoxides, N-oxides
		(NH_2)	primary amines
17	OH^+, $NH_3^{+\cdot}$	$M^{+\cdot}-17$ (OH)	acids (especially aromatic acids), hydroxylamines, N-oxides, nitro compounds, sulfoxides, tertiary alcohols
		(NH_3)	primary amines
18	$H_2O^{+\cdot}$, NH_4^+	$M^{+\cdot}-18$ (H_2O)	non-specific O-indicator abundant: alcohols, some acids, aldehydes, ketones, lactones, cyclic ethers
19	H_3O^+, F^+	$M^{+\cdot}-19$ (F)	fluorides F-indicator
20	$HF^{+\cdot}$, Ar^{++}, CH_2CN^{++}	$M^{+\cdot}-20$ (HF)	fluorides

21	$C_2H_2O^{++}$			
22	CO_2^{++}			
23	$Na^{+\cdot}$			
24	$C_2^{+\cdot}$			
25	C_2H^+	$M^{+\cdot}-25$	(C_2H)	terminal acetylenyl
26	$C_2H_2^{+\cdot}$, CN^+	$M^{+\cdot}-26$	(C_2H_2)	aromatics
			(CN)	nitriles
27	$C_2H_3^+$, $HCN^{+\cdot}$	$M^{+\cdot}-27$	(C_2H_3)	terminal vinyl, some ethyl esters and N-ethylamides, ethyl phos-phates
			(HCN)	aromatic N, nitriles
28	$C_2H_4^{+\cdot}$, $CO^{+\cdot}$, $N_2^{+\cdot}$, $HCNH^+$	$M^{+\cdot}-28$	(C_2H_4)	non-specific; abundant: cyclo-hex-enes, ethyl esters, propyl ketones, propyl-substituted aromatics
			(CO)	aromatic O, quinones, lactones, lactams, unsaturated cyclic ketones, allyl aldehydes
			(N_2)	diazo compounds
29	$C_2H_5^+$, CHO^+	$M^{+\cdot}-29$	(C_2H_5)	non-specific; abundant: ethyl
			(CHO)	phenols, furans, aldehydes
30	$CH_2O^{+\cdot}$, $CH_2NH_2^+$, NO^+, $C_2H_6^{+\cdot}$ $BF^{+\cdot}$, $N_2H_2^{+\cdot}$ N-indicator	$M^{+\cdot}-30$	(C_2H_6)	ethylalkanes, polymethyl compounds
			(CH_2O)	cyclic ethers, lactones, primary alcohols
			(NO)	nitro and nitroso compounds
31	CH_3O^+, $CH_3NH_2^{+\cdot}$, CF^+, $N_2H_3^+$ O-indicator	$M^{+\cdot}-31$	(CH_3O)	methyl esters, methyl ethers, primary alcohols
			(CH_3NH_2)	N-methylamines

MASS CORRELATION TABLE

Mass	Ion	Product ion and composition of the neutral particle lost. $M^{+\cdot}$ = molecular ion	Sub-structure or compound type
		(N_2H_3)	hydrazides
32	$O_2^{+\cdot}$, $CH_3OH^{+\cdot}$, $S^{+\cdot}$, $N_2H_4^{+\cdot}$	$M^{+\cdot}-32$ (CH_3OH)	methyl esters, methyl ethers
	O-indicator	(S)	sulfides
		(O_2)	cyclic peroxides
33	$CH_3OH_2^+$, SH^+, CH_2F^+	$M^{+\cdot}-33$ $(CH_3 + H_2O)$	non-specific O-indicator
		(SH)	non-specific S-indicator
		(CH_2F)	fluoromethyl
34	$SH_2^{+\cdot}$ S-indicator	$M^{+\cdot}-34$ (SH_2)	non-specific S-indicator
		$(OH + OH)$	nitro compounds
35	SH_3^+, Cl^+	$M^{+\cdot}-35$ (Cl)	chlorides
		$(OH + H_2O)$	nitro compounds 2 x O-indicator
36	$HCl^{+\cdot}$, C_3^+	$M^{+\cdot}-36$ (HCl)	chlorides
		$(H_2O + H_2O)$	2 x O-indicator
37	C_3H^+		
38	$C_3H_2^{+\cdot}$		
39	$C_3H_3^+$	$M^{+\cdot}-39$ (C_3H_3)	aromatics
40	$C_3H_4^{+\cdot}$, CH_2CN^+, $Ar^{+\cdot}$	$M^{+\cdot}-40$ (CH_2CN)	cyanomethyl
41	$C_3H_5^+$, $CH_3CN^{+\cdot}$	$M^{+\cdot}-41$ (C_3H_5)	alicyclics (especially polyalicyclics), alkenes
		(CH_3CN)	2-methyl-N-aromatics, N-methylanilines

m/z	Ions	Loss		Indication
42	$C_3H_6^{+\cdot}$, $C_2H_2O^{+\cdot}$, CON^+, $C_2H_4N^+$	$M^{+\cdot}-42$	(C_3H_6)	non-specific; abundant: propyl esters, butyl ketones, butyl aromatics, methyl cyclohexenes
			(C_2H_2O)	acetates (especially enolacetates), acetamides, cyclohexenones, αβ-unsaturated ketones
43	$C_3H_7^+$, $C_2H_3O^+$, $CONH^{+\cdot}$	$M^{+\cdot}-43$	(C_3H_7)	non-specific; abundant: propyl, cycloalkanes, cycloalkanones, cycloalkylamines, cycloalkanols, butyl aromatics
			(CH_3CO)	methyl ketones, acetates, aromatic methyl ethers
44	$CO_2^{+\cdot}$, $C_2H_6N^+$, $C_2H_4O^{+\cdot}$, $C_3H_8^{+\cdot}$ $CH_4Si^{+\cdot}$	$M^{+\cdot}-44$	(C_3H_8)	propyl alkanes
			(C_2H_6N)	N,N-dimethylamines, N-ethylamines
			(C_2H_4O)	cycloalkanols, cyclic ethers, ethylene ketals
			(CO_2)	anhydrides, lactones, carboxylic acids
45	$C_2H_5O^+$, CHS^+, $C_2H_7N^{+\cdot}$ <u>O-indicator</u>, <u>S-indicator</u>	$M^{+\cdot}-45$	(C_2H_5O)	ethyl esters, ethyl ethers, lactones ethyl sulfonates, ethyl sulfones
			(CHO_2)	carboxylic acids
			(C_2H_7N)	N,N-dimethylamines, N-ethylamines
46	$C_2H_5OH^{+\cdot}$, NO_2^+	$M^{+\cdot}-46$	(C_2H_6O)	ethyl esters, ethyl ethers, ethyl sulfonates
			($H_2O + C_2H_4$)	primary alcohols
			($H_2O + CO$)	carboxylic acids
			(NO_2)	nitro compounds
47	CH_3S^+, CCl^+, $C_2H_5OH_2^+$, $CH(OH)_2^+$ <u>S-indicator</u>, <u>2 x O-indicator</u> <u>P-indicator</u>	$M^{+\cdot}-47$	(CH_3S)	methyl sulfides

MASS CORRELATION TABLE

Mass	Ion	Product ion and composition of the neutral particle lost. $M^{+\cdot}$ = molecular ion	Sub-structure or compound type
48	$CH_3SH^{+\cdot}$, $CHCl^{+\cdot}$, $SO^{+\cdot}$	$M^{+\cdot}-48$ (CH_4S)	methyl sulfides
		(SO)	sulfoxides, sulfones, sulfonates
49	CH_2Cl^+, $CH_3SH_2^{+\cdot}$	$M^{+\cdot}-49$ (CH_2Cl)	chloromethyl
50	$C_4H_2^{+\cdot}$, $CH_3Cl^{+\cdot}$, $CF_2^{+\cdot}$	$M^{+\cdot}-50$ (CF_2)	trifluoromethyl aromatics, perfluoro alicyclics
51	$C_4H_3^+$, CHF_2^+		
52	$C_4H_4^{+\cdot}$		
53	$C_4H_5^+$		
54	$C_4H_6^{+\cdot}$, $C_2H_4CN^+$	$M^{+\cdot}-54$ (C_4H_6)	cyclohexene
		(C_2H_4CN)	cyanoethyl
55	$C_4H_7^+$, $C_3H_3O^+$	$M^{+\cdot}-55$ (C_4H_7)	non-specific; abundant: cycloalkanes, butyl esters, N-butylamides
56	$C_4H_8^{+\cdot}$, $C_3H_4O^{+\cdot}$	$M^{+\cdot}-56$ (C_4H_8)	butyl esters, N-butylamides, pentyl ketones, cyclohexenes, tetralins, pentylaromatics
		(C_3H_4O)	methylcyclohexenones, β-tetralones
57	$C_4H_9^+$, $C_3H_5O^+$, $C_3H_2F^+$	$M^{+\cdot}-57$ (C_4H_9)	non-specific
		(C_3H_5O)	ethyl ketones
58	$C_3H_6O^{+\cdot}$, $C_3H_8N^+$	$M^{+\cdot}-58$ (C_4H_{10})	alkanes
	N-indicator, O-indicator	(C_3H_6O)	α-methyl alkanals, methyl ketones, isopropylidene glycols
59	$C_3H_7O^+$, $C_2H_5NO^{+\cdot}$	$M^{+\cdot}-59$ (C_3H_7O)	propyl esters, propyl ethers
	O-indicator	($C_2H_3O_2$)	methyl esters

			(C_3H_9N)	amines, amides
60	$C_2H_4O_2^{+\cdot}$, $CH_2NO_2^+$, $C_2H_6NO^+$	$M^{+\cdot}-60$	(C_3H_8O)	propyl esters, propyl ethers
	O-indicator		$(C_2H_4O_2)$	acetates
			(CH_3OH+CO)	methyl esters
61	$C_2H_5O_2^+$, $C_2H_5S^+$	$M^{+\cdot}-61$	$(C_2H_5O_2)$	glycols, ethylene ketals
	S-indicator, 2 x O-indicator		(C_2H_5S)	ethyl sulfides
62	$C_2H_6O_2^{+\cdot}$, $C_2H_3Cl^{+\cdot}$	$M^{+\cdot}-62$	$(C_2H_6O_2)$	methoxymethyl ethers, ethylene glycols, ethylene ketals
			(C_2H_6S)	ethyl sulfides
63	$C_5H_3^+$, $C_2H_4Cl^+$, $COCl^+$	$M^{+\cdot}-63$	(C_2H_4Cl)	chloroethyl
			$(Cl+CO)$	acid chlorides
64	$C_5H_4^{+\cdot}$, $SO_2^{+\cdot}$, $S_2^{+\cdot}$	$M^{+\cdot}-64$	(SO_2)	sulfones, sulfonates
			(S_2)	disulfides
65	$C_5H_5^+$	$M^{+\cdot}-65$	(S_2H)	disulfides
66	$C_5H_6^{+\cdot}$	$M^{+\cdot}-66$	(C_5H_6)	cyclopentenes
67	$C_5H_7^+$, $C_4H_3O^+$	$M^{+\cdot}-67$	(C_4H_3O)	furyl ketones
68	$C_5H_8^{+\cdot}$, $C_4H_4O^{+\cdot}$, $C_3H_6CN^+$	$M^{+\cdot}-68$	(C_5H_8)	cyclohexenes, tetralins
			(C_4H_4O)	cyclohexenones, β-tetralones
69	$C_5H_9^+$, $C_4H_5O^+$, $C_3HO_2^+$, CF_3^+	$M^{+\cdot}-69$	(C_5H_9)	alicyclics, alkenes
			(CF_3)	trifluoromethyl
70	$C_6H_{10}^{+\cdot}$			alkanes, alkenes, cycloalkanes
	$C_4H_6O^{+\cdot}$			cycloalkanones
	$C_4H_8N^+$			pyrrolidines
71	$C_5H_{11}^+$			alkanes, larger alkyl groups
	$C_4H_7O^+$			alkanones, alkanals, tetrahydrofurans

MASS CORRELATION TABLE

Mass	Ion	Compound type	
72	$C_4H_8O^{+\cdot}$	alkanones, alkanals	O-indicator
	$C_4H_{10}N^+$	aliphatic amines	N-indicator
	C_6^+	perhalogenated benzenes	
73	$C_4H_9O^+$	alcohols, ethers, esters	O-indicator
	$C_3H_5O_2^+$	acids, esters, lactones	
	$C_3H_9Si^+$	trimethylsilyl compounds	
74	$C_4H_{10}O^{+\cdot}$	ethers	
	$C_3H_6O_2^{+\cdot}$	methyl esters of carboxylic acids, α-methyl carboxylic acids	
75	$C_3H_7O_2^+$	methyl acetals, glycols	2 x O-indicator
	$C_3H_7S^+$	sulfides, thiols	S-indicator
	$C_2H_7SiO^+$	trimethylsiloxyl compounds	
76	$C_6H_4^{+\cdot}$	aromatics	
77	$C_6H_5^+$	aromatics	
	$C_3H_6Cl^+$	chlorides	
78	$C_6H_6^{+\cdot}$	aromatics	
	$C_5H_4N^+$	pyridines	
	$C_3H_7Cl^{+\cdot}$	chlorides	
79	$C_6H_7^+$	aromatics with H-containing substituents	
	$C_5H_5N^{+\cdot}$	pyridines, pyrroles	
	Br^+	bromides	
80	$C_6H_8^{+\cdot}$	cyclohexenes, polycyclic alicycles	
	$C_5H_4O^{+\cdot}$	cyclopentenones	
	$HBr^{+\cdot}$	bromides	
	$C_5H_6N^+$	pyrroles, pyridines	
81	$C_6H_9^+$	cyclohexanes, cyclohexenyls, dienes	
	$C_5H_5O^+$	furans, pyrans	
82	$C_6H_{10}^{+\cdot}$	cyclohexanes	
	$C_5H_6O^{+\cdot}$	cyclopentenones, dihydropyrans	
	$C_5H_8N^+$	tetrahydropyridines	
	$C_4H_6N_2^{+\cdot}$	pyrazoles, imidazoles	

Mass	Ion	Compound type	
83	$C_6H_{11}^+$	alkenes, cycloalkanes, monosubstituted alkanes	
	$C_5H_7O^+$	cycloalkanones	
84	$C_5H_{10}N^+$	piperidines, N-methylpyrrolidines	
85	$C_6H_{13}^+$	alkanes	
	$C_5H_9O^+$	alkanones, alkanals, tetrahydropyrans, fatty acid derivatives	
86	$C_5H_{10}O^{+\cdot}$	alkanones, alkanals	
	$C_5H_{12}N^+$	aliphatic amines	N-indicator
87	$C_5H_{11}O^+$	alcohols, ethers, esters	O-indicator
	$C_4H_7O_2^+$	esters, acids	
88	$C_4H_8O_2^{+\cdot}$	ethyl esters of carboxylic acids, α-methyl-methyl esters, α-C_2-carboxylic acids	
89	$C_4H_9O_2^+$	diols, glycol ethers	2 x O-indicator
	$C_4H_9S^+$	sulfides	
90	$C_7H_6^{+\cdot}$	disubstituted aromatics	
91	$C_7H_7^+$	aromatics	
	$C_4H_8Cl^+$	alkyl chlorides	
92	$C_7H_8^{+\cdot}$	alkylbenzenes	
	$C_6H_6N^+$	alkylpyridines	
93	$C_6H_5O^+$	phenols, phenol derivatives	
	$C_6H_7N^{+\cdot}$	anilines	
	CH_2Br^+	bromides	
94	$C_6H_6O^{+\cdot}$	phenol esters, phenol ethers	
	$C_5H_4NO^+$	pyrryl ketones, pyridone derivatives	
95	$C_5H_3O_2^+$	furyl ketones	
96	$C_7H_{12}^{+\cdot}$	alicyclics	
97	$C_7H_{13}^+$	cycloalkanes, alkenes	
	$C_6H_9O^+$	cycloalkanones	
	$C_5H_5S^+$	alkylthiophenes	
98	$C_6H_{12}N^+$	N-alkylpiperidines	

MS

MASS CORRELATION TABLE

Mass	Ion	Compound type
99	$C_7H_{15}^+$	alkanes
	$C_6H_{11}O^+$	alkanones
	$C_5H_7O_2^+$	ethylene ketals
	$H_4PO_4^+$	alkyl phosphates
104	$C_8H_8^{+\cdot}$	tetralin derivatives, phenylethyl derivatives
	$C_7H_4O^{+\cdot}$	disubstituted α-ketobenzenes
105	$C_8H_9^+$	alkyl aromatics
	$C_7H_5O^+$	benzoyl derivatives
	$C_6H_5N_2^+$	diazophenyl derivatives
111	$C_5H_3OS^+$	thiophenoyl derivatives
115	$C_9H_7^+$	aromatics
	$C_6H_{11}O_2^+$	esters
	$C_5H_7O_3^+$	diesters
119	$C_9H_{11}^+$	alkyl aromatics
	$C_8H_7O^+$	tolyl ketones
	$C_2F_5^+$	perfluoroethyl derivatives
	$C_7H_5NO^{+\cdot}$	phenylcarbamates
120	$C_7H_4O_2^+$	γ-benzpyrones, salicylic acid derivatives
	$C_8H_{10}N^+$	pyridines, anilines
121	$C_8H_9O^+$	hydroxybenzene derivatives
	$C_7H_5O_2^+$	hydroxybenzene derivatives
127	$C_{10}H_7^+$	naphthalenes
	$C_6H_7O_3^+$	unsaturated diesters
	$C_6H_6NCl^{+\cdot}$	chlorinated N-aromatics
	I^+	iodides
128	$C_{10}H_8^{+\cdot}$	naphthalenes
	$C_6H_5OCl^{+\cdot}$	chlorinated hydroxybenzene derivatives
	$HI^{+\cdot}$	iodides
130	$C_9H_8N^+$	quinolines, indoles
	$C_9H_6O^{+\cdot}$	naphthoquinones

Mass	Ion	Compound type
131	$C_{10}H_{11}^+$	tetralins
	$C_5H_7S_2^+$	thioethylene ketals
	$C_3F_5^+$	perfluoroalkyl derivatives
135	$C_4H_8Br^+$	alkyl bromides
141	$C_{11}H_9^+$	naphthalenes
142	$C_{10}H_8N^+$	quinolines
149	$C_8H_5O_3^+$	phthalates
152	$C_{12}H_8^+$	diphenyl aromatics
165	$C_{13}H_9^+$	diphenylmethane derivatives
167	$C_8H_7O_4^+$	phthalates
205	$C_{12}H_{13}O_3^+$	phthalates
223	$C_{12}H_{15}O_4^+$	phthalates

Isotope Patterns of All Naturally Occurring Elements in the Periodic Table

Li 7	B 11	C 12	N 14	O 16	Ne 20	Mg 24	Si 28	S 32	Cl 35
Ar 40	K 39	Ca 40	Ti 48	V 51	Cr 52	Fe 56	Ni 58	Cu 63	Zn 64
Ga 69	Ge 74	Se 80	Br 79	Kr 84	Rb 85	Sr 88	Zr 90	Mo 98	Ru 102
Pd 106	Ag 107	Cd 114	In 115	Sn 120	Sb 121	Te 130	Xe 132	Ba 138	Ce 140
Nd 142	Sm 152	Eu 153	Gd 158	Dy 164	Er 166	Yb 174	Lu 175	Hf 180	W 184
Re 187	Os 192	Ir 193	Pt 195	Hg 202	Tl 205	Pb 208	U 238		

H	He	Be	F	Na	Al	P	Sc	Mn	Co	As	Y	Nb	Rh	I	Cs	La	Pr	Tb	Ho	Tm	Ta	Au	Bi	Th
1	4	9	19	23	27	31	45	55	59	75	89	93	103	127	133	139	141	159	165	169	181	197	209	232

Shown under the symbol of the element is the mass of the most abundant isotope; the lightest isotope is shown at the beginning of the mass scale. The row at the bottom summarizes all the monoisotopic - or almost monoisotopic - elements.

For the exact masses and abundances see p. M60 - M85.

Mass and Relative Abundance of the Isotopes of All Naturally Occurring Elements (for literature reference see p. M85)

Isotope	Mass	Abundance	Isotope	Mass	Abundance
^{1}H	1.0078	100	^{26}Mg	25.9826	13.938
^{2}H	2.0141	0.015			
			^{27}Al	26.9815	100
^{3}He	3.0160	$\sim 10^{-4}$			
^{4}He	4.0026	100	^{28}Si	27.9769	100
			^{29}Si	28.9765	5.063
^{6}Li	6.0151	8.108	^{30}Si	29.9738	3.361
^{7}Li	7.0160	100			
			^{31}P	30.9738	100
^{9}Be	9.0122	100			
			^{32}S	31.9721	100
^{10}B	10.0129	24.844	^{33}S	32.9715	0.789
^{11}B	11.0093	100	^{34}S	33.9679	4.431
			^{36}S	35.9671	0.022
^{12}C	12.0000	100			
^{13}C	13.0034	1.112	^{35}Cl	34.9689	100
			^{37}Cl	36.9659	31.978
^{14}N	14.0031	100			
^{15}N	15.0001	0.367	^{36}Ar	35.9675	0.338
			^{38}Ar	37.9627	0.063
^{16}O	15.9949	100	^{40}Ar	39.9624	100
^{17}O	16.9991	0.038			
^{18}O	17.9991	0.200	^{39}K	38.9637	100
			^{40}K	39.9640	0.013
^{19}F	18.9984	100	^{41}K	40.9618	7.217
^{20}Ne	19.9924	100	^{40}Ca	39.9626	100
^{21}Ne	20.9938	0.298	^{42}Ca	41.9586	0.667
^{22}Ne	21.9914	10.187	^{43}Ca	42.9588	0.139
			^{44}Ca	43.9555	2.152
^{23}Na	22.9898	100	^{46}Ca	45.9537	0.004
^{24}Mg	23.9850	100	^{48}Ca	47.9524	0.193
^{25}Mg	24.9858	12.660			
			^{45}Sc	44.9559	100

MS

ISOTOPES, MASSES AND ABUNDANCES

Isotope	Mass	Abundance	Isotope	Mass	Abundance
^{46}Ti	45.9526	10.840	^{68}Zn	67.9249	38.683
^{47}Ti	46.9518	9.892	^{70}Zn	69.9254	1.235
^{48}Ti	47.9479	100			
^{49}Ti	48.9479	7.453	^{69}Ga	68.9257	100
^{50}Ti	49.9448	7.317	^{71}Ga	70.9248	66.389
^{50}V	49.9472	0.251	^{70}Ge	69.9243	56.164
^{51}V	50.9440	100	^{72}Ge	71.9217	75.068
			^{73}Ge	72.9234	21.370
^{50}Cr	49.9460	5.191	^{74}Ge	73.9212	100
^{52}Cr	51.9405	100	^{76}Ge	75.9214	21.370
^{53}Cr	52.9407	11.338			
^{54}Cr	53.9389	2.817	^{75}As	74.9216	100
^{55}Mn	54.9381	100	^{74}Se	73.9225	1.815
			^{76}Se	75.9192	18.145
^{54}Fe	53.9396	6.324	^{77}Se	76.9199	15.323
^{56}Fe	55.9349	100	^{78}Se	77.9174	47.379
^{57}Fe	56.9354	2.399	^{80}Se	79.9165	100
^{58}Fe	57.9333	0.305	^{82}Se	81.9167	18.952
^{59}Co	58.9332	100	^{79}Br	78.9184	100
			^{81}Br	80.9163	97.278
^{58}Ni	57.9353	100			
^{60}Ni	59.9308	38.231	^{78}Kr	77.9204	0.614
^{61}Ni	60.9311	1.655	^{80}Kr	79.9164	3.947
^{62}Ni	61.9284	5.259	^{82}Kr	81.9135	20.351
^{64}Ni	63.9280	1.333	^{83}Kr	82.9141	20.175
			^{84}Kr	83.9115	100
^{63}Cu	62.9296	100	^{86}Kr	85.9106	30.351
^{65}Cu	64.9278	44.571			
			^{85}Rb	84.9117	100
^{64}Zn	63.9292	100	^{87}Rb	86.9092	38.571
^{66}Zn	65.9261	57.407			
^{67}Zn	66.9272	8.436			

Isotope	Mass	Abundance
^{84}Sr	83.9134	0.678
^{86}Sr	85.9093	11.940
^{87}Sr	86.9089	8.477
^{88}Sr	87.9056	100
^{89}Y	88.9059	100
^{90}Zr	89.9048	100
^{91}Zr	90.9057	21.905
^{92}Zr	91.9051	33.372
^{94}Zr	93.9063	33.683
^{96}Zr	95.9084	5.403
^{93}Nb	92.9065	100
^{92}Mo	91.9068	61.500
^{94}Mo	93.9052	38.334
^{95}Mo	94.9059	65.976
^{96}Mo	95.9047	69.126
^{97}Mo	96.9060	39.577
^{98}Mo	97.9055	100
^{100}Mo	99.9075	39.909
^{96}Ru	95.9074	17.468
^{98}Ru	97.9047	5.949
^{99}Ru	98.9057	40.190
^{100}Ru	99.9041	39.873
^{101}Ru	100.9052	53.797
^{102}Ru	101.9039	100
^{104}Ru	103.9051	59.177
^{103}Rh	102.9041	100

Isotope	Mass	Abundance
^{102}Pd	101.9051	3.732
^{104}Pd	103.9037	40.761
^{105}Pd	104.9046	81.705
^{106}Pd	105.9032	100
^{108}Pd	107.9039	96.817
^{110}Pd	109.9045	42.883
^{107}Ag	106.9050	100
^{109}Ag	108.9047	92.905
^{106}Cd	105.9060	4.351
^{108}Cd	107.9040	3.098
^{110}Cd	109.9030	43.474
^{111}Cd	110.9040	44.553
^{112}Cd	111.9030	83.989
^{113}Cd	112.9046	42.534
^{114}Cd	113.9036	100
^{116}Cd	115.9050	26.070
^{113}In	112.9043	4.493
^{115}In	114.9039	100
^{112}Sn	111.9051	3.086
^{114}Sn	113.9029	2.160
^{115}Sn	114.9033	1.235
^{116}Sn	115.9018	45.370
^{117}Sn	116.9029	23.765
^{118}Sn	117.9016	75.000
^{119}Sn	118.9032	26.543
^{120}Sn	119.9021	100
^{122}Sn	121.9032	14.198
^{124}Sn	123.9051	17.284

MS

ISOTOPES, MASSES AND ABUNDANCES

Isotope	Mass	Abundance
^{121}Sb	120.9036	100
^{123}Sb	122.9039	74.520
^{120}Te	119.9045	0.284
^{122}Te	121.9028	7.692
^{123}Te	122.9042	2.686
^{124}Te	123.9028	14.249
^{125}Te	124.9044	21.124
^{126}Te	125.9032	56.065
^{128}Te	127.9047	93.757
^{130}Te	129.9067	100
^{127}I	126.9044	100
^{124}Xe	123.9061	0.372
^{126}Xe	125.9042	0.335
^{128}Xe	127.9035	7.100
^{129}Xe	128.9048	98.141
^{130}Xe	129.9035	15.242
^{131}Xe	130.9051	78.810
^{132}Xe	131.9042	100
^{134}Xe	133.9054	38.662
^{136}Xe	135.9072	33.085
^{133}Cs	132.9051	100
^{130}Ba	129.9063	0.148
^{132}Ba	131.9051	0.141
^{134}Ba	133.9043	3.371
^{135}Ba	134.9056	9.194
^{136}Ba	135.9044	10.954
^{137}Ba	136.9056	15.662
^{138}Ba	137.9050	100
^{138}La	137.9068	0.090
^{139}La	138.9061	100
^{136}Ce	135.9071	0.215
^{138}Ce	137.9057	0.283
^{140}Ce	139.9053	100
^{142}Ce	141.9090	12.523
^{141}Pr	140.9074	100
^{142}Nd	141.9075	100
^{143}Nd	142.9096	44.895
^{144}Nd	143.9099	87.726
^{145}Nd	144.9122	30.593
^{146}Nd	145.9127	63.362
^{148}Nd	147.9165	21.231
^{150}Nd	149.9207	20.789
^{144}Sm	143.9117	11.654
^{147}Sm	146.9146	56.767
^{148}Sm	147.9146	42.481
^{149}Sm	148.9169	52.256
^{150}Sm	149.9170	27.820
^{152}Sm	151.9195	100
^{154}Sm	153.9220	84.962
^{151}Eu	150.9196	91.571
^{153}Eu	152.9209	100
^{152}Gd	151.9195	0.805
^{154}Gd	153.9207	8.776
^{155}Gd	154.9226	59.581
^{156}Gd	155.9221	82.407
^{157}Gd	156.9239	63.003
^{158}Gd	157.9241	100
^{160}Gd	159.9271	88.003

Isotope	Mass	Abundance
^{159}Tb	158.9250	100
^{156}Dy	155.9238	0.213
^{158}Dy	157.9240	0.355
^{160}Dy	159.9248	8.298
^{161}Dy	160.9266	67.021
^{162}Dy	161.9265	90.426
^{163}Dy	162.9284	88.298
^{164}Dy	163.9288	100
^{165}Ho	164.9303	100
^{162}Er	161.9288	0.417
^{164}Er	163.9293	4.792
^{166}Er	165.9304	100
^{167}Er	166.9321	68.304
^{168}Er	167.9324	79.762
^{170}Er	169.9355	44.345
^{169}Tm	168.9344	100
^{168}Yb	167.9339	0.409
^{170}Yb	169.9349	9.591
^{171}Yb	170.9365	44.969
^{172}Yb	171.9366	68.868
^{173}Yb	172.9383	50.692
^{174}Yb	173.9390	100
^{176}Yb	175.9427	39.937
^{175}Lu	174.9409	100
^{176}Lu	175.9427	2.669
^{174}Hf	173.9403	0.455
^{176}Hf	175.9417	14.773
^{177}Hf	176.9435	52.841
^{178}Hf	177.9439	76.989
^{179}Hf	178.9460	39.034
^{180}Hf	179.9468	100
^{180}Ta	179.9475	0.012
^{181}Ta	180.9491	100
^{180}W	179.9470	0.424
^{182}W	181.9483	85.752
^{183}W	182.9503	46.625
^{184}W	183.9510	100
^{186}W	185.9543	93.251
^{185}Re	184.9530	59.744
^{187}Re	186.9560	100
^{184}Os	183.9526	0.049
^{186}Os	185.9539	3.854
^{187}Os	186.9560	3.902
^{188}Os	187.9560	32.439
^{189}Os	188.9583	39.268
^{190}Os	189.9586	64.390
^{192}Os	191.9614	100
^{191}Ir	190.9609	59.490
^{193}Ir	192.9633	100
^{190}Pt	189.9560	0.030
^{192}Pt	191.9614	2.337
^{194}Pt	193.9628	97.337
^{195}Pt	194.9648	100
^{196}Pt	195.9650	74.852
^{198}Pt	197.9675	21.302
^{197}Au	196.9666	100

ISOTOPES, MASSES AND ABUNDANCES

Isotope	Mass	Abundance
^{196}Hg	195.9658	0.506
^{198}Hg	197.9668	34.064
^{199}Hg	198.9683	57.336
^{200}Hg	199.9683	77.909
^{201}Hg	200.9703	44.519
^{202}Hg	201.9706	100
^{204}Hg	203.9735	22.934
^{203}Tl	202.9723	41.892
^{205}Tl	204.9745	100
^{204}Pb	203.9731	2.672
^{206}Pb	205.9745	45.992
^{207}Pb	206.9759	42.176
^{208}Pb	207.9766	100
^{209}Bi	208.9804	100
^{232}Th	232.0382	100
^{234}U	234.0409	0.006
^{235}U	235.0439	0.725
^{238}U	238.0508	100

The relative abundances were compiled by the IUPAC Commission on Atomic Weights and Isotopic Abundances: N.E. Holden, R.L. Martin, I.L. Barnes, Pure & Appl. Chem. 55, 1119-1136 (1983).

Calculation of Isotope Distributions

The characteristic abundance patterns resulting from the combination of more than one polyisotopic element can be calculated using the following scheme:

A →		m_{a_1}	$m_{a_1}+1$	$m_{a_1}+2$	$m_{a_1}+3$	$m_{a_1}+4$	$m_{a_1}+5$ ...
B ↓		$a_1=1$	a_2	a_3	a_4		
m_{b_1}	$b_1=1$	$1\cdot 1$	$1\cdot a_2$	$1\cdot a_3$	$1\cdot a_4$		
$m_{b_1}+1$	b_2		$b_2\cdot 1$	$b_2\cdot a_2$	$b_2\cdot a_3$	$b_2\cdot a_4$	
$m_{b_1}+2$	b_3			$b_3\cdot 1$	$b_3\cdot a_2$	$b_3\cdot a_3$	$b_3\cdot a_4$
$m_{b_1}+3$ ⋮		1	b_2+a_2	$(b_2\cdot a_2)$ $+a_3$ $+b_3$	$(b_2\cdot a_3)$ $+(b_3\cdot a_2)$ $+a_4$	$(b_2\cdot a_4)$ $+(b_3\cdot a_3)$	$b_3\cdot a_4$
		$m_{a_1}+m_{b_1}$	+1	+2	+3	+4	+5

An element A consisting of the isotopes of mass m_{a_1}, m_{a_2}, m_{a_3}, m_{a_4} of the natural abundances a_1, a_2, a_3, a_4 are combined with element B with isotopes of mass m_{b_1}, m_{b_2}, m_{b_3} of the natural abundance b_1, b_2, b_3.

The relative abundances of the lightest isotopes m_{a_1} and m_{b_1} are normalized to make their abundances $a_1 = b_1 = 1$. The normalized abundances of element A are entered according to their increasing mass in the first horizontal row and those of element B in the first vertical column. Each entry in a row is then multiplied with that of the column, moving from left to right. The sum of each column represents the abundance distribution of A and B by increasing mass.

The resulting distribution can be entered as a "super element" to be combined with additional elements or "super elements" using the same computational scheme.

Example: What is the isotope pattern of $ZnBr_2$?
Zn consists of isotopes of mass 64, 66, 67, 68 and 70 (see p. M65 and M55) of the abundance distribution (rounded off) 1 : 0.57 : 0.08 : 0.38 : 0.01. Br_2 is entered as a "super element" consisting of the isotopes of mass 158, 160 and 162 (see p. M105 and M100) and an abundance distribution (rounded off) 1 : 2 : 1.

	Zn →	64	65	66	67	68	69	70	71	72	73	74
Br_2 ↓		1		0.57	0.08	0.38		0.01				
158	1	1		0.57	0.08	0.38		0.01				
159												
160	2			2		1.14	0.16	0.76		0.02		
161												
162	1					1		0.57	0.08	0.38		0.01
		1		2.57	0.08	2.52	0.16	1.34	0.08	0.40		0.01

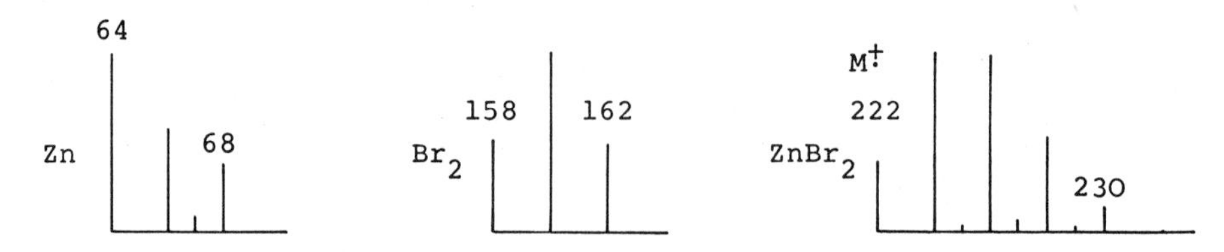

Isotope Patterns of Various Combinations of Cl and Br

Note: The signals are separated by 2 mass units.

For exact abundances see p. M105, M110.

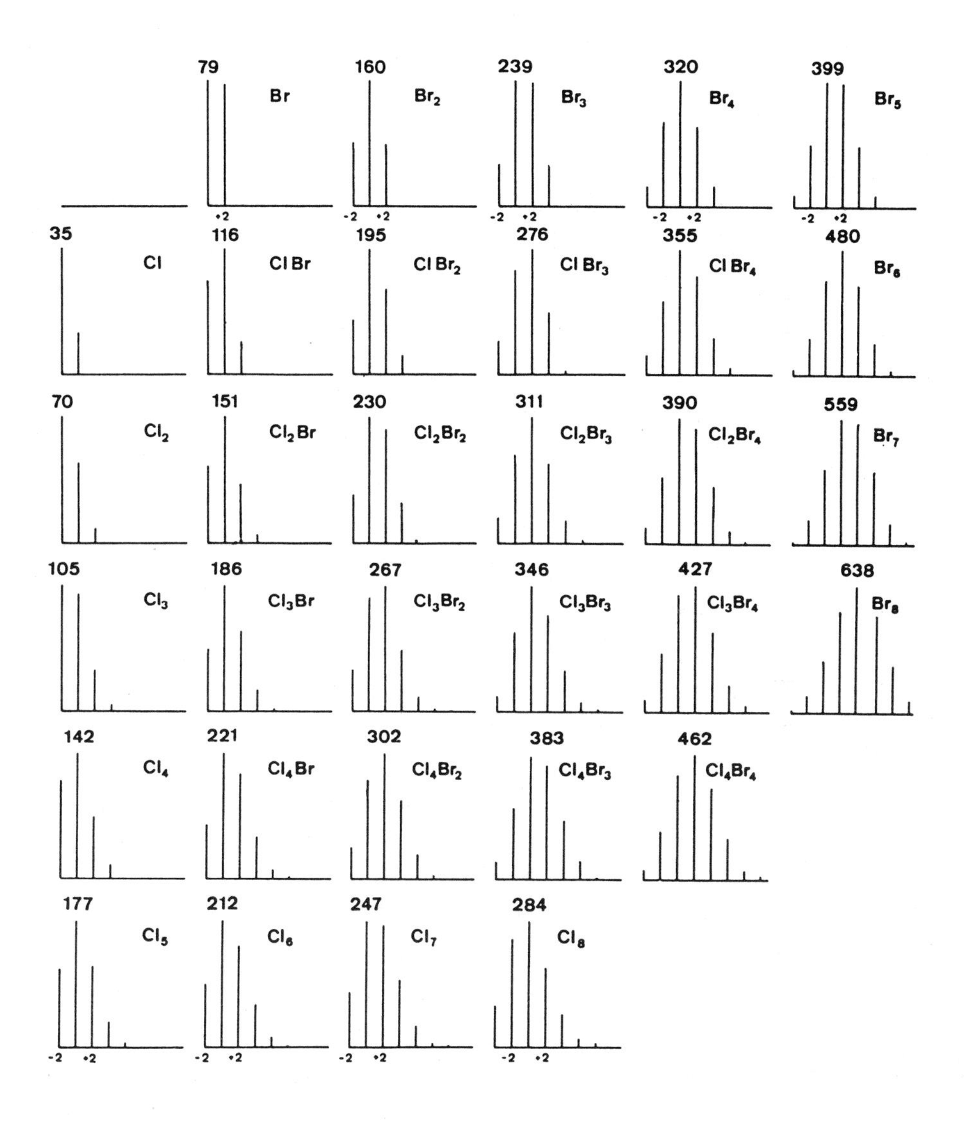

HALOGENS, ISOTOPE DISTRIBUTIONS

Isotopic Abundances of Various Combinations of Chlorine and Bromine

Halogens	Mass	Rel. Abundance	Halogens	Mass	Rel. Abundance
Cl_1	35	100	Br_1	79	100
	37	31.98		81	97.88
Cl_2	70	100	Br_2	158	51.09
	72	63.96		160	100
	74	10.23		162	48.93
Cl_3	105	100	Br_3	237	34.05
	107	95.93		239	100
	109	30.67		241	97.89
	111	3.27		243	31.94
Cl_4	140	77.96	Br_4	316	17.40
	142	100		318	68.09
	144	47.82		320	100
	146	10.19		322	65.26
	148	0.82		324	15.96
Cl_5	175	62.53	Br_5	395	10.43
	177	100		397	51.09
	179	63.94		399	100
	181	20.45		401	97.94
	183	3.28		403	47.89
	185	0.21		405	9.38
Cl_6	210	52.12	Br_6	474	5.32
	212	100		476	31.26
	214	79.95		478	76.62
	216	34.08		480	100
	218	8.21		482	73.38
	220	1.05		484	28.73
	222	0.06		486	4.68
Cl_1Br_1	114	76.70	Cl_1Br_4	351	14.26
	116	100		353	60.41
	118	24.46		355	100
				357	79.93
Cl_1Br_2	193	43.83		359	30.39
	195	100		361	4.25
	197	69.83			
	199	13.66	Cl_1Br_5	430	8.02
				432	41.85
Cl_1Br_3	272	26.15		434	89.50
	274	85.22		436	100
	276	100		438	61.10
	278	48.90		440	19.12
	280	7.86		442	2.35

Halogens	Mass	Rel. Abundance
Cl_2Br_1	149	61.35
	151	100
	153	45.67
	155	6.38
Cl_2Br_2	228	38.35
	230	100
	232	89.63
	234	31.89
	236	3.90
Cl_2Br_3	307	20.49
	309	73.38
	311	100
	313	63.78
	315	18.71
	317	2.03
Cl_2Br_4	386	11.92
	388	54.36
	390	100
	392	94.03
	394	47.21
	396	11.82
	398	1.15
Cl_3Br_1	184	51.12
	186	100
	188	65.22
	190	17.73
	192	1.74
Cl_3Br_2	263	31.35
	265	92.01
	267	100
	269	50.01
	271	11.70
	273	1.03
Cl_3Br_3	342	16.50
	344	64.58
	346	100
	348	77.78
	350	31.90
	352	6.58
	354	0.54

Halogens	Mass	Rel. Abundance
Cl_4Br_1	219	43.79
	221	100
	223	83.86
	225	33.42
	227	6.93
	229	0.48
Cl_4Br_2	298	24.14
	300	78.63
	302	100
	304	63.54
	306	21.54
	308	3.73
	310	0.26
Cl_4Br_3	377	13.63
	379	57.78
	381	100
	383	91.19
	385	47.31
	387	14.03
	389	2.22
	391	0.13
Cl_5Br_1	254	37.60
	256	98.11
	258	100
	260	52.18
	262	14.89
	264	2.22
	266	0.12
Cl_5Br_2	333	19.19
	335	68.85
	337	100
	339	76.56
	341	33.64
	343	8.56
	345	1.17
	347	0.06

Indications of Structural Type

Certain sequences of abundance maxima in the lower mass range and the masses of unique signals are often characteristic of a particular compound type. The abundance distribution of such ion series is

Mass values \ Compound types	Alkanes	Alkenes, Mono-cyclo-alkanes	Alkynes, Dienes, Cyclo-alkenes	Poly-cyclic alicycles	Alkanones, Alkanals	Alcohols, Alkyl ethers	Cyclic alcohols	Cyclo-alkanones	Aliphat. acids, Esters, Lactones	Alkyl-amines
12 + (14)n m/z 26,40,54,68, 82,96,110...		C_nH_{2n-2}	C_nH_{2n-2}	C_nH_{2n-2}			C_nH_{2n-2}			
13 + (14)n m/z 27,41,55,69, 83,97,111...	C_nH_{2n-1}	C_nH_{2n-1}	C_nH_{2n-1}	C_nH_{2n-1}		C_nH_{2n-1}	C_nH_{2n-1}	$C_nH_{2n-3}O$ C_nH_{2n-1}	C_nH_{2n-1}	
14 + (14)n m/z 28,42,56,70, 84,98,112...	C_nH_{2n}	C_nH_{2n}		C_nH_{2n}		C_nH_{2n}		$C_nH_{2n-2}O$		
15 + (14)n m/z 29,43,57,71, 85,99,113...	C_nH_{2n+1}	C_nH_{2n+1}	C_nH_{2n+1}	C_nH_{2n+1}	C_nH_{2n+1} $C_nH_{2n-1}O$		$C_nH_{2n-1}O$			
16 + (14)n m/z 30,44,58,72, 86,100,114...					$C_nH_{2n}O$					C_nH_{2n+2}
17 + (14)n m/z 31,45,59,73, 87,101,115...						$C_nH_{2n+1}O$			$C_nH_{2n+1}O$ $C_nH_{2n-1}O_2$	
18 + (14)n m/z 32,46,60,74, 88,102,116...									$C_nH_{2n}O_2$	
19 + (14)n m/z 33,47,61,75, 89,103,117...						$C_nH_{2n+3}O$			$C_nH_{2n+1}O_2$	
20 + (14)n m/z 34,48,62,76, 90,104,118...										
21 + (14)n m/z 35,49,63,77, 91,105,119...										
22 + (14)n m/z 36,50,64,78, 92,106,120...				C_nH_{2n-6}						
23 + (14)n m/z 37,51,65,79, 93,107,121...				C_nH_{2n-5}						
24 + (14)n m/z 38,52,66,80, 94,108,122...				C_nH_{2n-4}						
25 + (14)n m/z 39,53,67,81, 95,109,123...			C_nH_{2n-3}	C_nH_{2n-3}						
11,12,13,14+(13)n m/z 39,52±1,64±1, 78±1,91±1,...										

in general smooth. Abrupt abundance changes (maxima and minima) are therefore always of structural significance. The ion or ion series which is most indicative of a particular compound type is underlined in the table.

Aliphat. amides	Thiols, Sulfides	Glycols, Glycol ethers	Alkyl chlorides	Acid chlorides	Phenols, Aryl ethers	Aryl ketones	Aromat. hydrocarbons	Alkyl-benzenes	Alkyl-anilines	Compound types / Mass values
			C_nH_{2n+1}							12 + (14)$_n$ m/z 26,40,54,68, 82,96,110...
	C_nH_{2n-1}	C_nH_{2n-1}	C_nH_{2n-1}							13 + (14)$_n$ m/z 27,41,55,69 83,97,111...
	C_nH_{2n}		C_nH_{2n}							14 + (14)$_n$ m/z 28,42,56,70, 84,98,112...
		C_nH_{2n+1}		C_nH_{2n+1} $C_nH_{2n-1}O$						15 + (14)$_n$ m/z 29,43,57,71, 85,99,113...
$\underline{C_nH_{2n+2}N}$ $C_nH_{2n}NO$										16 + (14)$_n$ m/z 30,44,58,72, 86,100,114...
		$C_nH_{2n+1}O$								17 + (14)$_n$ m/z 31,45,59,73, 87,101,115...
										18 + (14)$_n$ m/z 32,46,60,74, 88,102,116...
	$\underline{C_nH_{2n+1}S}$	$\underline{C_nH_{2n+1}O_2}$								19 + (14)$_n$ m/z 33,47,61,75, 89,103,117...
	$C_nH_{2n+2}S$	$C_nH_{2n+2}O_2$						C_8H_8 + C_nH_{2n}		20 + (14)$_n$ m/z 34,48,62,76, 90,104,118...
			$\underline{C_nH_{2n}Cl}$	COCl		C_7H_5O + C_nH_{2n}		C_7H_7 + C_nH_{2n}		21 + (14)$_n$ m/z 35,49,63,77, 91,105,119...
									C_6H_6N + C_nH_{2n}	22 + (14)$_n$ m/z 36,50,64,78, 92,106,120...
					$C_nH_{n-1}O$ + C_nH_{2n}					23 + (14)$_n$ m/z 37,51,65,79, 93,107,121...
					C_nH_nO					24 + (14)$_n$ m/z 38,52,66,80, 94,108,122...
										25 + (14)$_n$ m/z 39,53,67,81, 95,109,123...
					$C_nH_{n\pm1}$	$C_nH_{n\pm1}$	$\underline{C_nH_{n\pm1}}$	$C_nH_{2n\pm1}$		11,12,13,14+(13)$_n$ m/z 39,52±1,64±1, 78±1,91±1,...

Indications of the Presence of Heteroatoms

In unit-resolution mass spectra one often observes characteristic abundances of ions due to isotope patterns, specific masses of fragment ions and characteristic mass differences (Δm) between the molecular ion ($M^{+\cdot}$) and fragment ions (F^+) or between pairs of fragment ions. It should be noted that the mass differences between fragment ions can be used reliably only if their relationship is supported by the observation of the corresponding signal ("metastable peak", m^*) due to this unimolecular decomposition.

Indication of O: Δm 17 from $M^{+\cdot}$, in N-free compounds
Δm 18 from $M^{+\cdot}$
Δm 18 from F^+ (m^*), particularly in aliphatic compounds
Δm 28, 29 from $M^{+\cdot}$ for aromatic compounds
Δm 28 from F^+ (m^*) for aromatic compounds
m/z 15, relatively abundant
m/z 19
m/z 31, 45, 59, 73 ... + $(14)_n$
m/z 32, 46, 60, 74 ... + $(14)_n$
m/z 33, 47, 61, 75 ... + $(14)_n$ for 2 x O, in the absence of S
m/z 69 for aromatic compounds meta-disubstituted by oxygen

Indication of N: $M^{+\cdot}$ odd-numbered
large number of even-numbered fragment ions
Δm 17 from $M^{+\cdot}$ or F^+ (m^*), in O-free compounds
Δm 27 from $M^{+\cdot}$ or F^+ (m^*) for aromatic compounds or nitriles
Δm 30, 46 for nitro compounds
m/z 30, 44, 58, 72 ... + $(14)_n$ for aliphatic compounds

Indication of S: Isotope peak $M^{+\cdot} + 2 \geqslant 5\% M^{+\cdot}$
Δm 33, 34, 47, 48, 64, 65 from $M^{+\cdot}$
Δm 34, 48, 64 from F^{+} (m*)
m/z 33, 34, 35
m/z 45, in O-free compounds
m/z 47, 61, 75, 89 ... + $(14)_n$, unless compound contains 2 oxygens
m/z 48, 64 for S-oxides

Indication of F: Δm 19, 20, 50 from $M^{+\cdot}$
Δm 20 from F^{+} (m*)
m/z 20
m/z 57 without m/z 55 in aromatics

Indication of Cl: Isotope peak $[M+2]^{+\cdot} \geqslant 33\% M^{+\cdot}$
Δm 35, 36 from $M^{+\cdot}$
Δm 36 from F^{+} (m*)
m/z 35/37, 36/38, 49/51*

Indication of Br: Isotope peak $[M+2]^{+\cdot} \geqslant 98\% M^{+\cdot}$
Δm 79, 80 from $M^{+\cdot}$
Δm 80 from F^{+} (m*)
m/z 79/81, 80/82*

Indication of I: Isotope peak $M^{+\cdot} + 1$ of very low abundance at relatively high mass
Δm 127 from $M^{+\cdot}$
Δm 127, 128 from F^{+} (m*)
m/z 127, 128, 254*

Indication of P: m/z 47, in compounds free of S or two oxygens
m/z 99 without isotope peak at m/z 100 in alkylphosphates

*produce doublets with halogen-free ions at the same mass which can be easily recognized even at moderate resoulution

MOLECULAR ION

Rules for the Determination of Relative Molecular Weight (M_r)

- The molecular ion ($M^{+\cdot}$) is defined as that ion the mass of which is the sum of all elements present in the molecule, using the most abundant isotope in each case.
- $M^{+\cdot}$ is always accompanied by isotope peaks. Their relative abundance depends on the number and kind of the elements present and their natural isotopic distribution. The abundance of [$M^{+\cdot}$ + 1] indicates the maximum number of carbon atoms (C_{max}) according to the following relationship:

$$C_{max} = \frac{[M^{+\cdot} + 1] \times 100}{[M^{+\cdot}] \times 1.1} \, .$$

  [$M^{+\cdot}$ + 2] and higher masses indicate the number and kind of elements that have a relatively abundant isotope two mass units heavier (such as S, Si, Cl, Br).
- $M^{+\cdot}$ is always an even number if the molecule contains only elements for which the atomic mass and valence are both even-numbered or both odd-numbered (such as H, C, O, S, Si, P, F, Cl, Br, I). In the presence of other elements $M^{+\cdot}$ becomes an odd number unless these elements are present in an even number (this holds for N, ^{13}C, ^{2}H).
- $M^{+\cdot}$ can only form fragment ions of a mass which differs from that of the molecular ion by a chemically logical value (Δm). In this context chemically illogical differences are: $\Delta m = 3$ (in the absence of $\Delta m = 1$) to $\Delta m = 14$, $\Delta m = 21$ (in the absence of $\Delta m = 1$) to $\Delta m = 24$, $\Delta m = 37$, 38 and all Δm less than the mass of an element of characteristic isotope pattern in cases where the same isotope pattern is not retained in the fragment ion.
- $M^{+\cdot}$ of a compound must contain all elements (and the maximum number of each) that are shown to be present in the fragment ions.
- $M^{+\cdot}$ is the ion with lowest appearance potential.
- $M^{+\cdot}$ exhibit the same effusion rate as can be determined from the fragment ions (this requires that the sample flows into the ion source through a molecular leak).

- The abundance of $M^{+\cdot}$ ($[M^{+\cdot}]$) is proportional to the sample pressure in the ion source.
- In the absence of a signal for $M^{+\cdot}$ the molecular weight must have a value which shows a logical and reasonable mass difference Δm to all observed fragment ions.
- MH^+ ($M^{+\cdot}$ + 1) is often observed in the mass spectra of polar compounds.
- The abundance of MH^+ changes in proportion to the square of the sample pressure in the ion source.
- For chemical ionization, field ionization, field desorption and particle desorption (FAB, SIMS, ^{252}Cf) MH^+ (or $M^{+\cdot}$ - $H\cdot$) dominates over $M^{+\cdot}$.

Signals Due to a Unimolecular Fragmentation Step ("Metastable Peaks", m*)

Mass spectra recorded in magnetic sector instruments often exhibit weak, broad peaks which are due to the fragmentation of ions after they have left the ion source. They are generally called "metastable peaks", m*, and are very useful in the interpretation process. If the spectrum is recorded in the conventional way one observes almost exclusively products resulting from the processes of the type $m_1^+ \rightarrow m_2^+ + (m_1 - m_2)$ and $m_1^{++} \rightarrow m_2^{++} + (m_1 - m_2)$. The ion m_2^+ is recorded at the apparent mass m* which is related to m_1^+ according to the equation

$$m^* = \frac{m_2^2}{m_1}$$

and thus defines the origin of m_2^+. The assignment of m_1 and m_2 to a particular m* is done by trial and error. The calculation is a simple operation on any pocket calculator based on the rearranged equation $m_2^+ = \sqrt{m^* \times m_1}$. Multiplication of the apparent mass of the "metastable" peak(m*) with that of a peak (m_1) present in the spectrum and taking the square root of the product results in the mass of the fragment ion (m_2) that may have been produced by unimolecular decomposition of m_1 (as long as the mass difference between m_1 and m_2 makes chemical sense). As a general approximation one can use the fact that on a linear mass scale and for small mass differences the distance between m_1^+ and m_2^+ are approximately the same as those between m_2^+ and m*.

In bar graph representations of mass spectra the position of metastable peaks is often indicated as small arrows along the x-axis; in fragmentation schemes those processes that are substantiated by a metastable peak are indicated by an asterisk beneath the arrow representing that process ($m_1 \underset{*}{\rightarrow} m_2$); and for mass spectra in tabular form they are listed as $m^* = m_1^+ \rightarrow m_2^+ + (m_1 - m_2)$ or appropriate shorter versions (see the examples on p. M150).

Example:

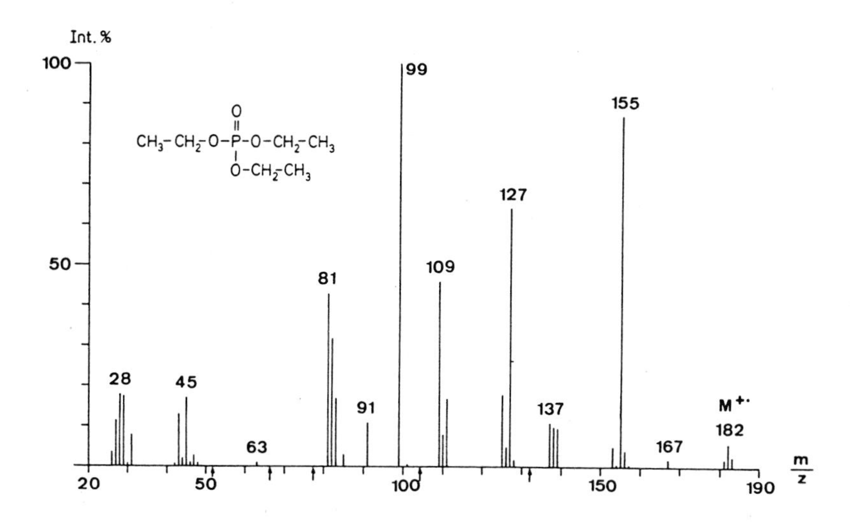

m^*	m_1^+		m_2^+		$(m_1 - m_2)$
132.0	182	⟶	155	+	27
104.1	155	⟶	127	+	28
77.2	127	⟶	99	+	28
66.3	99	⟶	81	+	18
51.7	127	⟶	81	+	46

MS

SOLVENT SPECTRA

The label [50] along the intensity scale indicates that it ends at 50% relative intensity and is subdivided in 5% steps. In those cases the abundance of the base peak has to be doubled to bring it to 100%.

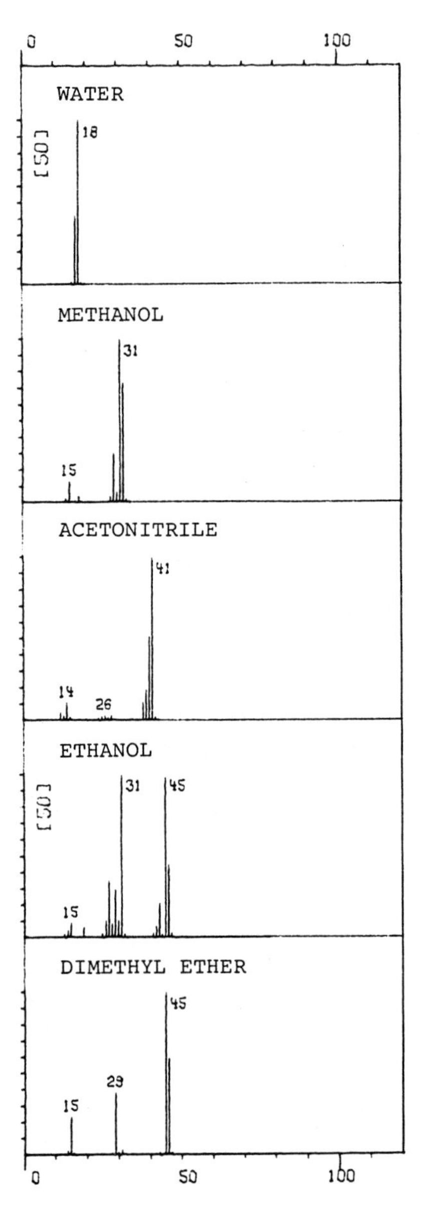

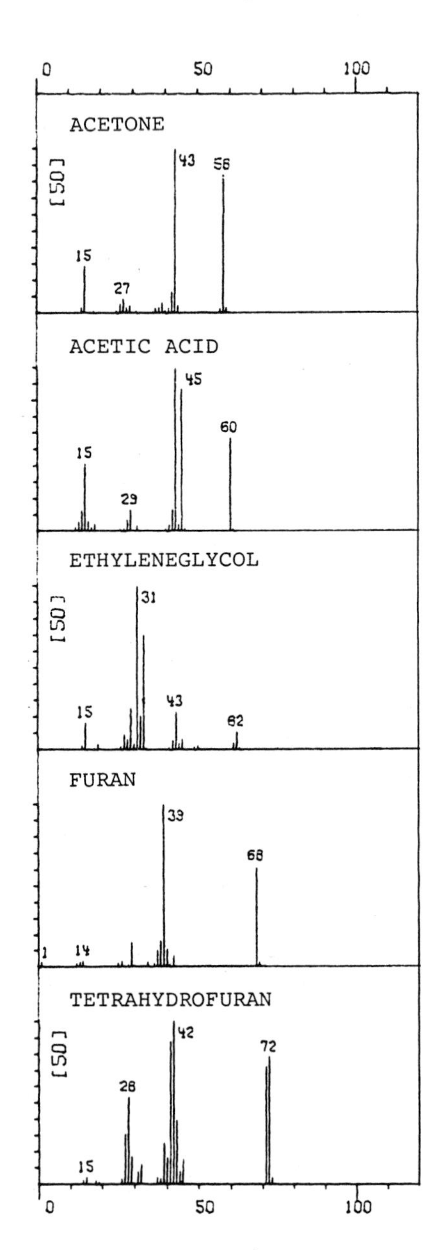

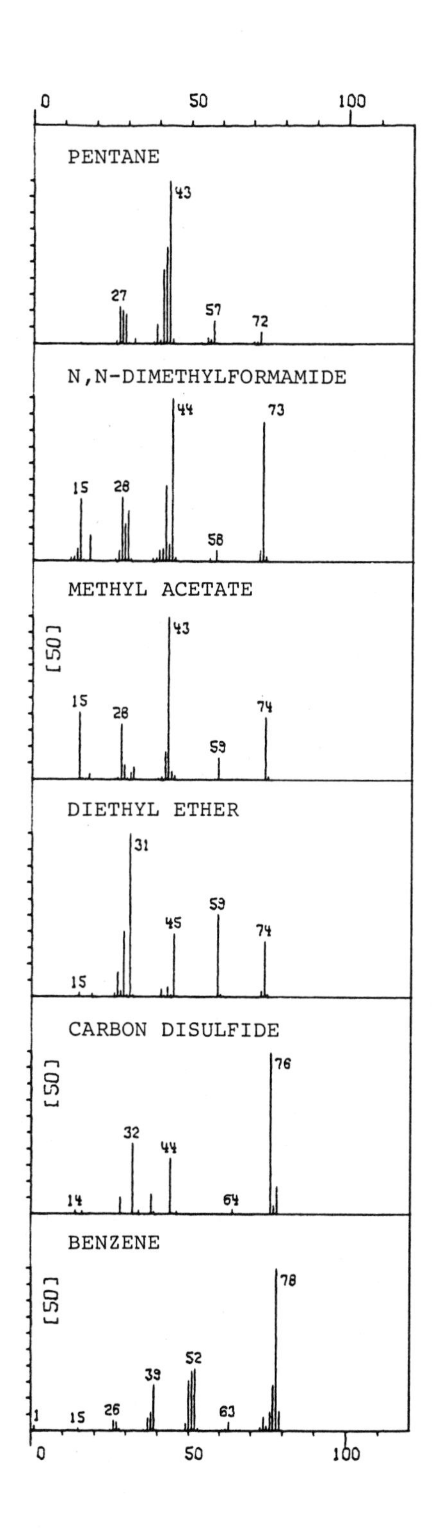
0
50
100
PENTANE
43
27
57
72
N,N-DIMETHYLFORMAMIDE
44
73
15
28
58
METHYL ACETATE
[50]
43
15
28
74
59
DIETHYL ETHER
31
59
45
74
15
CARBON DISULFIDE
[50]
76
32
44
14
64
BENZENE
[50]
78
52
39
1
15
26
63
0
50
100

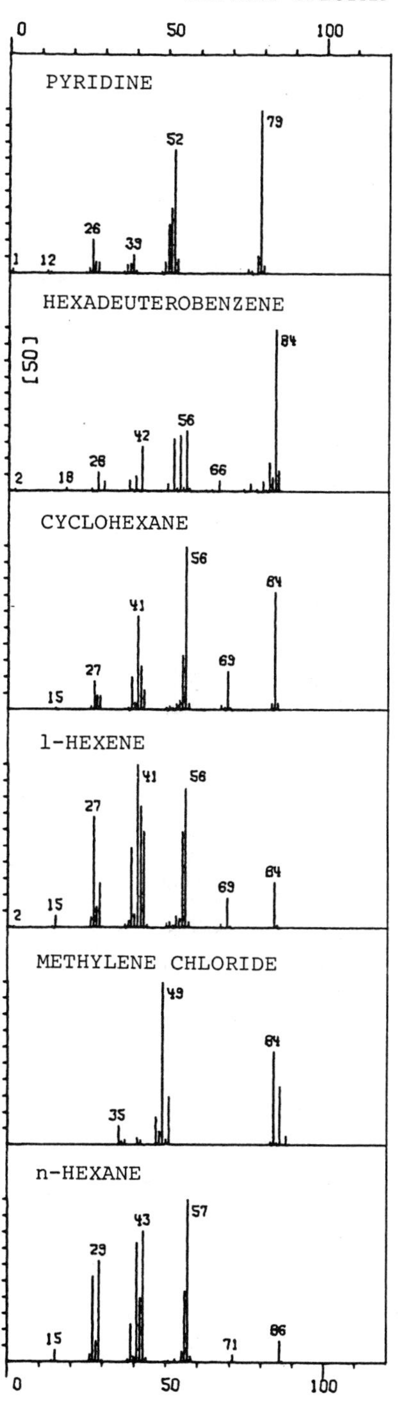
0
50
100
PYRIDINE
79
52
26
39
1
12
HEXADEUTEROBENZENE
[50]
84
56
42
28
2
18
66
CYCLOHEXANE
56
84
41
69
27
15
1-HEXENE
41
56
27
84
69
2
15
METHYLENE CHLORIDE
49
84
35
n-HEXANE
57
43
29
15
71
86
0
50
100

MS

SOLVENT SPECTRA

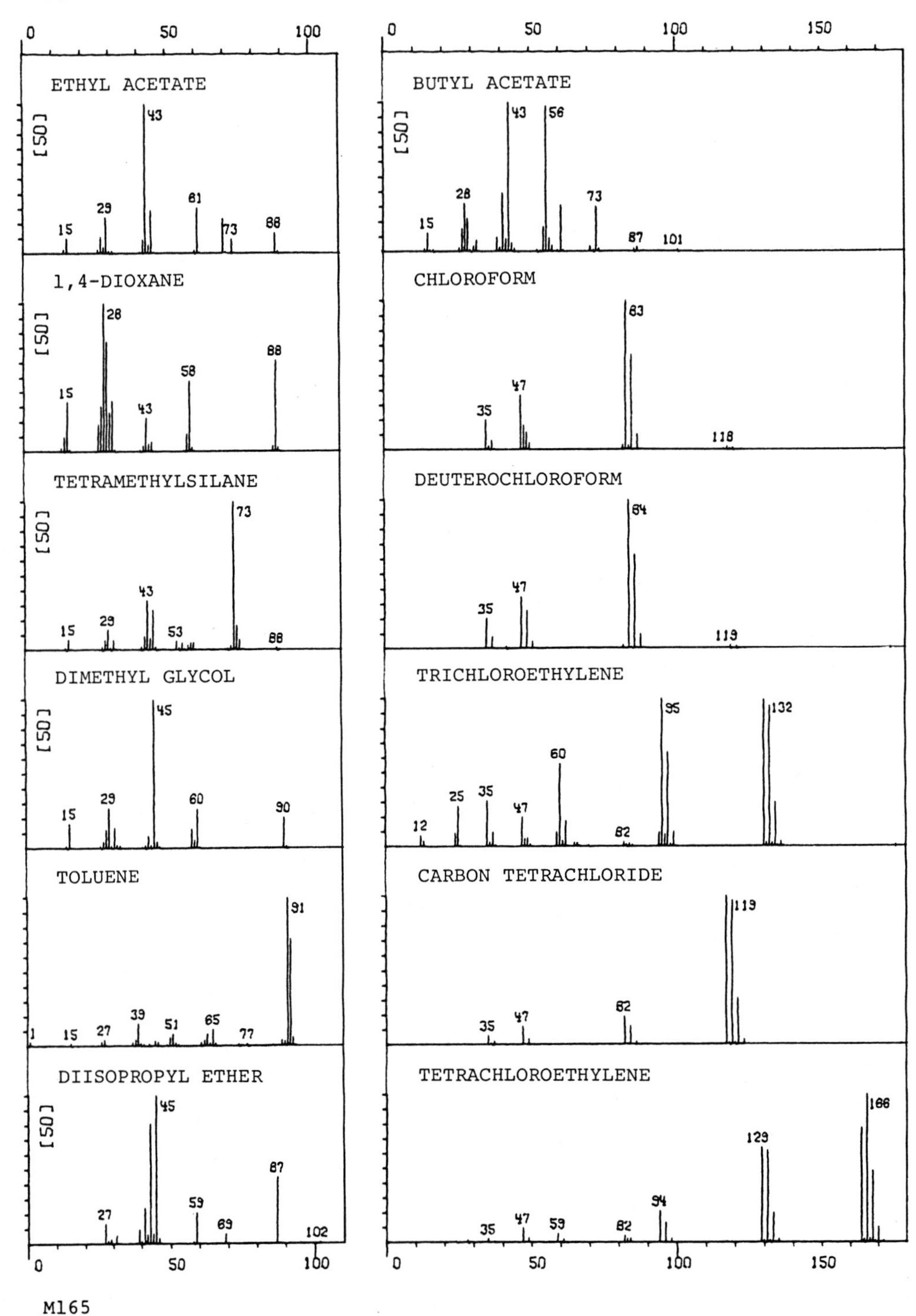

M165

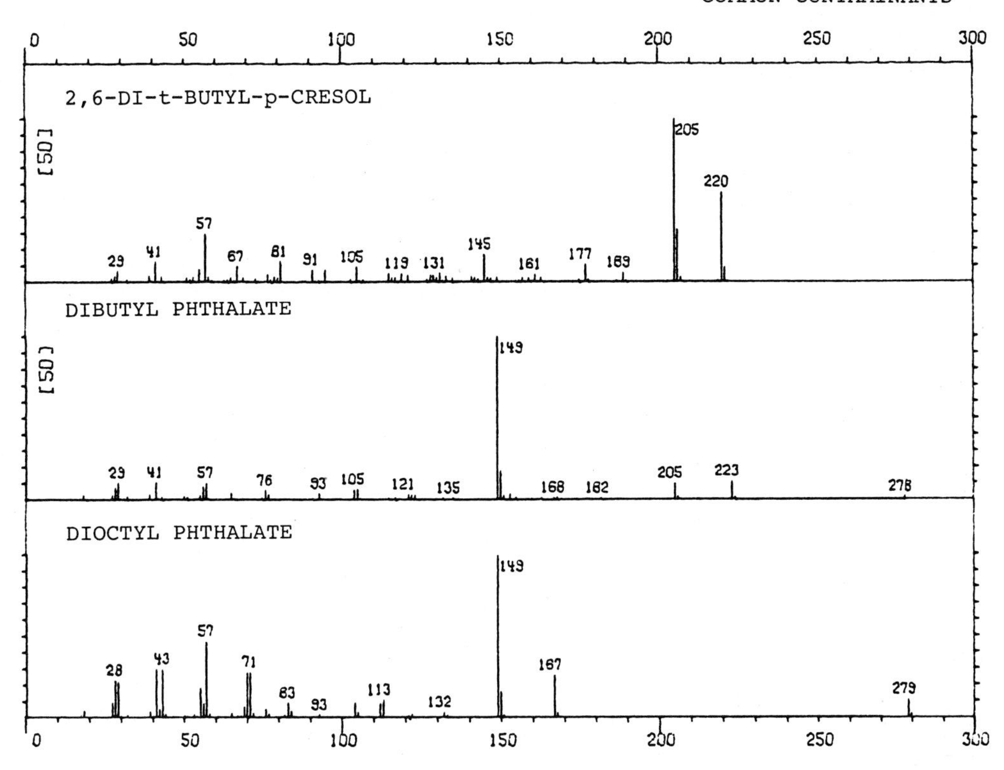
0 50 100 150 200 250 300
2,6-DI-t-BUTYL-p-CRESOL
[50]
29 41 57 67 81 91 105 119 131 145 161 177 189 205 220
DIBUTYL PHTHALATE
[50]
29 41 57 76 93 105 121 135 149 168 182 205 223 278
DIOCTYL PHTHALATE
28 43 57 71 83 93 113 132 149 167 279

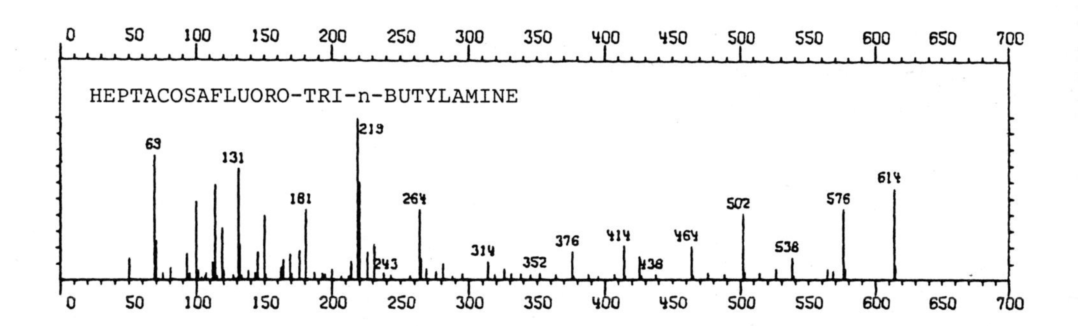
0 50 100 150 200 250 300 350 400 450 500 550 600 650 700
HEPTACOSAFLUORO-TRI-n-BUTYLAMINE
69 131 181 219 243 264 314 352 376 414 438 464 502 538 576 614

MS

MATRICES FOR FAB

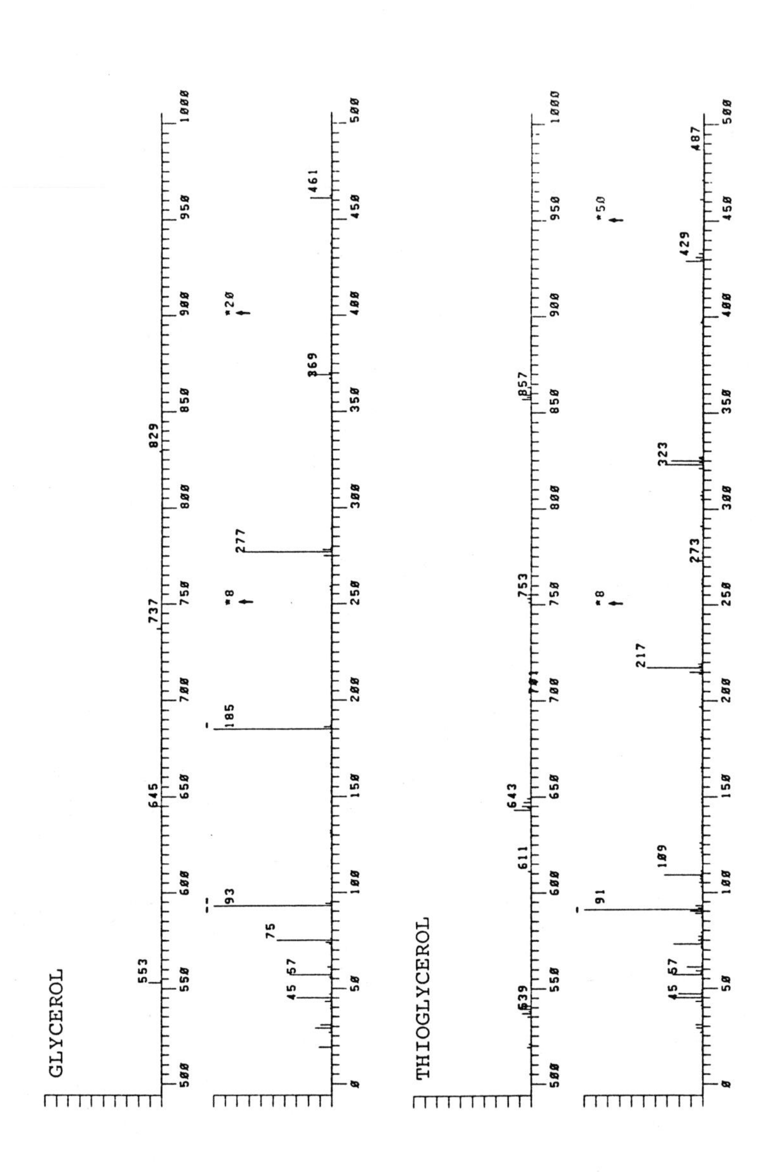

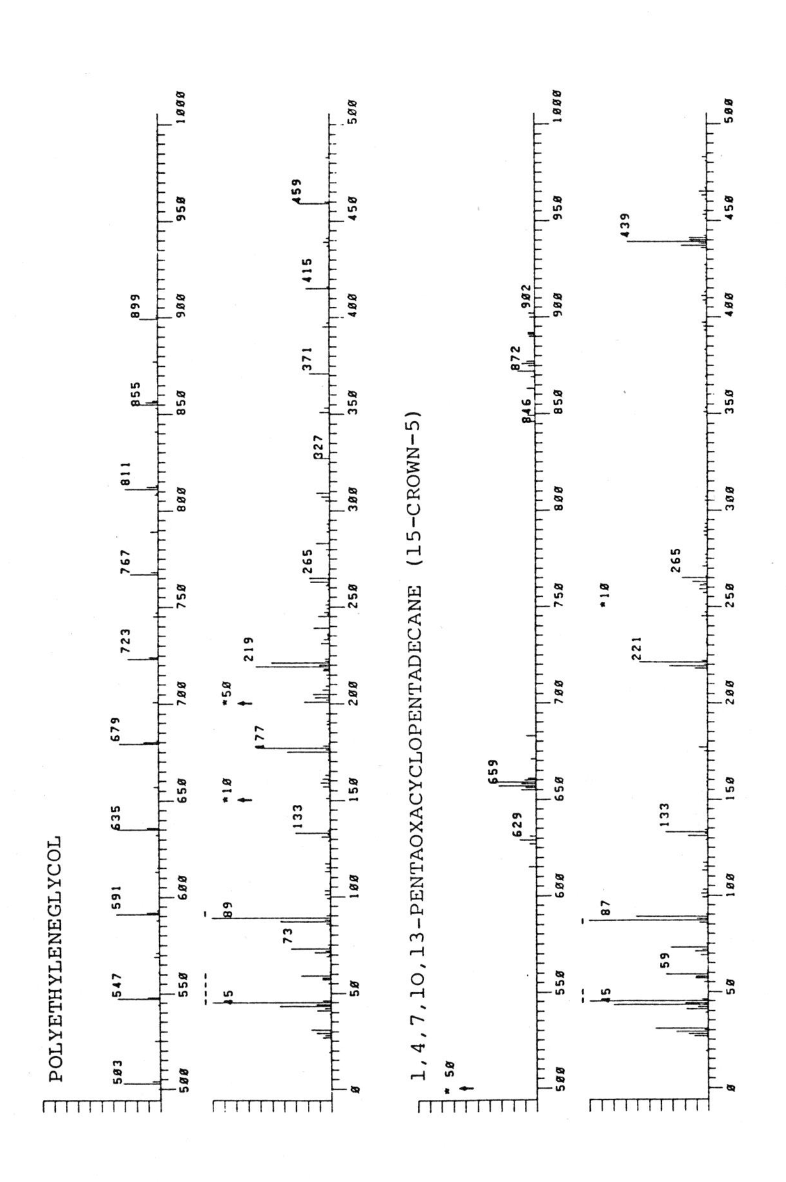
POLYETHYLENEGLYCOL
503
547
591
635
679
723
767
811
855
899
45
73
89
133
*10
177
*50
219
265
327
371
415
459
1,4,7,10,13-PENTAOXACYCLOPENTADECANE (15-CROWN-5)
* 50
629
659
846
872
902
45
59
87
133
221
*10
265
439

MS

MATRICES FOR FAB

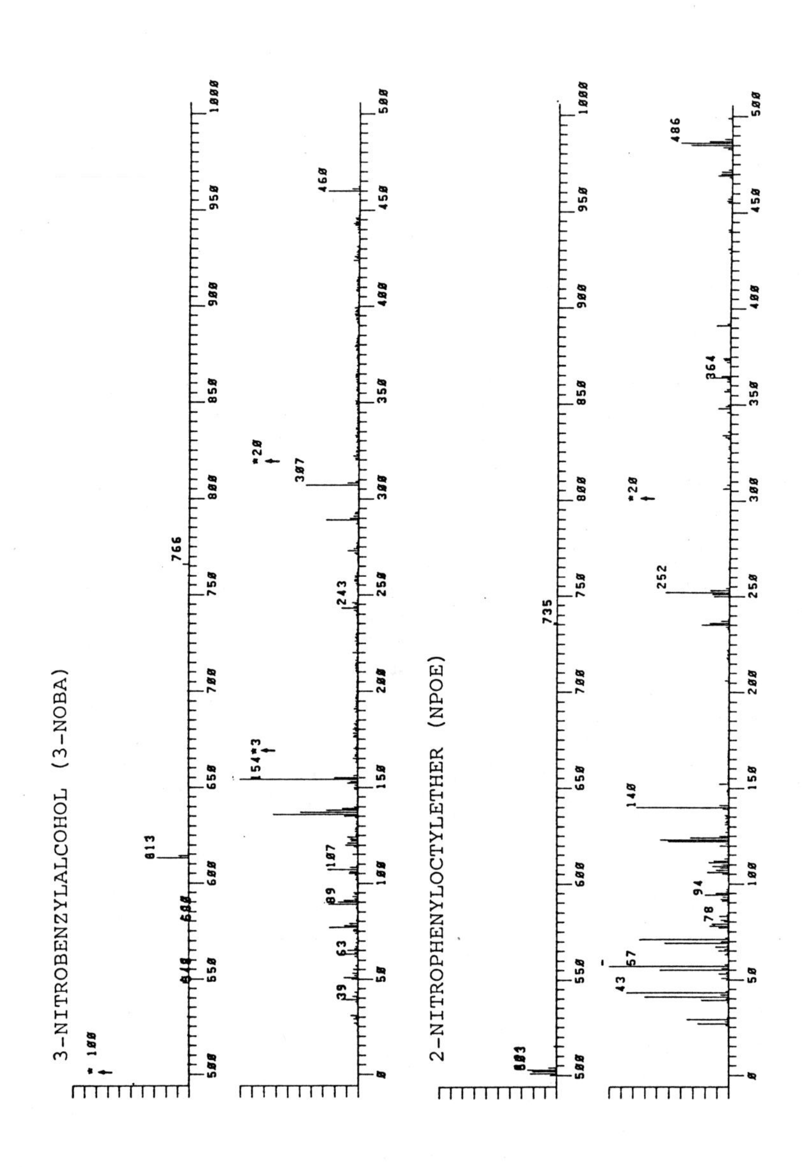

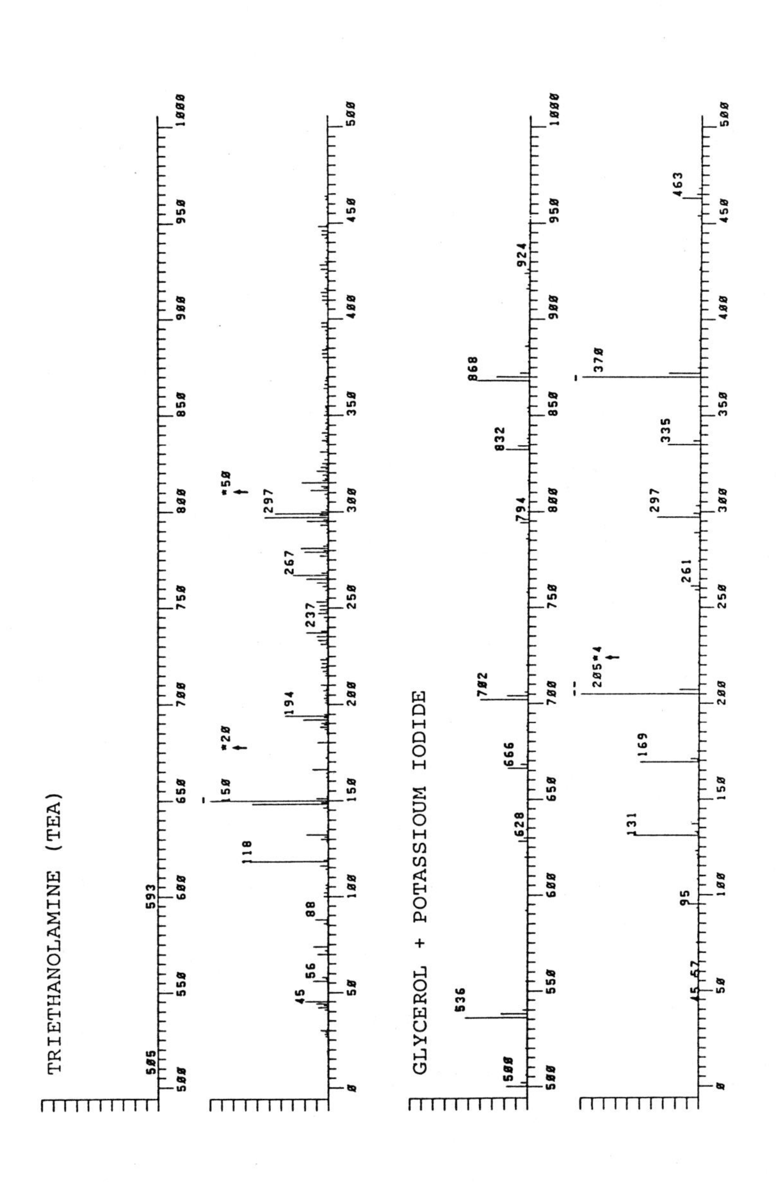
TRIETHANOLAMINE (TEA)
505
593
45
56
88
118
150
*20
194
237
267
297
*50
GLYCEROL + POTASSIOUM IODIDE
500
536
628
666
702
794
832
868
924
45
57
95
131
169
205*4
261
297
335
370
463

Typical fragmentation processes and characteristics of mass spectra of monofunctional compound classes

---

Hydrocarbons

n-Alkanes : $M^{+\cdot}$ of medium abundance; typical ion series $[C_nH_{2n+1}]^+$, m/z 29, 43, 57 ... $+(14)_n$, accompanied by $[C_nH_{2n}]^{+\cdot}$ and $[C_nH_{2n-1}]^+$ ions of lower abundance. Abundance maxima at n = 3, 4, smoothly decreasing to a minimum at m/z $[M-15]^+$.

iso-Alkanes : Abundance of $M^{+\cdot}$ decreases with increasing degree of branching; $M^{+\cdot}$ absent for highly branched compounds. Ion series same as for n-alkanes; abundance due to cleavage at branching points according to

$$[R\text{-}CH(R'')\text{-}R']^{+\cdot} \xrightarrow{-R''\cdot} [R\text{-}CH\text{-}R']^+$$
$$[R\text{-}CH(R'')\text{-}R']^{+\cdot} \xrightarrow{-RH''} [R\text{-}C{=}R']^{+\cdot}$$

for the ions of the series $[C_nH_{2n+1}]^+$ and $[C_nH_{2n}]^{+\cdot}$

Alkenes : $M^{+\cdot}$ abundant; decreasing with increasing degree of branching. Ions of the $[C_nH_{2n-1}]^+$ series accompanied by $[C_nH_{2n}]^{+\cdot}$ and $[C_nH_{2n+1}]^+$ of lower abundance. Position of the double bond only recognizable if substituted which leads to even mass maxima of the type $[C_nH_{2n}]^{+\cdot}$ formed via

$$\left[\text{R, R', H, R'' (cyclic H-transfer)}\right]^{+\cdot} \xrightarrow{-\,R''\text{-CH=CH}_2} \left[R\text{-}C(H)(R')\text{-CH=CH}_2\right]^{+\cdot}$$

Cycloalkanes : Similar to alkenes. For polycyclic compounds of increasing mass, increasing tendency to exhibit intensity maxima corresponding to ions of higher unsaturation

$$[C_nH_{2n-1}]^+ \longrightarrow [C_nH_{2n-3}]^+ \longrightarrow [C_nH_{2n-5}]^+ \ \ldots$$

MS

TYPICAL FRAGMENTATION PROCESSES

Cyclohexenes : Site of the double bond can be deduced from maxima at even mass $[C_nH_{2n-2}]^{+\cdot}$ and/or $[C_nH_{2n}]^{+\cdot}$ derived by retro-Diels-Alder cleavage:

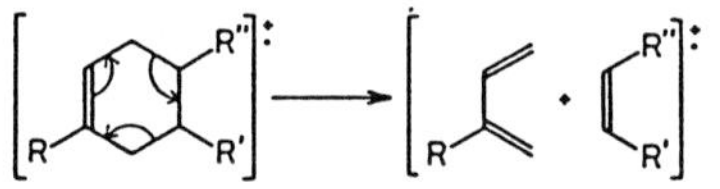

Alkynes : $M^{+\cdot}$ often of low abundance; generally more $[M-1]^+$; mixture of ion series $[C_nH_{2n-3}]^+$, $[C_nH_{2n-1}]^+$, $[C_nH_{2n+1}]^+$ with even mass ions in between.

Aromatics : $M^{+\cdot}$ abundant; typical ion series $[C_nH_{n\pm1}]^+$, for polycyclic compounds gradually changing to more highly unsaturated ions; large number of doubly charged ions. ions.

Alkyl-aromatics : Abundant ions due to cleavage of a benzyl bond with and without H-transfer to aromatic ring. Polysubstituted compounds may eliminate entire side chain as an olefin.

Ketones

saturated aliphatic : $M^{+\cdot}$ relatively abundant; typical ion series $[C_nH_{2n-1}O]^+$ or $[C_nH_{2n+1}]^+$, m/z 29, 43, 57 ... $+(14)_n$ like iso-alkanes, abundance maxima due to cleavage at the carbonyl group to form acyl ion and their decarbonylation. Even-mass maxima two dalton higher than for iso-alkanes due to olefin elimination (McLafferty arrangement):

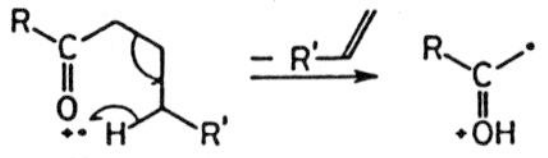

Sometimes the presence of oxygen can be deduced from weak signals at $[M-18]^+$, m/z 31, 45.

aromatic aliphatic : $M^{+\cdot}$ abundant; dominating benzoyl ion due to cleavage at the benzoyl bond; decarbonylation leads to less abundant phenyl ion. Even-mass maxima due to olefin elimination via McLafferty rearrangement (see above).

saturated alicyclic : $M^{+\cdot}$ abundant; typical ion series $[C_nH_{2n-3}O]^+$, $[C_nH_{2n-1}]^+$, with maxima due to alkyl loss after ring opening next to the carbonyl group and H-transfer:

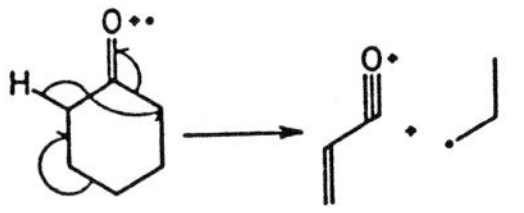

Prominent even mass maxima by elimination of substituents at the 2,6-positions as olefin via sterically favored McLafferty rearrangements:

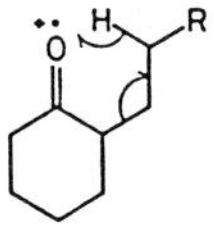

Aldehydes

saturated aliphatic : $M^{+\cdot}$ and ion series similar to ketones. McLafferty rearrangement generally with reversed charge distribution giving rise to $[C_nH_{2n}]^{+\cdot}$ ions. Elimination of water from $M^{+\cdot}$ to $[M-18]^{+\cdot}$ occasionally very pronounced, particularly for aliphatic compounds.

aromatic : $M^{+\cdot}$ abundant; pronounced loss of H to form benzoyl ion, followed by decarbonylation.

TYPICAL FRAGMENTATION PROCESSES

Alcohols

saturated aliphatic : $M^{+\cdot}$ often of low abundance or absent; elimination of water leads to olefin-like spectra with additional "oxygen-type" peaks $[C_nH_{2n+1}O]^+$, m/z 31, 45, 59 ... $+(14)_n$; within this series abundance maxima due to α-cleavage such as

$$R-CH(\ddot{O}H)-R' \longrightarrow R-CH=\overset{+}{O}H$$

In the region below the molecular ion typical series of peaks at $[M-15]^+$, $[M-18]^{+\cdot}$, $[M-33]^+$.

alicyclic : $M^{+\cdot}$ often of low abundance, but always detectable. Cleavage at the C-atom bearing the OH-group followed by a reaction analogous to that shown for cyclic ketones leads to maxima of the ion series $[C_nH_{2n-1}O]^+$. In addition, ion series $[C_nH_{2n-1}]^+$ and $[C_nH_{2n-3}]^+$.

phenols : $M^{+\cdot}$ abundant. Other abundant ions formed by loss of CO (Δm 28) and/or CHO (Δm 29) from $M^{+\cdot}$ and fragments.

Ethers

saturated aliphatic : $M^{+\cdot}$ generally detectable. Alkyl ions $[C_nH_{2n+1}]^+$ with maxima due to cleavage of the C-O bond; olefin ion series $[C_nH_{2n}]^{+\cdot}$ and $[C_nH_{2n-1}]^+$ due to elimination of an alcohol moiety; oxygen containing ion series $[C_nH_{2n+1}O]^+$ with maxima due to cleavage of the C-C bond next to oxygen.

saturated cyclic : $M^{+\cdot}$ generally detectable; maxima due to loss of substituents at the carbon atoms next to oxygen, often followed by elimination of water.

furans : $M^{+\cdot}$ abundant; maxima due to benzyl-type cleavage, followed by loss of CO.

Aryl-alkyl ethers : $M^{+\cdot}$ abundant; maxima due to olefin elimination to form a phenol ion, followed by decarbonylation (Δm 28). Methyl ethers lose CH_3 followed by decarbonylation to give $[M-43]^+$; elimination of formaldehyde (Δm 30).

aromatic : $M^{+\cdot}$ abundant; decarbonylation with recombination of the aromatic moieties.

Amines

saturated aliphatic : $M^{+\cdot}$ (odd mass) of low abundance or absent, often protonated to $[M+H]^+$; even mass ion series $[C_nH_{2n+2}N]^+$, m/z 30, 44, 58, 72 ... $+(14)_n$ with dominant maxima due to α-cleavage

$$R'R''\dot{N}^{+}-CH_2-R \longrightarrow R'R''\overset{+}{N}=CH_2$$

followed by elimination of olefins to $R'-\overset{+}{N}H=CH_2$ and $\overset{+}{N}H_2=CH_2$, m/z 30.

Cycloalkyl amines : $M^{+\cdot}$ generally detectable; after ring opening by cleavage at the N-bearing C-atom followed by loss of alkyl (as shown for cyclic ketones) ions of the type $[C_nH_{2n}N]^+$ are formed; elimination of NH_3.

saturated cyclic : $M^{+\cdot}$ generally detectable; ion series $[C_nH_{2n}N]^+$ and $[C_nH_{2n+1}N]^{+\cdot}$, maxima due to loss of the substitutent at the N-bearing C-atom (often also of $H^{\cdot}$).

Pyridines, Pyrroles, Indoles : $M^{+\cdot}$ abundant; maxima due to benzyl-type cleavage of substitutents; 2-alkyl substituted pyridines may lose the side chain beyond the α-carbon as an olefin via a McLafferty rearrangement or undergo cleavage of the β-γ or γ-δ bonds and lose a shorter alkyl group. HCN (Δm 27) elimination from fragments.

Alkyl-anilines : $M^{+\cdot}$ abundant; cleavage of alkyl substituents at N for aliphatic amines, HCN elimination from fragments.

<u>Carboxylic acids</u>

saturated aliphatic : $M^{+\cdot}$ generally detectable; easily protonated to $[M+H]^+$; ion series $[C_nH_{2n+1}]^+$, $[C_nH_{2n-1}]^+$, $[C_nH_{2n-1}O_2]^+$, in long chain acids the latter exhibiting maxima at n = 3, 7, 11, 15 ...; even mass maxima of the ion series $[C_nH_{2n}O_2]^{+\cdot}$ due to McLafferty rearrangement

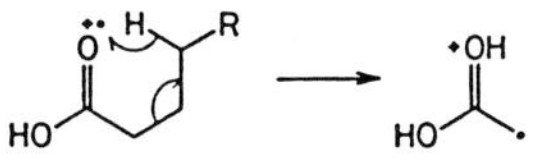

Loss of HO (Δm 17) and H_2O (Δm 18) from $M^{+\cdot}$, followed by decarbonylation.

aromatic : $M^{+\cdot}$ abundant; pronounced loss of $HO^{\cdot}$ (Δm 17) from $M^{+\cdot}$ to form benzoyl ion, followed by decarbonylation; H_2O elimination (Δm 18) if H-bearing ortho-substituent present; some acids decarboxylate (Δm 44).

<u>Anhydrides</u>

linear : $M^{+\cdot}$ of low abundance or absent; abundant acyl ions due to cleavage next to carbonyl.

cyclic : $M^{+\cdot}$ of low abundance, easily protonated to $[M+H]^+$; maxima due to carboxylation (Δm 44) followed by decarbonylation.

Esters

saturated aliphatic : $M^{+\cdot}$ often of low abundance, easily protonated to $[M+H]^+$; ion series $[C_nH_{2n-1}]^+$, $[C_nH_{2n-1}O_2]^+$, the latter in the case of methyl esters with maxima at n = 4, 8, 12, 16 ...; maximum in the $[C_nH_{2n+1}]^+$ series for the alkyl group at the ester-oxygen; except for methyl esters even mass maxima due to olefin elimination at both sides of the -COO- moiety via McLafferty rearrangements; leading at the acid side to $[C_nH_{2n}O_2]^{+\cdot}$ as in the case of carboxylic acids, at the alcohol side to the corresponding acid $[C_nH_{2n}O_2]^{+\cdot}$ and/or the corresponding olefin $[C_nH_{2n}]^{+\cdot}$. For esters of higher alcohols $[C_nH_{2n+1}O_2]^+$ ions due to the protonated acid (formed by double H-rearrangements) predominate; loss of alkoxyl to form acylium ions and elimination of alcohol followed by decarbonylation; α-substituted esters may lose the substituent and then CO (Δm 28) via alkoxyl rearrangement; β-substituted esters may eliminate ketene in an analogous reaction (Δm 42).

Esters of aromatic acids : $M^{+\cdot}$ often of low abundance; loss of alkoxy group to form benzoyl ion followed by decarbonylation dominates; olefin elimination to $[M]^{+\cdot}$ of corresponding acid, and/or further H-transfer to protonated acid; if H-bearing ortho-substituent present, alcohol elimination is a competing reaction. In the case of alkyl phthalates (other than dimethyl phthalate), first alkenyl elimination to the protonated ester acid, followed by olefin elimination from the other ester group, and finally water elimination to the protonated anhydride ion which forms the base peak at m/z 149.

Lactones

aliphatic : $M^{+\cdot}$ often of low abundance, easily protonated to $[M+H]^+$; maxima due to loss of substituent at the O-bearing C-atom, followed by decarboxylation (Δm 44), decarbonylation (Δm 28), ketene elimination; decarboxylation of $M^{+\cdot}$ rarely significant.

aromatic : $M^{+\cdot}$ abundant; maxima due to two consecutive decarbonylations.

Amides

saturated aliphatic : $M^{+\cdot}$ (odd mass) observable, easily protonated; characteristic even mass ion series $[C_nH_{2n}NO]^+$, maxima due to olefin elimination on the acid side to form the corresponding acetamide (via McLafferty rearrangement), on the amine side to the ion of the desalkyl amide, often via double H-rearrangement to the protonated desalkyl amide ion; competitive reactions are cleavage at the carbonyl group and of the C-C bond attached to N, respectively.

aromatic : Amides of aromatic acids exhibit maxima due to amide bond cleavage to form the benzoyl ion, followed by decarbonylation; derivatives of aromatic amines with aliphatic acids eliminate a ketene to form the amine ion.

lactams : $M^{+\cdot}$ often observable, more abundant than for corresponding lactones; maxima due to cleavage of the C-C bond at the N-bearing C-atom; competing reactions are the cleavage of the CO-N bond followed by loss of CO or by further cleavage of the C-C bond next to N forming an iminium ion.

Sulfur compounds

Behavior analogous to the corresponding O-compounds; $M^{+\cdot}$ generally more abundant; typical loss of $SH^{\cdot}$ (Δm 33) and H_2S (Δm 34) and formation of H_3S^+ (m/z 35), CS^+ (m/z 45) as non-specific S-indicators; for sulfoxides loss of SO (Δm 48), for sulfones loss of SO and SO_2 (Δm 64) is characteristic; saturated aliphatic sulfides and thiols show maxima in the ion series $[C_nH_{2n+1}S]^+$ m/z 47, 61, 75 ... $+(14)_n$ due to cleavage of the C-C bond next to S; the natural isotope ratio of 4.5% ^{34}S relative to ^{32}S results in a diagnostically useful isotope pattern differing in mass by 2 dalton.

Halides

saturated aliphatic : $M^{+\cdot}$ decreasing with increasing molecular weight, branching and number of heteroatoms, absent for many polyhalogenated compounds; an important fragmentation is the loss of a halogen radical (for Cl and F often more H-Hal elimination) and hydrogen halide from fragments. For longer aliphatic chains maximum $[C_4H_8Hal]^+$ with Cl and Br. Cl and Br exhibit characteristic isotope patterns which facilitate identification (see p. M100, M105, M110).

aromatic : $M^{+\cdot}$ abundant; some loss of halogen radicals from $M^{+\cdot}$, consecutive in the case of polyhalogenated compounds; otherwise elimination of hydrogen halide from fragment ions. Trifluoromethyl groups eliminate CF_2 (Δm 50).

Nitriles

saturated aliphatic : $M^{+\cdot}$ of low abundance or absent, easily protonated to $[M+H]^+$ and $H^{\cdot}$ eliminating to $[M-H]^+$; ion series

$[C_nH_{2n-1}]^+$, $[C_nH_{2n-1}N]^{+\cdot}$ m/z 41, 55, 69, 83 ... $+(14)_n$, with longer aliphatic chains also even mass ion series $[C_nH_{2n-2}N]^+$.

aromatic : $M^{+\cdot}$ abundant; most significant fragmentation is the elimination of HCN (Δm 27).

Nitrocompounds

saturated aliphatic : $M^{+\cdot}$ of low abundance, often absent; uncharacteristic spectra, most frequent fragmentation is the elimination of HNO_2 (Δm 47) and HNO (Δm 31).

aromatic : $M^{+\cdot}$ abundant; often typical group of ions below $M^{+\cdot}$ are $[M-16]^+$, $[M-30]^+$ and $[M-46]^+$ due to loss of $O^\cdot$, $NO^\cdot$ and $NO_2{}^\cdot$ as well as $[M-58]^+$ by loss of ($NO^\cdot$ + CO); loss of NO_2 (Δm 46) from fragment ions.

Aromatic diazo compounds

$M^{+\cdot}$ abundant; significant peaks due to cleavages at the diazo group, followed by loss of N_2 (Δm 28) to give a more abundant secondary fragment (can be used to differentiate from acyl ions which generally produce an aryl ion that is less abundant than the acyl ion).

Phosphorous compounds

alkyl phosphates : $M^{+\cdot}$ observable; maxima due to alkenyl loss from $M^{+\cdot}$ via double H-rearrangement, followed by olefin elimination down to protonated phosphoric acid, m/z 99; PO^+ (m/z 47), $H_2PO_2{}^+$ (m/z 65), $H_2PO_3{}^+$ (m/z 81) often as non-specific P-indicators.

phosphines and phosphine oxides:

saturated aliphatic : $M^{+\cdot}$ observable; maxima of the ion series $[C_nH_{2n+3}P]^+$, m/z 48, 62, 76, 90 ... due to olefin eliminations.

aromatic : $M^{+\cdot}$ abundant, easily loses $H^\cdot$ to $[M-1]^+$; maxima due to loss of an aryl group, followed by H_2-elimination to form the phosphafluorenyl ion:

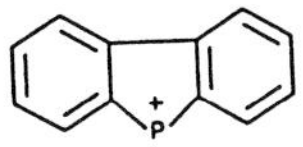

Silicon compounds

trialkyl silylethers : $M^{+\cdot}$ often of low abundance or absent, easily protonated to $[M+H]^+$; maxima due to loss of alkyl attached to Si (preferential loss of larger groups), to cleavage of the C-C bond adjacent to O followed by olefin elimination to form fragments of the type $[C_nH_{2n+3}OSi]^+$, m/z 75, 89, 103, 117 ..., and to loss of alkoxyl, followed by olefin eliminations to fragments of the type $[C_nH_{2n+3}Si]^+$, m/z 45, 59, 73, 87 ....
Occasionally maxima at even mass by elimination of trialkylsilanol. The R_2Si-OR' cation has the tendency to attack in an electrophilic manner free electron pairs or π-electron centers even over long distances, causing the expulsion of neutral fragments from the interior of the molecule via a rearrangement:

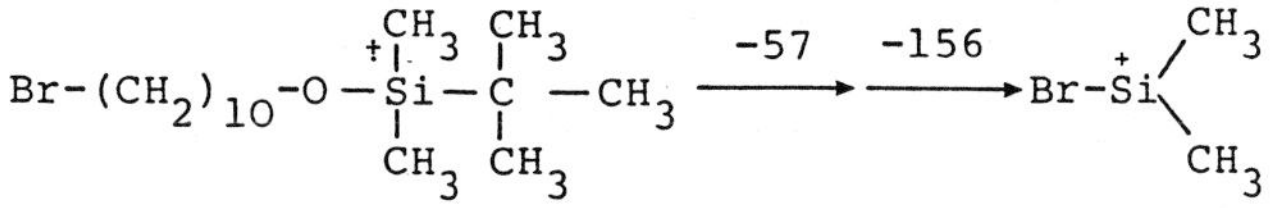

$$Br-(CH_2)_{10}-O-\overset{+}{Si}(CH_3)_2-C(CH_3)_2-CH_3 \xrightarrow{-57} \xrightarrow{-156} Br-\overset{+}{Si}(CH_3)_2$$

Correlation Between Wavelength of Absorbed Radiation and Observed Color

Absorbed Light		Observed (transmitted) color
Wavelength nm	Corresponding color	
400	violet	yellow-green
425	indigo blue	yellow
450	blue	orange
490	blue-green	red
510	green	purple
530	yellow-green	violet
550	yellow	indigo blue
590	orange	blue
640	red	blue-green
730	purple	green

UV/VIS-Absorption of Simple Chromophores

Chromophore	Compound	Transition	λ_{max} [nm]	ε_{max}	Solvent
C-C	CH_3-CH_3	$\sigma \rightarrow \sigma^*$	135	strong	gas
C-H	CH_4	$\sigma \rightarrow \sigma^*$	122	strong	gas
C-O	CH_3OH	$n \rightarrow \sigma^*$	177	200	hexane
	CH_3-O-CH_3	$n \rightarrow \sigma^*$	184	2500	gas
C-N	$(C_2H_5)_2NH$	$n \rightarrow \sigma^*$	193	2500	hexane
	$(CH_3)_3N$	$n \rightarrow \sigma^*$	199	4000	hexane
C-S	CH_3-SH	$n \rightarrow \sigma^*$	195	1800	gas
		$n \rightarrow \sigma^*$	235	180	gas
	$C_2H_5-S-C_2H_5$	$n \rightarrow \sigma^*$	194	4500	gas
		$n \rightarrow \sigma^*$	225	1800	gas
S-S	$C_2H_5-S-S-C_2H_5$	$n \rightarrow \sigma^*$	194	5500	hexane
		$n \rightarrow \sigma^*$	250	380	
C-Cl	CH_3Cl	$n \rightarrow \sigma^*$	173	200	hexane
C-Br	$n-C_3H_7Br$	$n \rightarrow \sigma^*$	208	300	hexane
C-I	CH_3I	$n \rightarrow \sigma^*$	259	400	hexane
C=C	$CH_2=CH_2$	$\pi \rightarrow \pi^*$	162.5	15000	heptane
	$(CH_3)_2C=C(CH_3)_2$	$\pi \rightarrow \pi^*$	196.5	11500	heptane
C=O	$(CH_3)_2-C=O$	$n \rightarrow \sigma^*$	166	16000	gas
		$\pi \rightarrow \pi^*$	189	900	hexane
		$n \rightarrow \pi^*$	279	15	hexane
	$CH_3-\overset{O}{\overset{\Vert}{C}}-OH$	$n \rightarrow \pi^*$	200	50	gas
	$CH_3-\overset{O}{\overset{\Vert}{C}}-OC_2H_5$	$n \rightarrow \pi^*$	210	50	gas
	$CH_3-\overset{O}{\overset{\Vert}{C}}-ONa$	$n \rightarrow \pi^*$	210	150	water
	$CH_3-\overset{O}{\overset{\Vert}{C}}-NH_2$	$n \rightarrow \pi^*$	220	63	water
	$CH_2-C(=O)-NH-C(=O)-CH_2$ (cyclic; succinimide)		191	15200	aceto-nitrile

Chromophore	Compound	Transition	λ_{max} [nm]	ε_{max}	Solvent
C=N	$H_2N-C(=NH)-NH_2 \cdot HCl$		265	15	water
	$(CH_3)_2C=NOH$		193	2000	ethanol
	$(CH_3)_2-C=NONa$		265	200	ethanol
N=N	$CH_3-N=N-CH_3$		340	16	ethanol
N=O	$(CH_3)_3C-NO$		300 665	100 20	ether
	$(CH_3)_3C-NO_2$		276	27	ethanol
	$n-C_4H_9-O-NO$		218 313-384	1050 20-40	 ethanol
	$C_2H_5-O-NO_2$		260	15	ethanol
C=S	$CH_3-C(=S)-CH_3$		460	weak	
	=S		495	weak	ethanol
C≡C	$HC\equiv CH$		173	6000	gas
	$n-C_5H_{11}-C\equiv C-CH_3$		177.5 196 222.5	10000 2000 160	hexane
C≡N	$CH_3-C\equiv N$		<190		
X=C=Y	$CH_2=C=CH_2$		170 227	4000 630	
	$(C_2H_5)_2C=C=O$		227 375	360 20	
	$C_2H_5-N=C=N-C_2H_5$		230 270	4000 25	
	$C_2H_5-N=C=S$		250	1200	hexane

UV-Absorption of α,β-Unsaturated Carbonyl Compounds

(extended Woodward rules to estimate the position of the $\pi \rightarrow \pi^*$ transition)

```
δ   β   X
|   |   |
-C=C-C=C-C=O
   |   |
   γ   α
```

Parent system:

(acyclic enone, X on carbonyl C)	X : alkyl	215 nm
	X : H	207 nm
	X : OH, O-alkyl	193 nm
(cyclohexenone)		215 nm
(cyclopentenone)		202 nm

Increments: for each additional conjugated double bond +30 nm

for each exocyclic double bond (C=C) + 5 nm

C=C

for each homoannular diene system (homoannular arrangement of double bonds) +39 nm

For each substituent at the π-electron system	(Increment in nm) α	β	γ	δ	and beyond
C-substituent	10	12	18	18	18
OH	35	30		50	50
$OCOCH_3$	6	6	6	6	6
O-alkyl	35	30	17	31	31
S-alkyl		85			
Cl	15	12			
Br	25	30			
$N(alkyl)_2$		95			

Solvent corrections:

water	+ 8 nm
ethanol, methanol	0 nm
chloroform	- 1 nm
dioxane	- 5 nm
diethyl ether	- 7 nm
hexane, cyclohexane	-11 nm

To estimate the position of the absorption maximum in a particular solvent the appropriate increments are added to the base value for the parent system. For cross-conjugated systems the value for the chromophore absorbing at the longest wavelength should be calculated.

Example:

Estimating the absorption maximum for

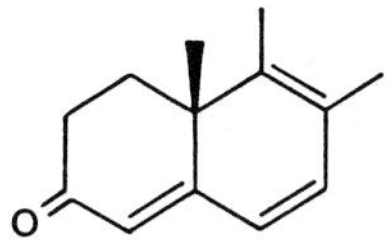

in ethanolic solution.

	nm
base value	215
2 additional conjugated double bonds	60
exocyclic double bond	5
homoannular diene system	39
C-substituent in β	12
3 additional C-substituents	54
solvent correction	0
estimated:	385 nm (ethanol)
determined:	388 nm (ethanol)

UV-Absorption of Dienes and Polyenes

(Woodward-Fieser rules to estimate the position of the $\pi \rightarrow \pi^*$ transition)

Parent system:

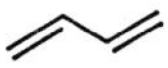	acyclic (and in non-condensed ring systems)	217 nm
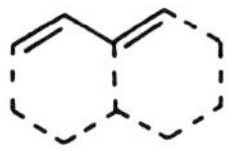	heteroannular	214 nm
	homoannular	253 nm

Increments:

for each additional conjugated double bond		+30 nm
for each exocyclic double bond (C=C)		+ 5 nm
for each substituent:	C-substituent	+ 5 nm
	$OCOCH_3$	+ 0 nm
	O-alkyl	+ 6 nm
	S-alkyl	+30 nm
	Cl, Br	+ 5 nm
	$N(alkyl)_2$	+60 nm
	solvent correction	∿ 0 nm

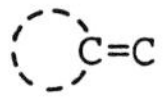

Example:

Estimating the absorption maximum for

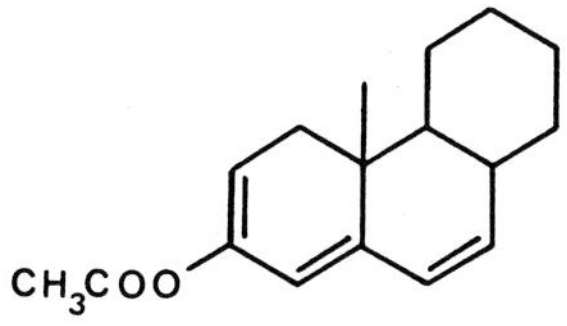

CH_3COO

	nm
base value (homoannular)	253
additional conjugated double bond	30
exocyclic double bond	5
3 C-substituents	15
estimated:	303 nm
determined:	306 nm

UV-Absorption of Aromatic Carbonyl Compounds

{Scott rules to estimate the position of the K-band (in ethanol as the solvent)}

Parent system:

C_6H_5-CO-R	R : alkyl, alicyclyl	246 nm
C_6H_5-CO-H		250 nm
C_6H_5-CO-OH		230 nm
C_6H_5-CO-OR		230 nm

Increments (in nm) per substituent in

	ortho	meta	para
alkyl, alicyclyl	3	3	10
OH, O-alkyl	7	7	25
O^-	11	20	78
Cl	0	0	10
Br	2	2	15
NH_2	13	13	58
$NHCOCH_3$	20	20	45
$N(CH_3)_2$	20	20	85

Example:

Estimating the position of the absorption maximum (K-band) for

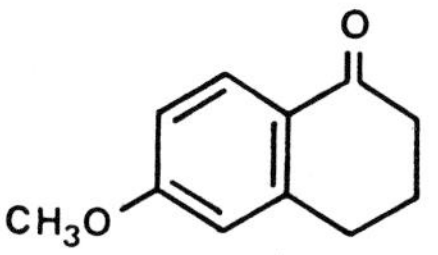

CH_3O

	nm
base value:	246
alicyclyl in o-position	3
O-alkyl in p-position	25
estimated:	274 nm
determined:	276 nm

UV-Absorption of Aromatic Compounds

Substituent R (solvent)	Transitions							
	$\pi \rightarrow \pi^*$ (allowed)		$\pi \rightarrow \pi^*$ ("forbidden")		$\pi \rightarrow \pi^*$ (substituent delocalized by aromatic system); K-band		$n \rightarrow \pi^*$ (substituent with non-bonding electron pair); R-band	
	λ_{max} [nm]	ε	λ_{max} [nm]	ε	λ_{max} [nm]	ε	λ_{max} [nm]	ε
	∼180-230	∼2000-10000	∼250-290	∼100-2000	∼220-250	∼10000-30000	∼275-350	∼10-100
-H (cyclohexane)	198	8000	255	230				
$-CH_3$ (hexane)	208	7900	262	230				
-OH (water)	211	6200	270	1450				
$-O^-$ (water)	235	9400	287	2600				
$-NH_2$ (water)	230	8600	280	1430				
$-NH_3^+$ (water)	203	7500	254	160				
$-NO_2$ (hexane)	{208, 213}	{9800, 8100}	270	800	251	9000	322	150
-Cl (ethanol)	210	7500	257	170				
$-CH=CH_2$ (ethanol)			282	450	244	12000		
-C≡CH (hexane)			278	650	236	12500		
$-COCH_3$ (ethanol)			278	1100	243	13000	319	50
-CHO (hexane)			280	1400	242	14000	∼330	∼60
-COOH (water)	202	8000	270	800	230	10000		
-C≡N (water)			271	1000	22	13000		

UV/ VIS

REFERENCE SPECTRA

$CH_3-CH=CH-CH=CH_2$

(in heptane)

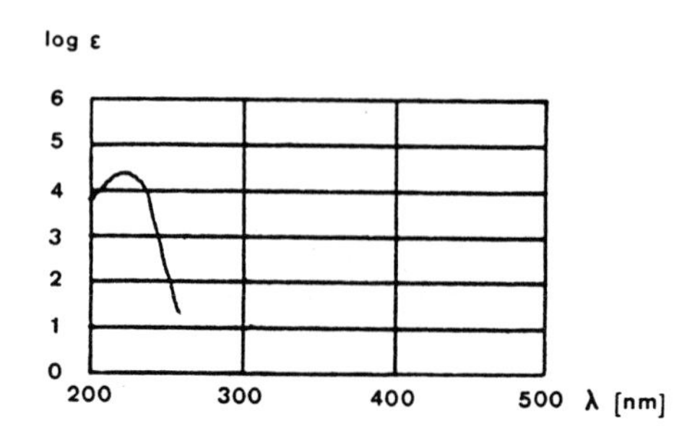

$CH_3(H)C=C(H)COCH_3$

(in ethanol)

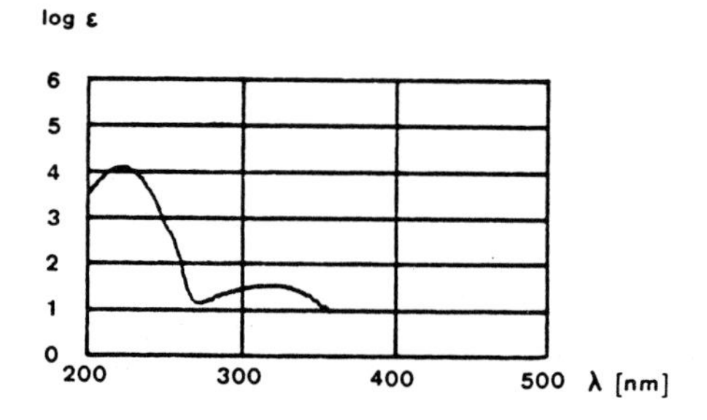

$CH_3(H)C=C(H)COOH$

(in water)

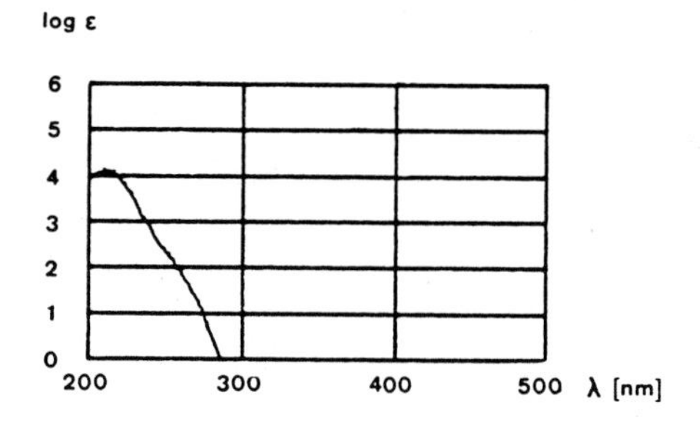

CH_3-NO_2

(in hexane)

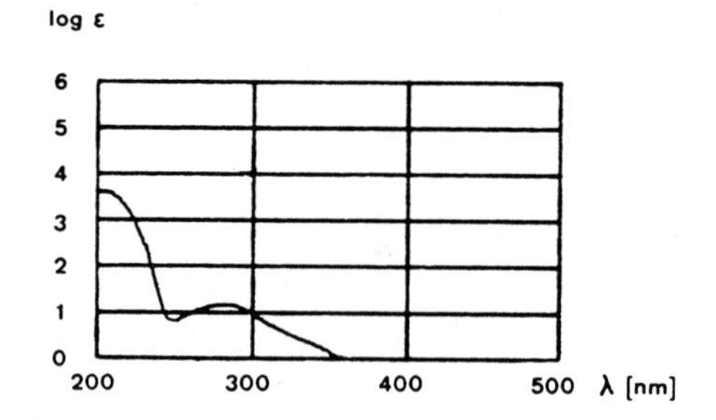

$CH_3CH_2COCH_3$

(in heptane)

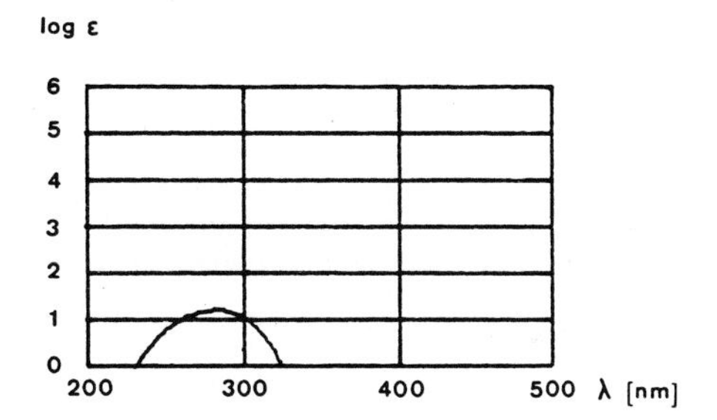

CH_3 / H / C=C / H / CHO

(in hexane)

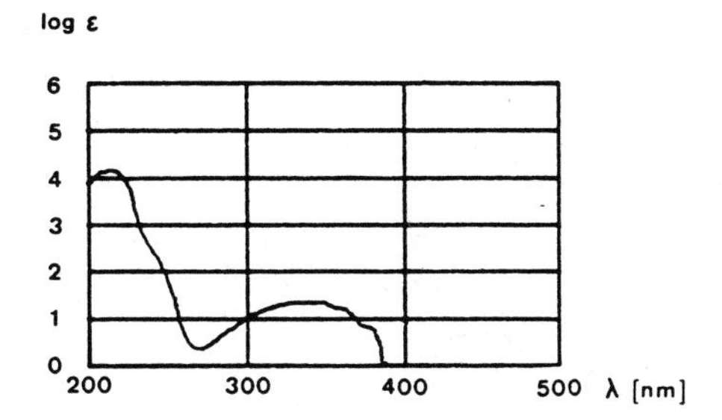

CH_3 / H / C=C / H / COO^-

(in water)

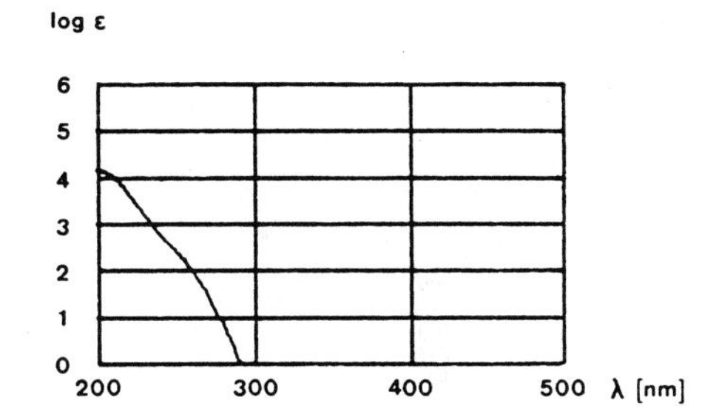

$(CH_3)_3C{-}O{-}NO$

(in hexane)

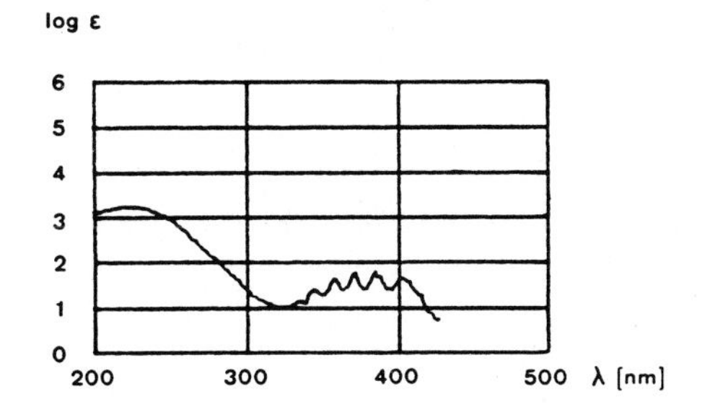

CH_3-CH_2-SH

(in heptane)

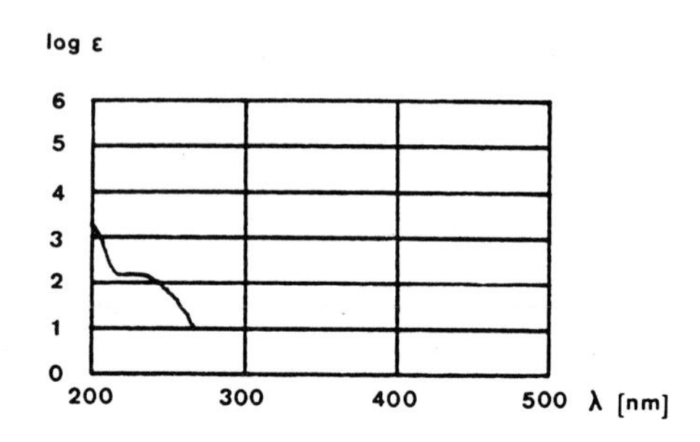

$CH_3-S-S-CH_3$

(in ethanol)

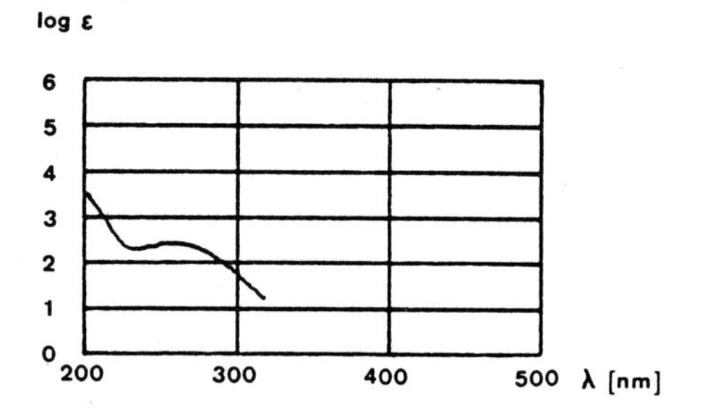

$(CH_3)_3C-Br$

(in heptane)

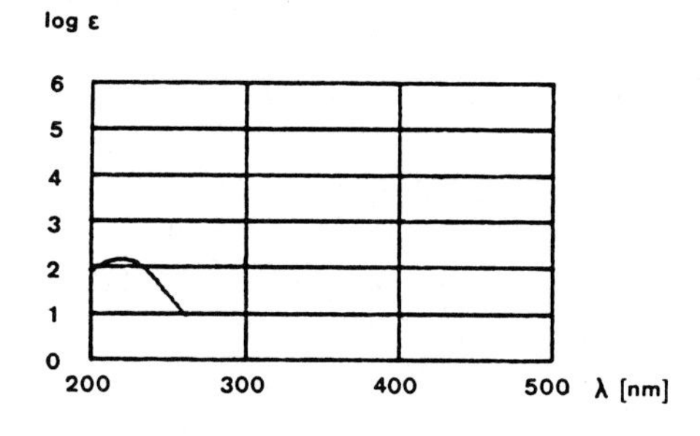

Cl

(in heptane)

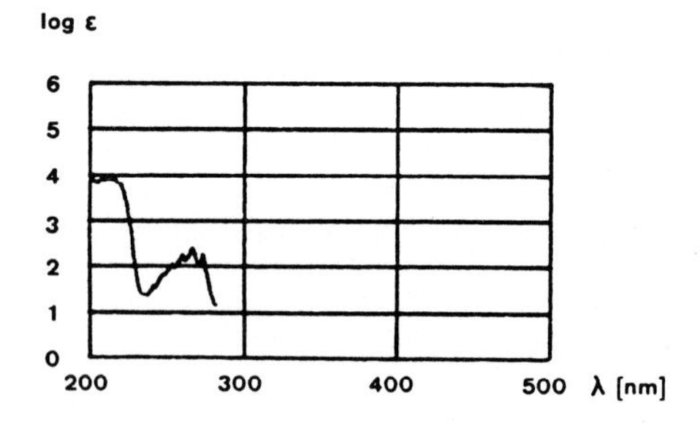

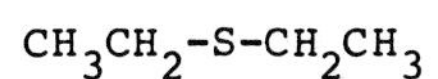

(in heptane)

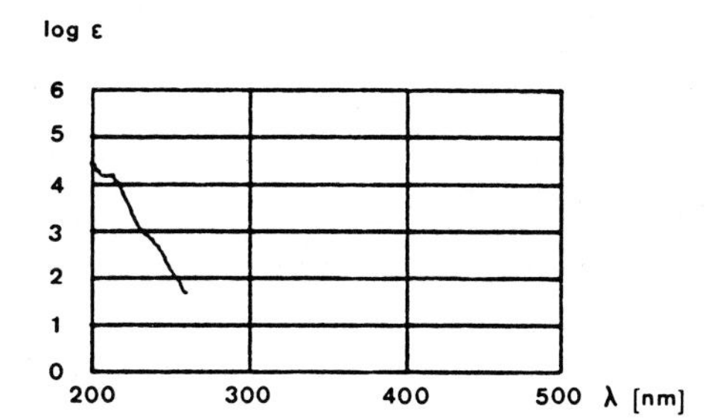

$(CH_3CH_2)_3N$

(in heptane)

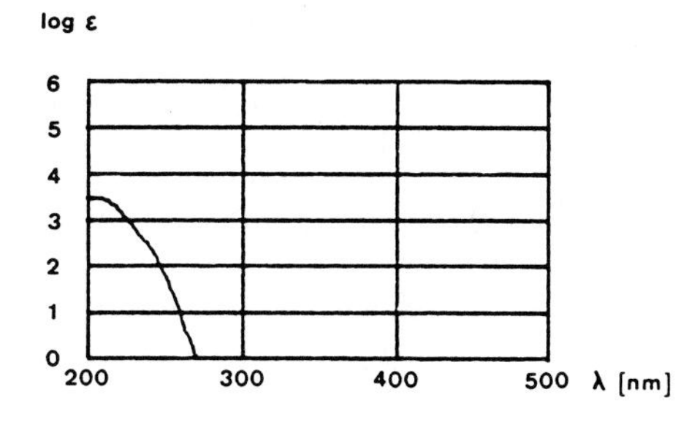

$(CH_3)_2CH-I$

(in heptane)

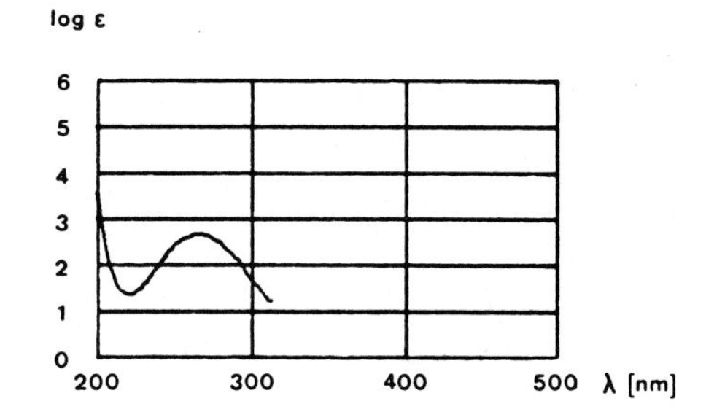

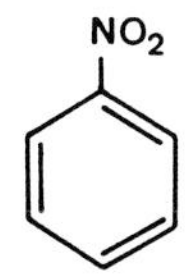

(in petroleum ether)

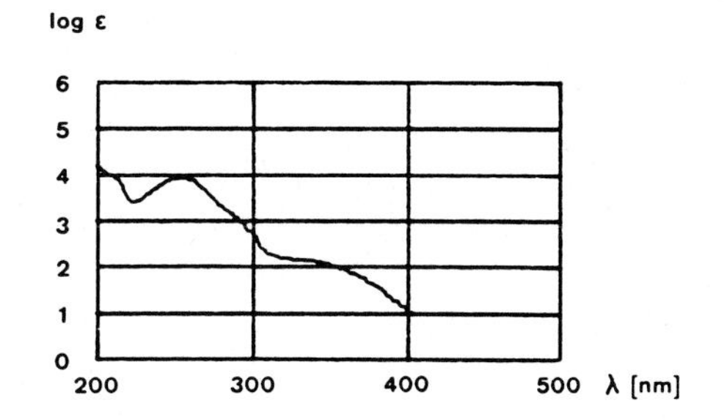

NH_2

(in water)

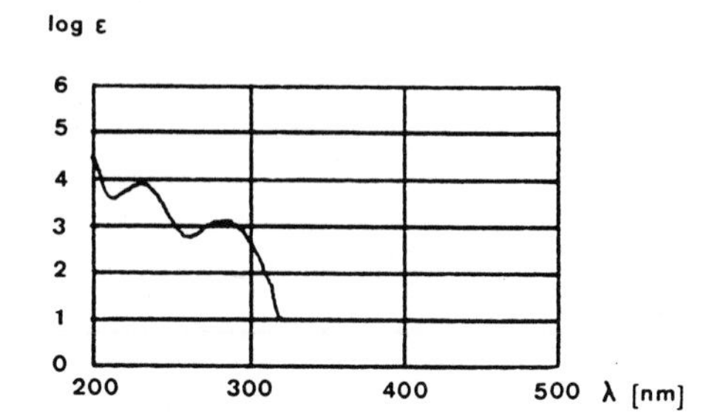

$N(CH_2CH_3)_2$

(in heptane)

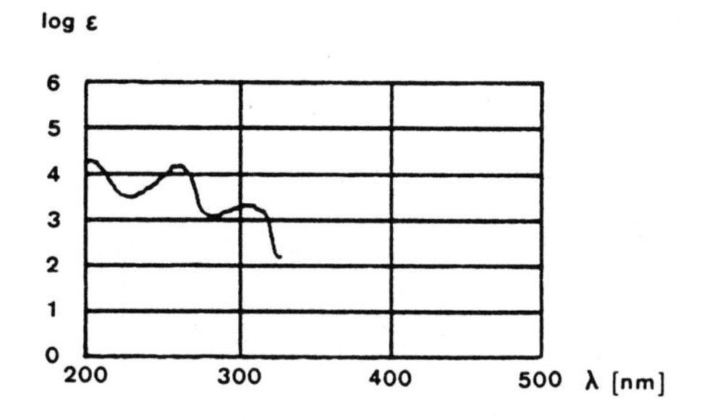

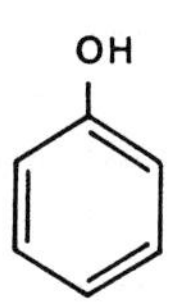

(in water)

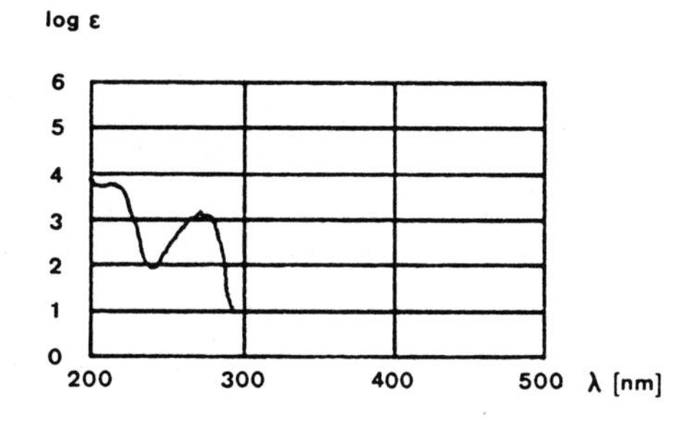

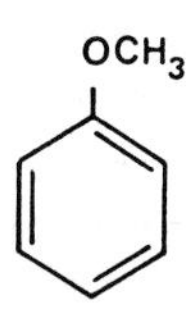

(in isooctane)

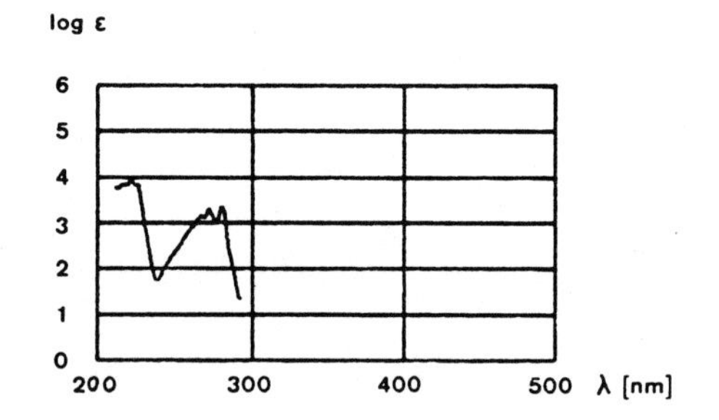

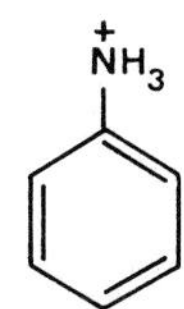

(in water)

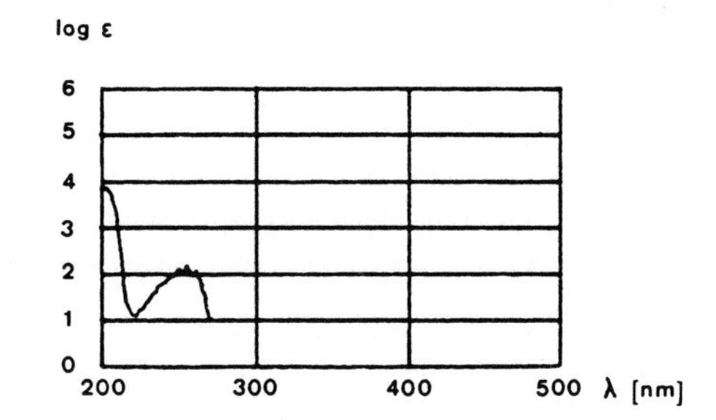

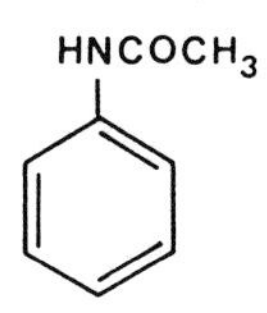

(in cyclohexane)

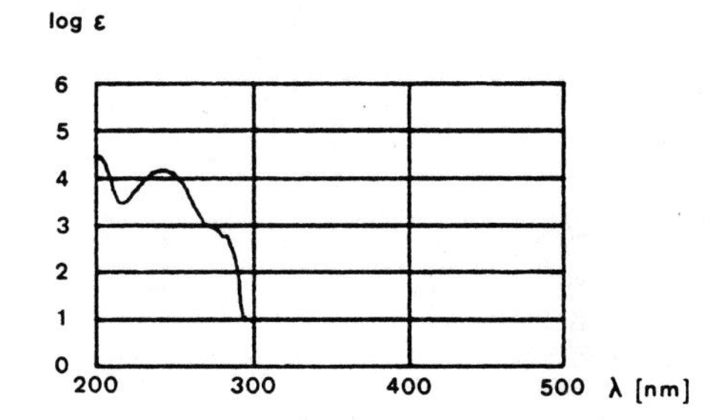

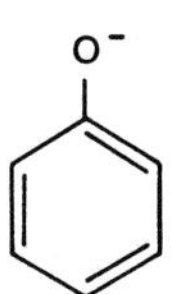

(in water)

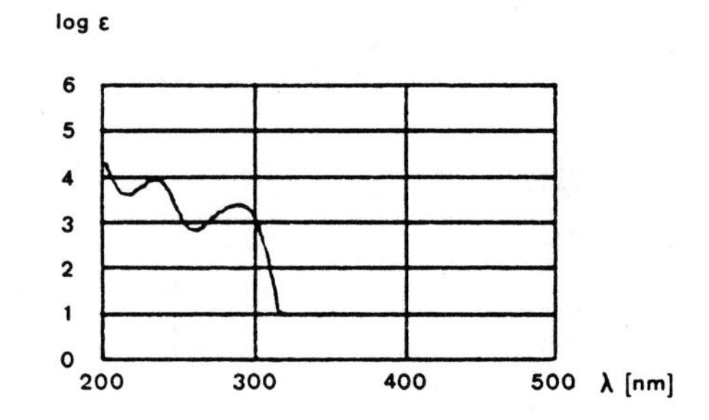

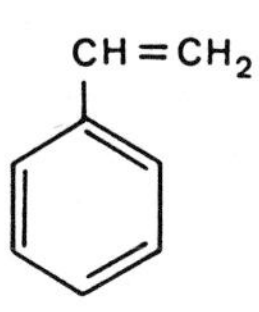

(in heptane)

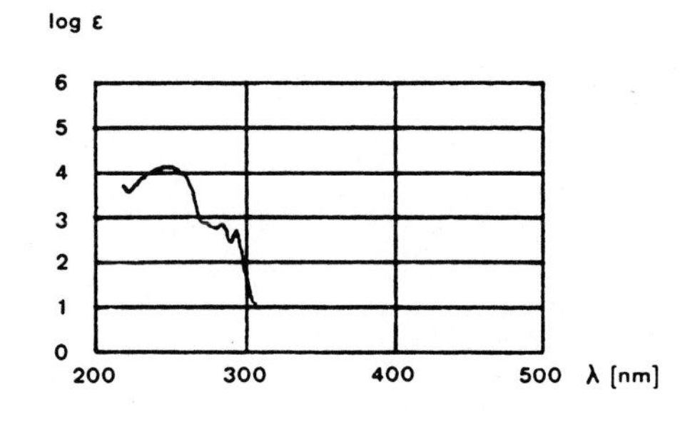

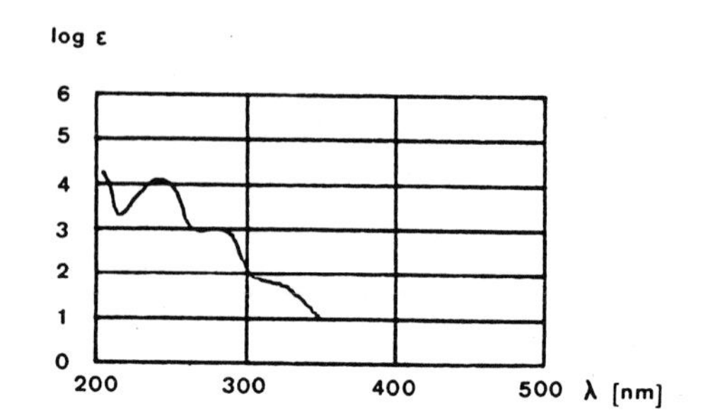

COCH$_3$

(in ethanol)

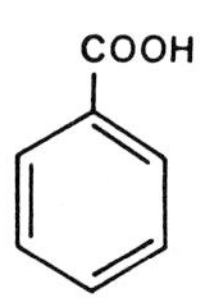

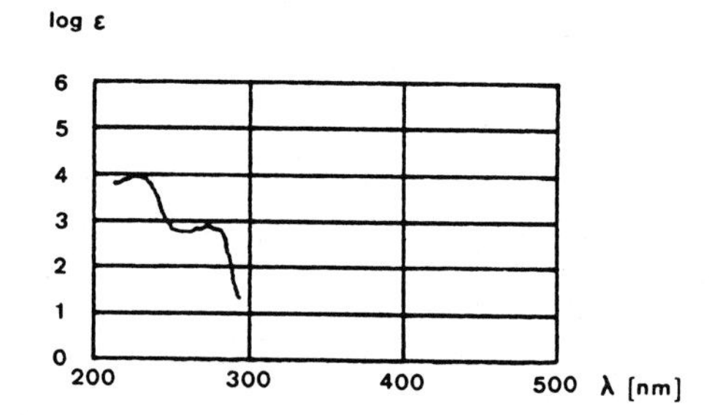

COOH

(in ethanol)

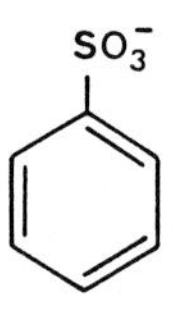

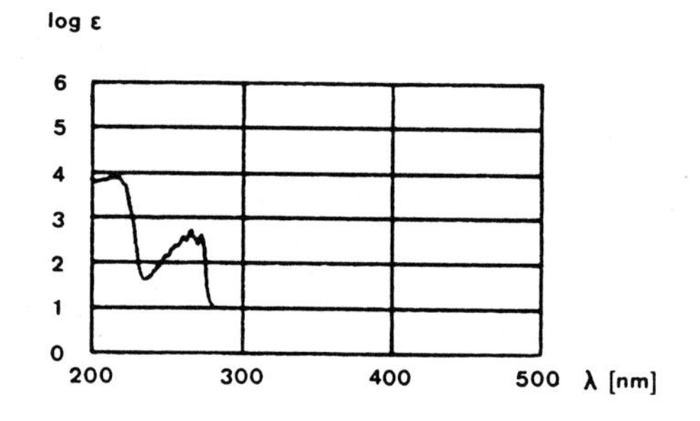

SO_3^-

(in water)

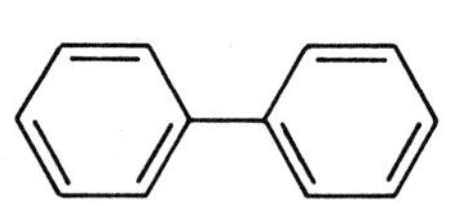

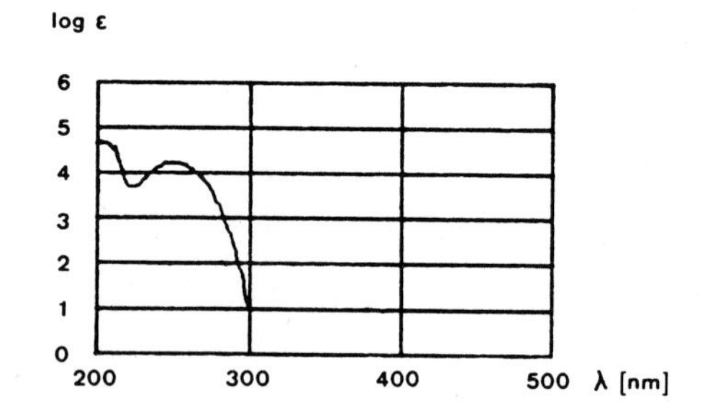

(in petroleum ether)

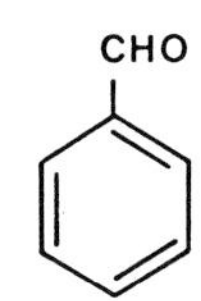

(in heptane)

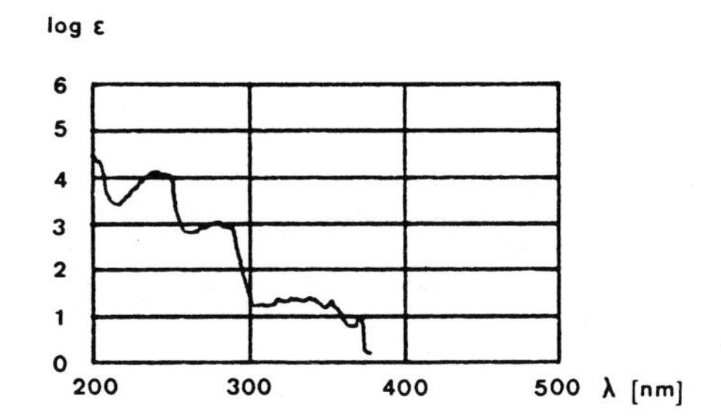

COO^-

(in water)

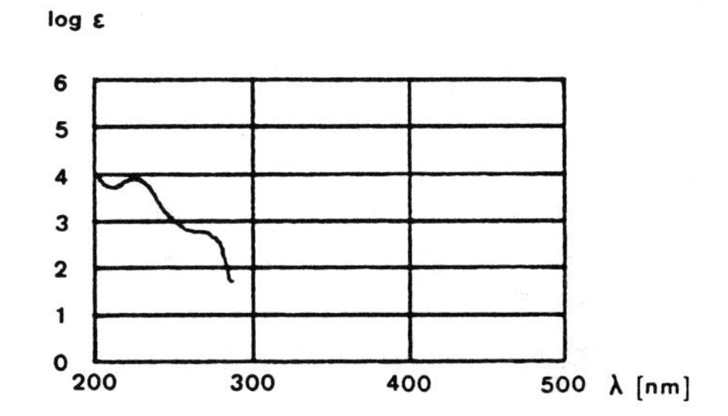

SCH_3

(in heptane)

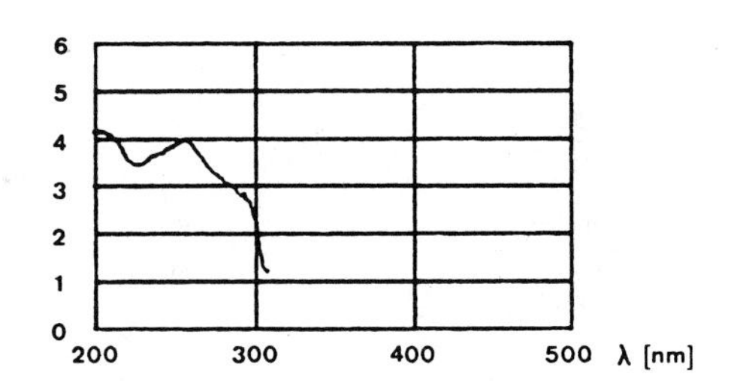

CH_3 CH_3

CH_3 CH_3

(in petroleum ether)

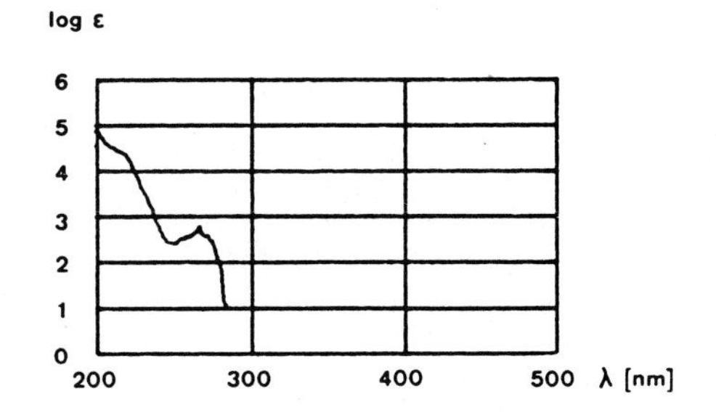

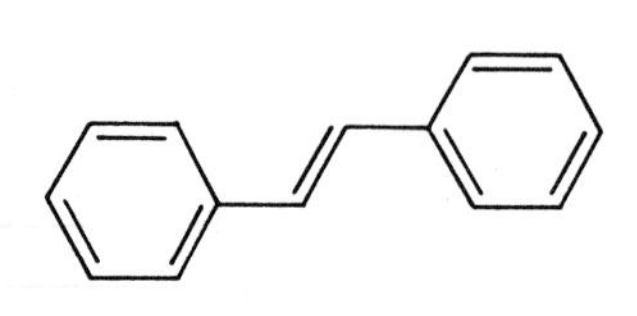

(in ethanol)

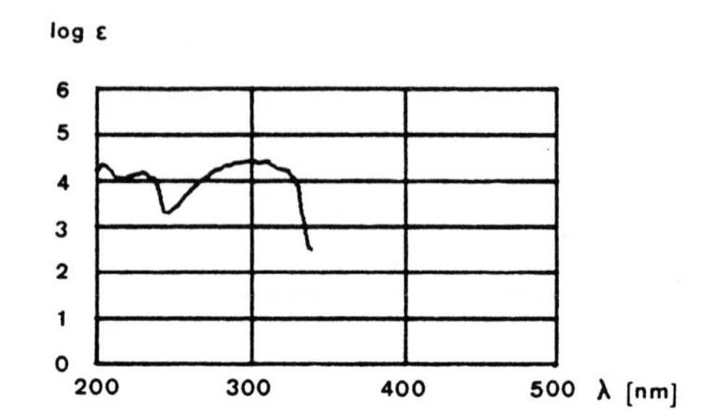

(in hexane)

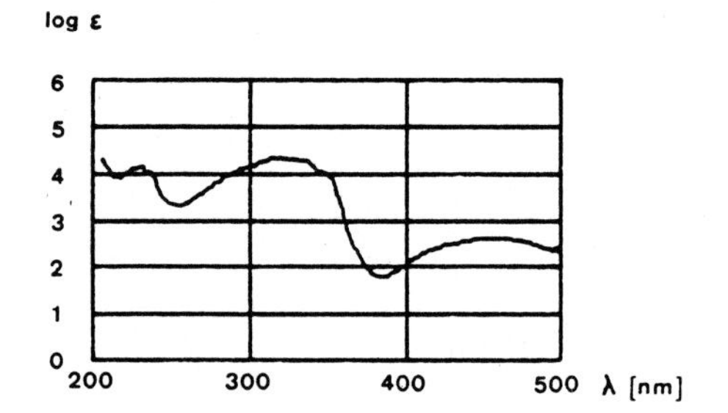

(in ethanol)

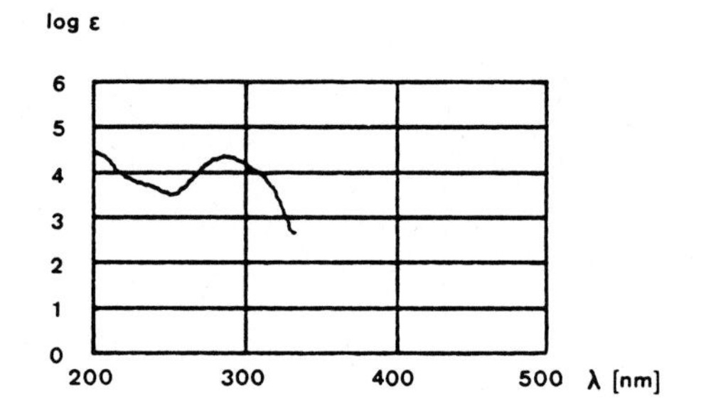

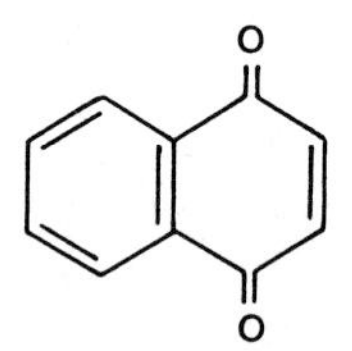

(in methanol)

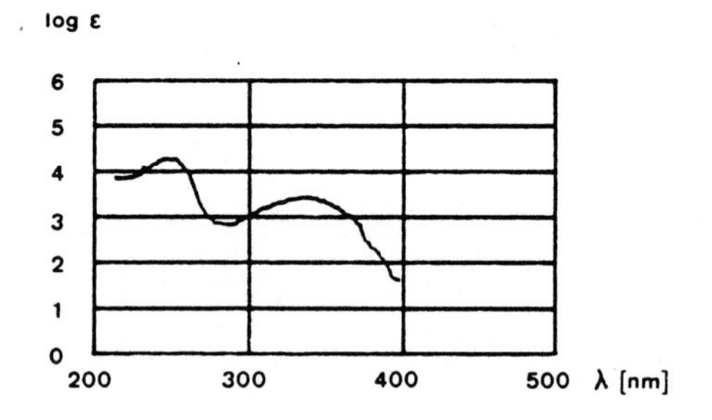

CO

(in ethanol)

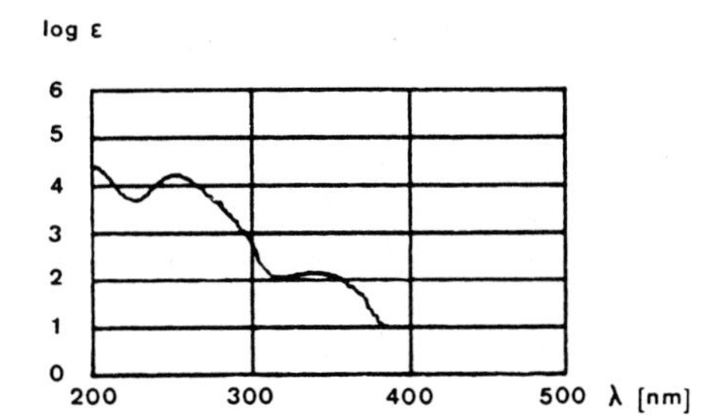

O

(in ethanol)

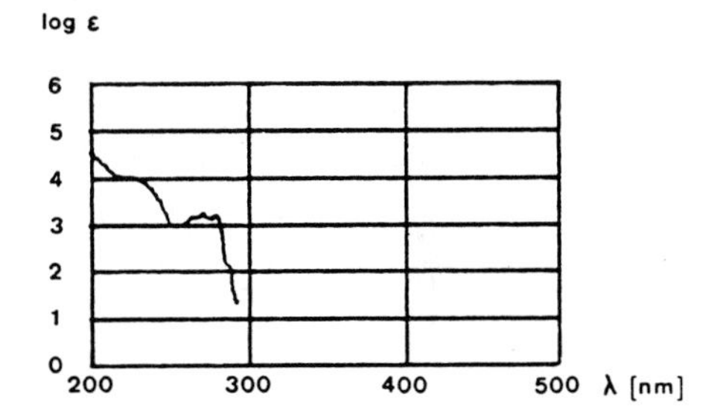

S

(in ethanol)

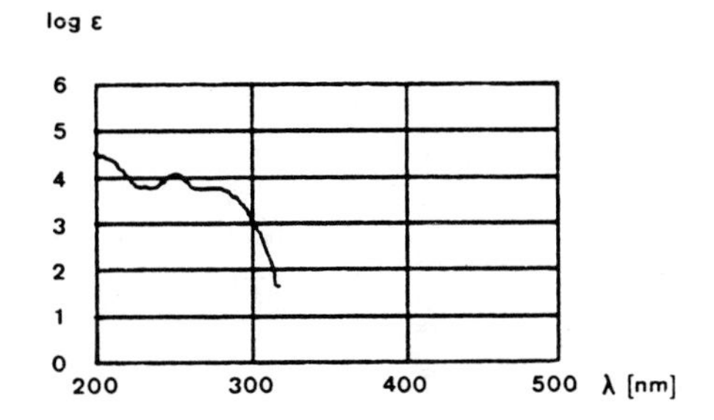

O
O

(in methanol)

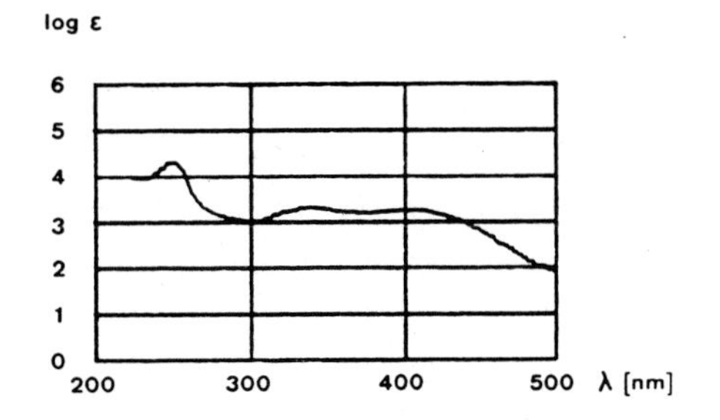

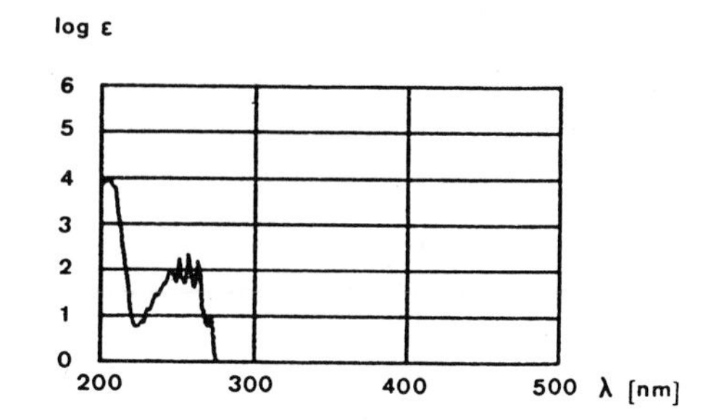

(in hexane)

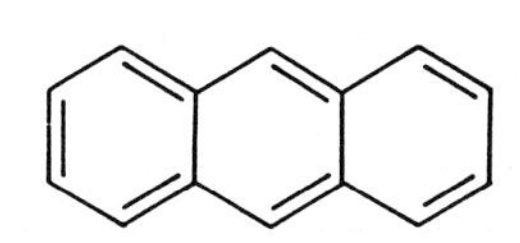

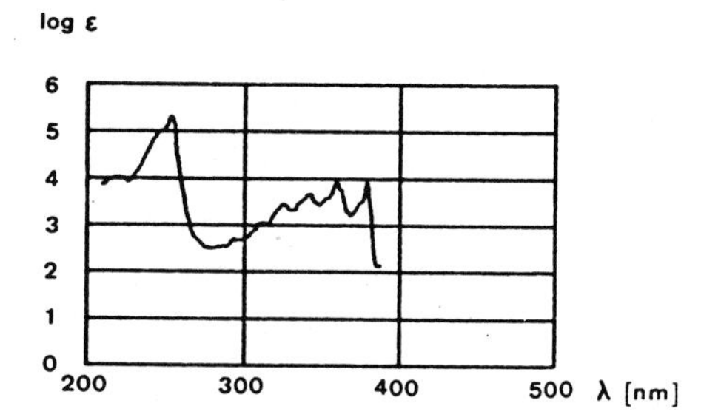

(in hexane)

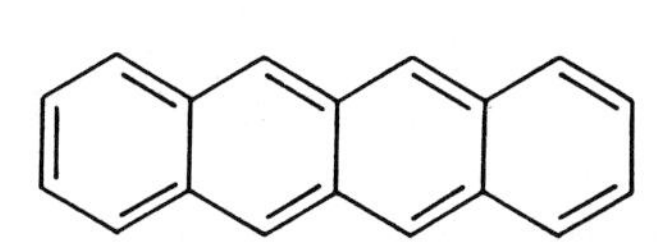

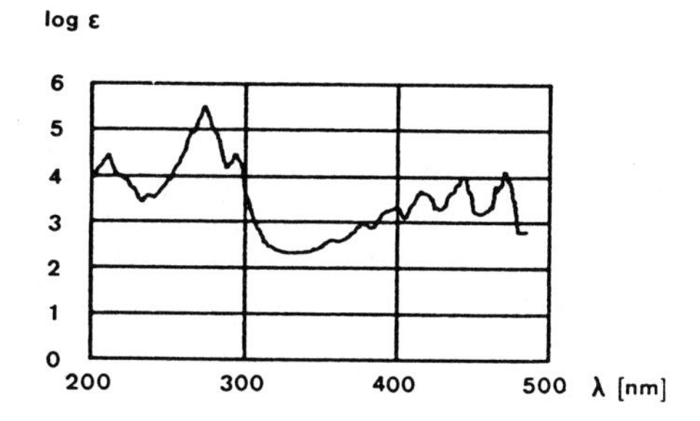

(in heptane)

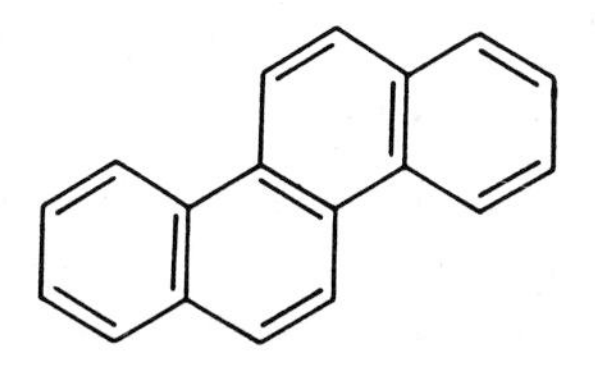

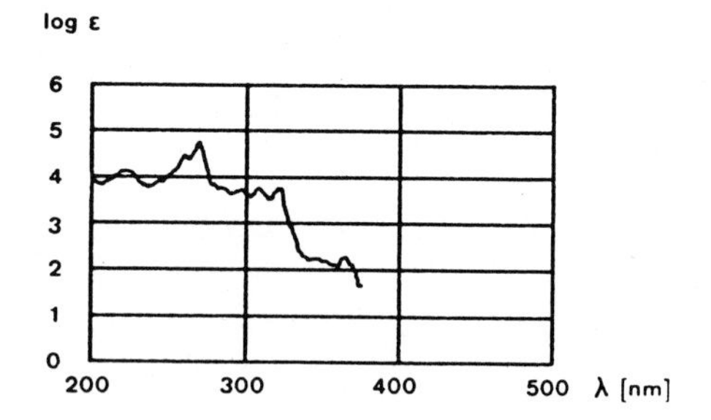

(in heptane)

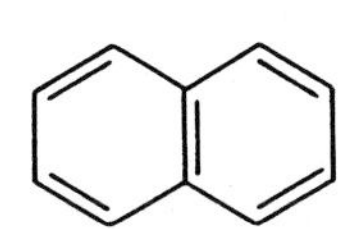

(in hexane)

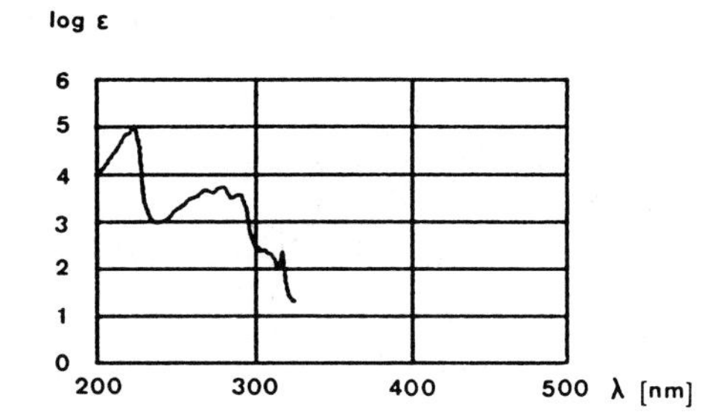

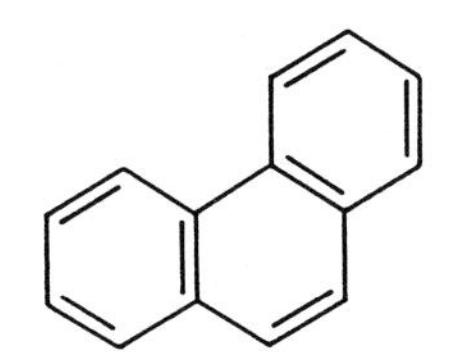

(in hexane)

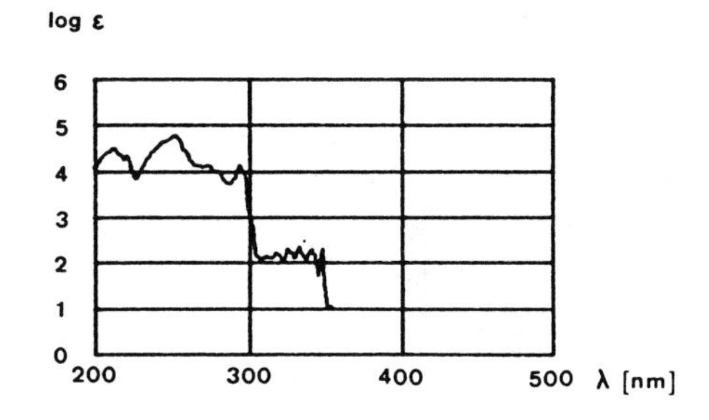

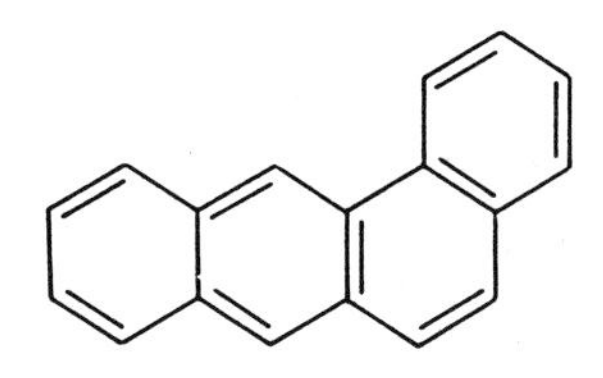

(in heptane)

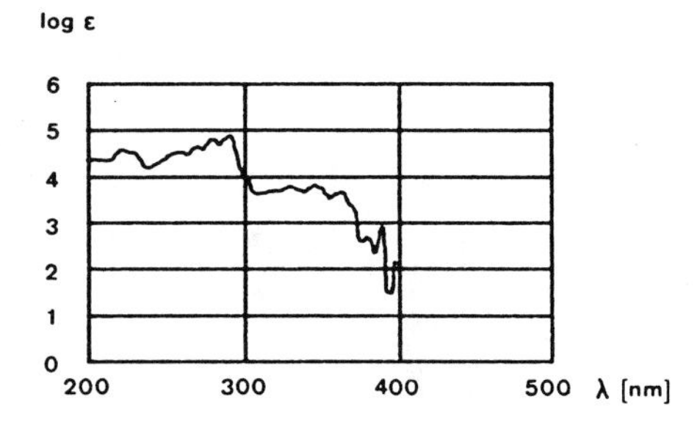

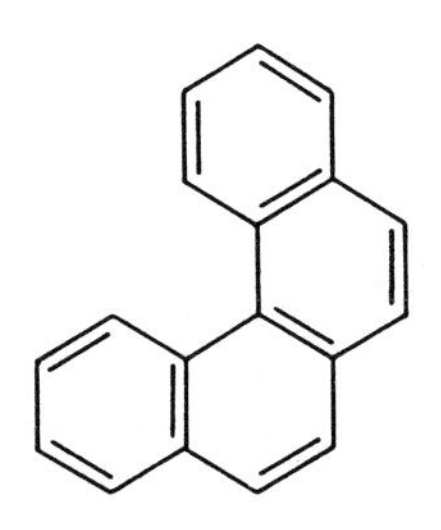

(in heptane)

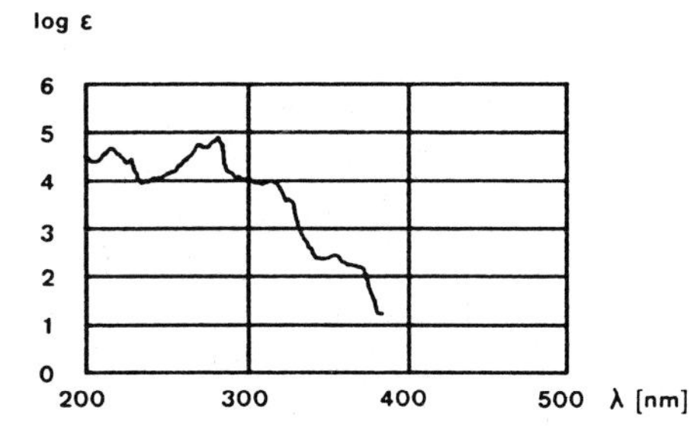

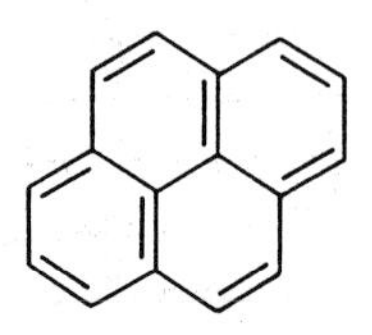

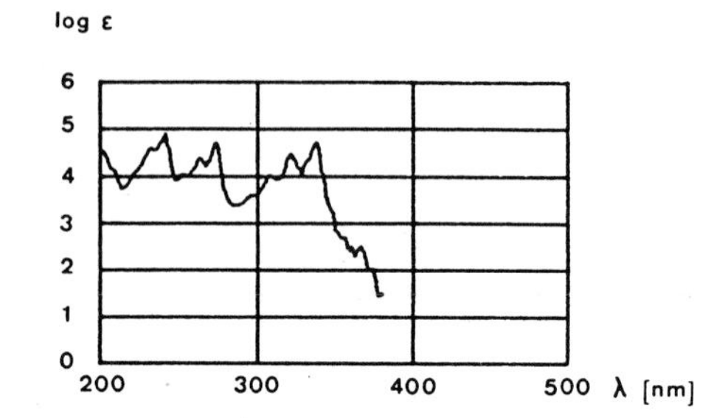

(in petroleum ether)

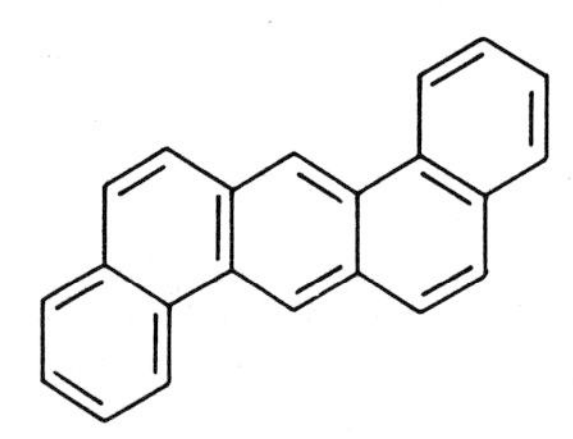

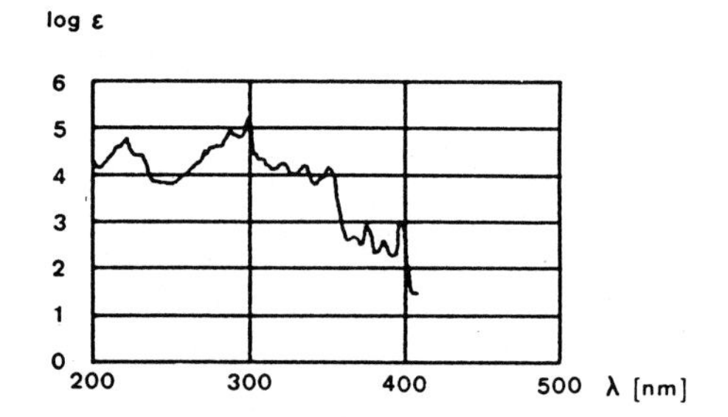

(in heptane)

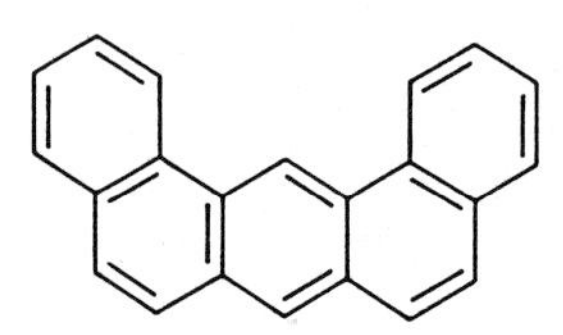

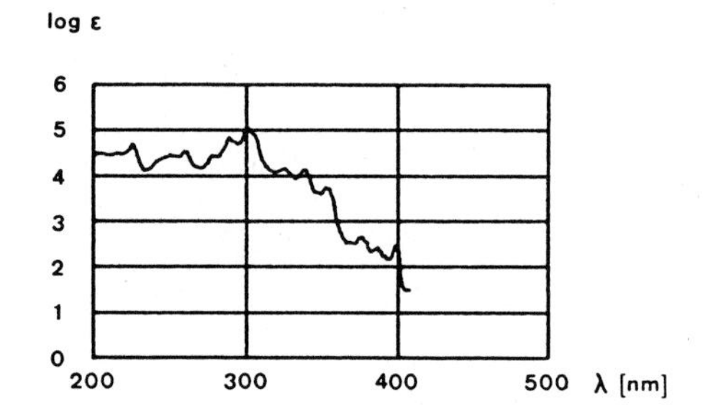

(in heptane)

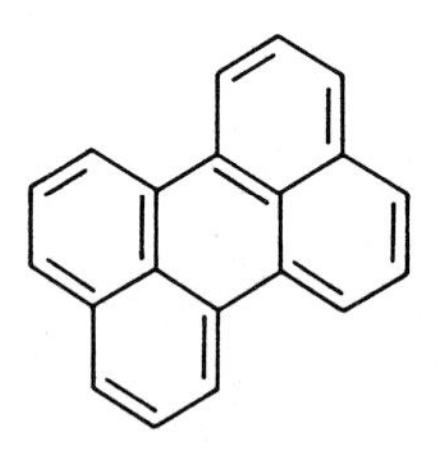

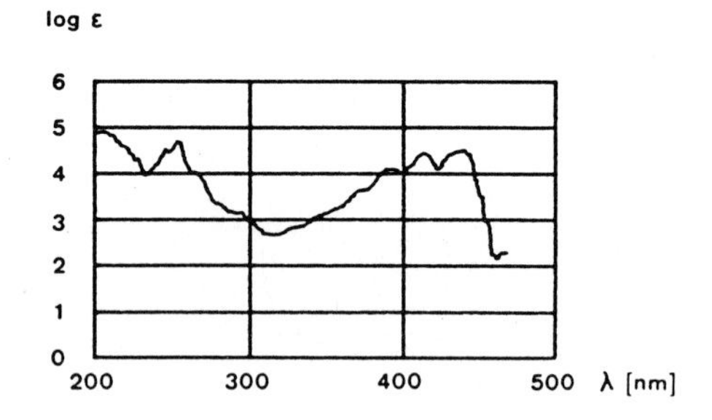

(in petroleum ether)

(in heptane)

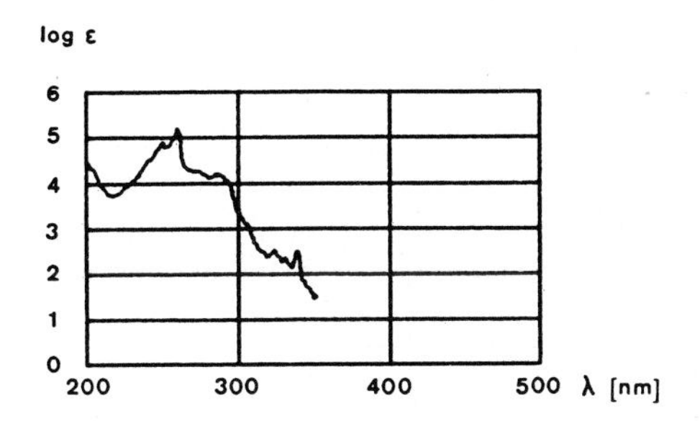

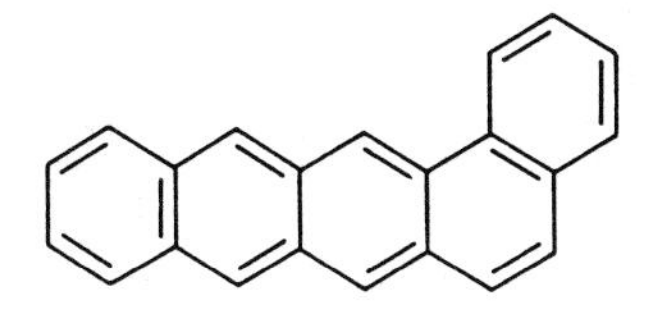

(in heptane)

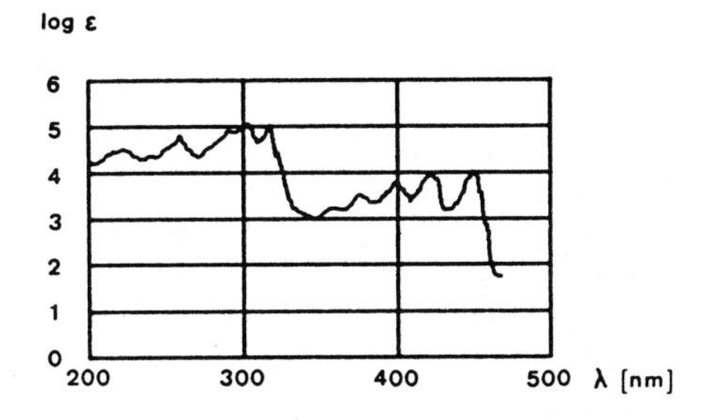

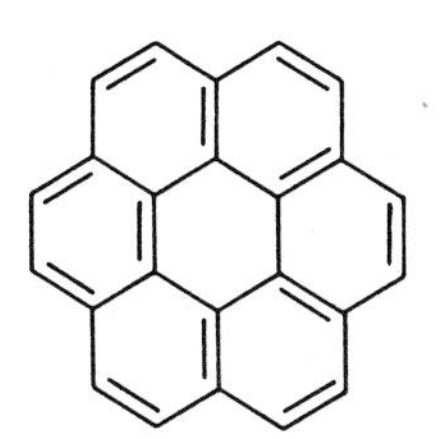

(in petroleum ether)

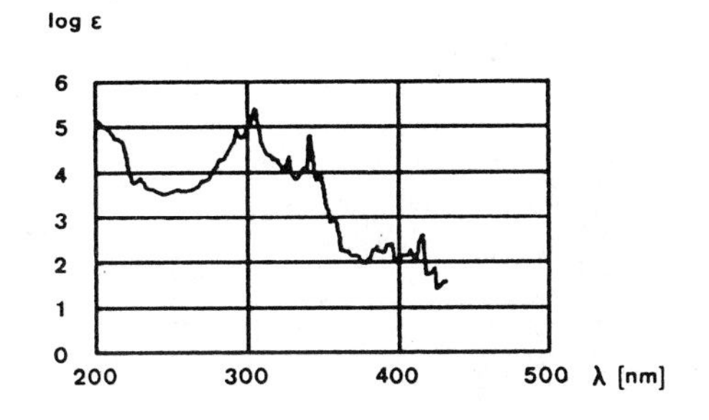

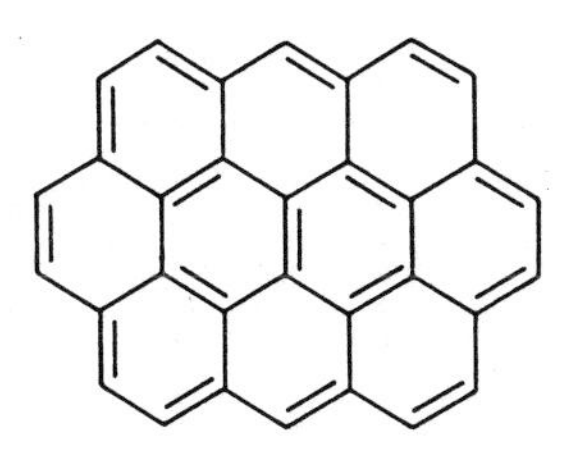

(in petroleum ether)

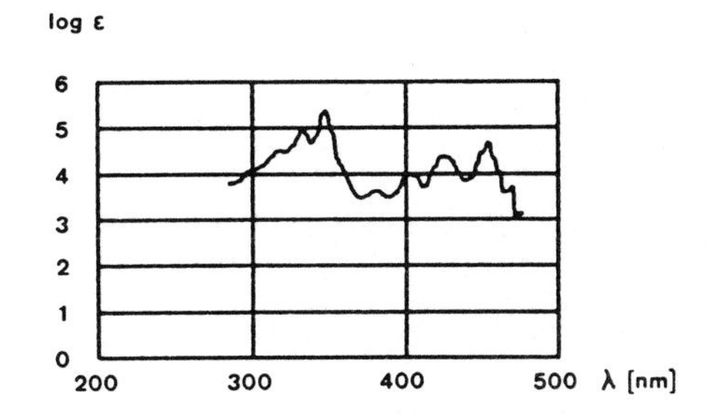

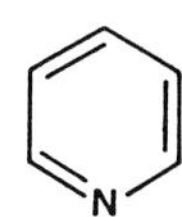

(in petroleum ether)

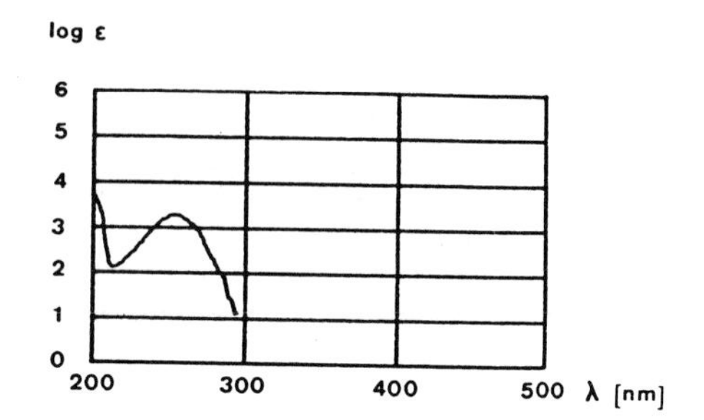

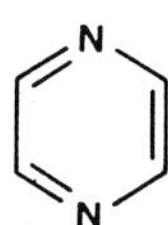

(in petroleum ether)

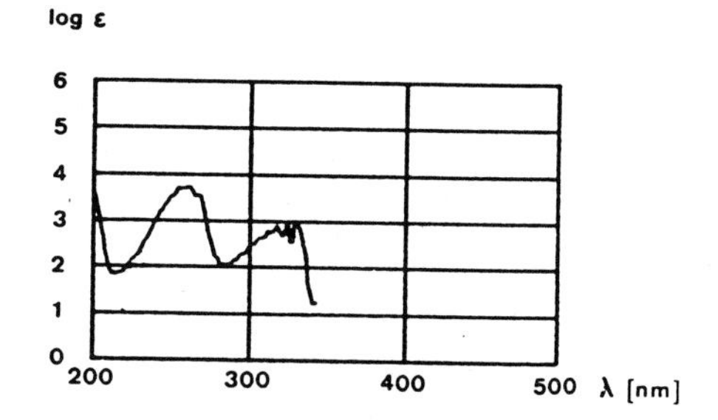

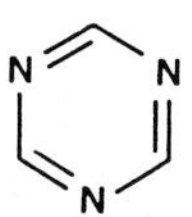

(in petroleum ether)

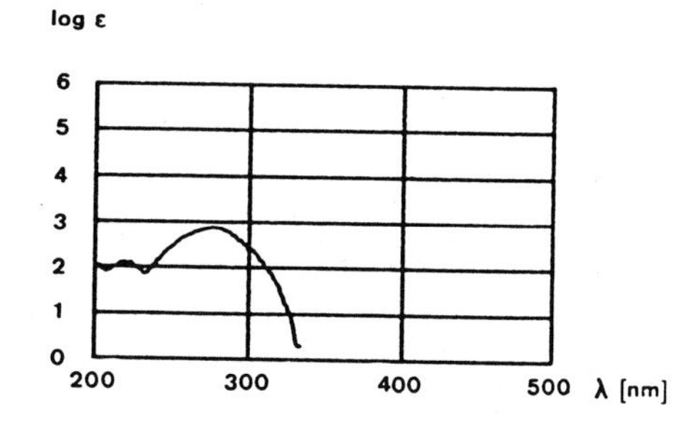

(in petroleum ether)

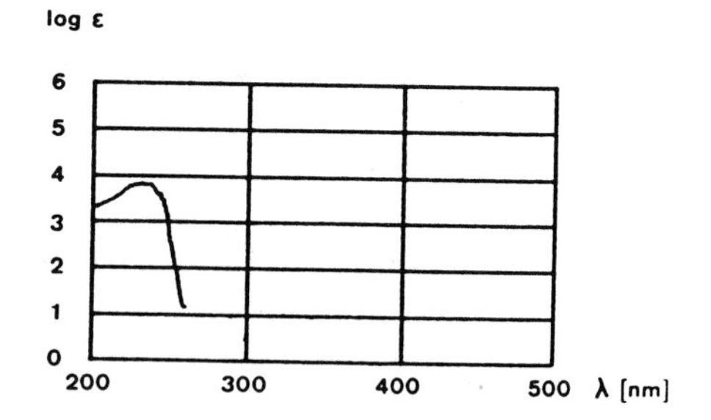

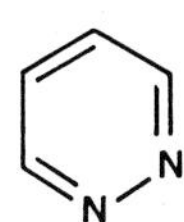

(in petroleum ether)

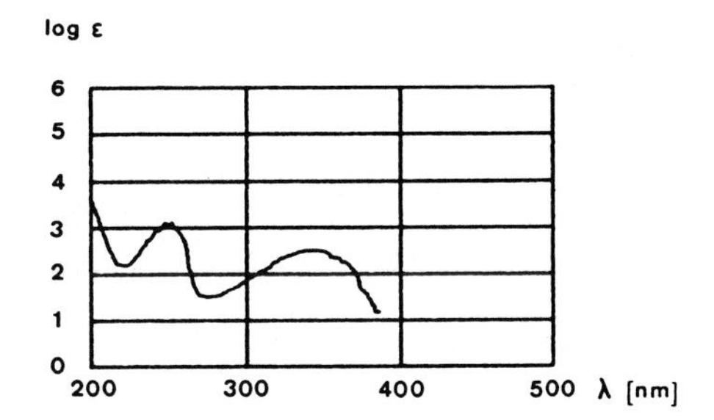

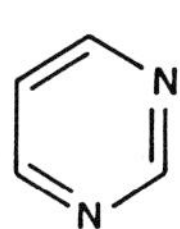

(in petroleum ether)

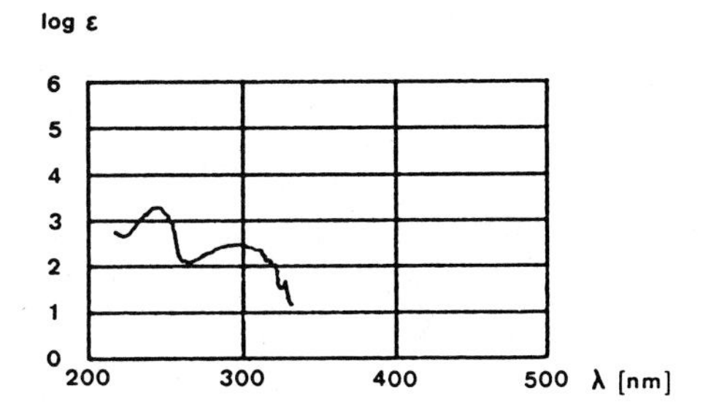

(in petroleum ether)

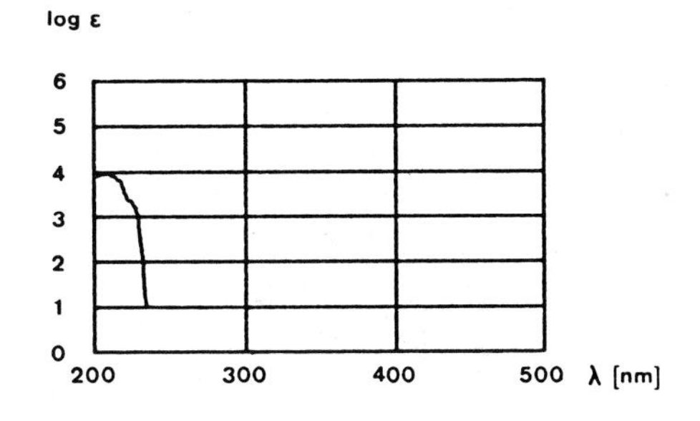

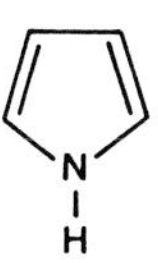

(in petroleum ether)

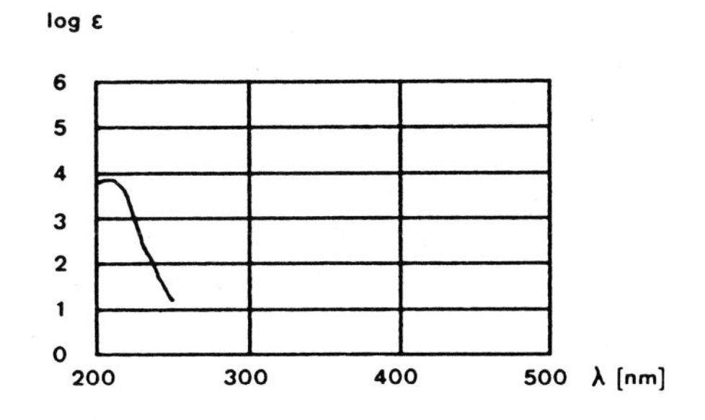

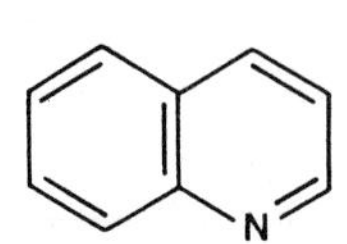

(in methanol)

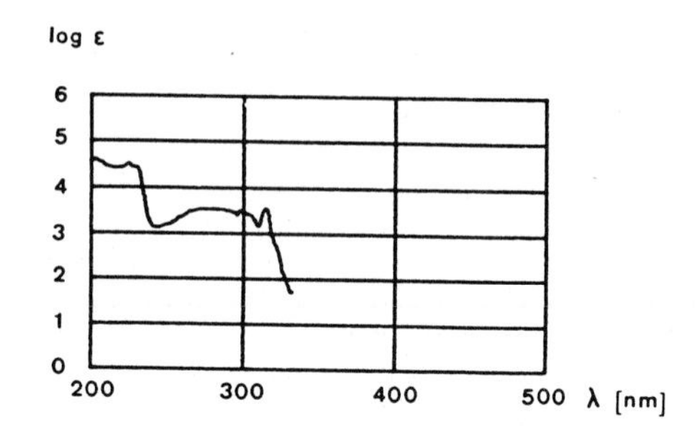

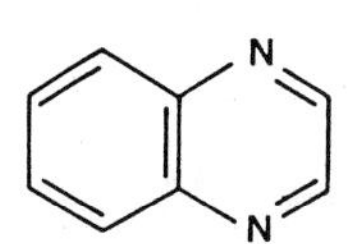

(in heptane)

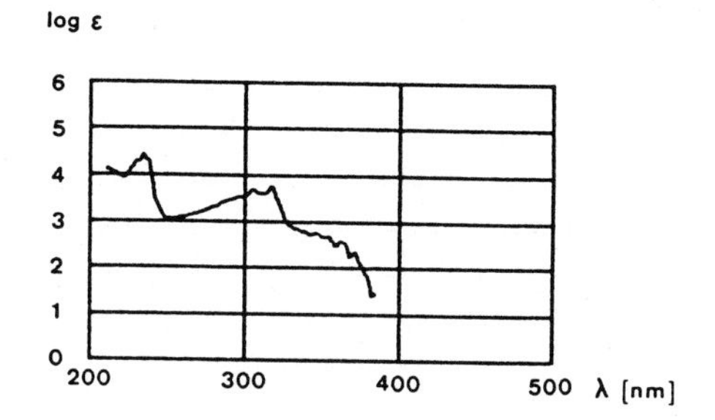

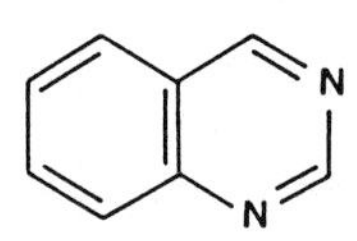

(in cyclohexane)

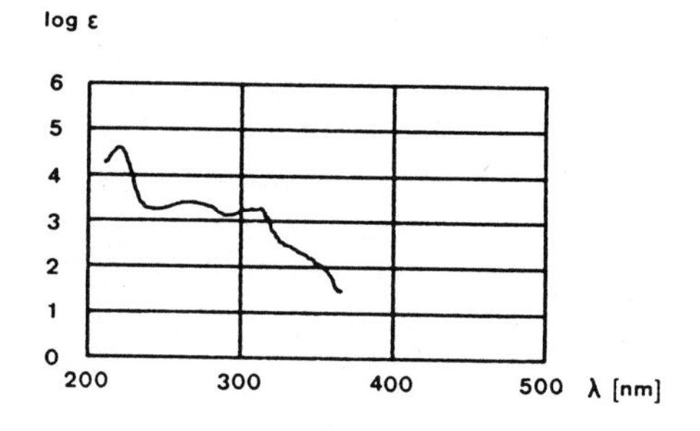

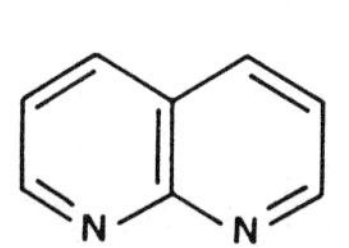

(in cyclohexane)

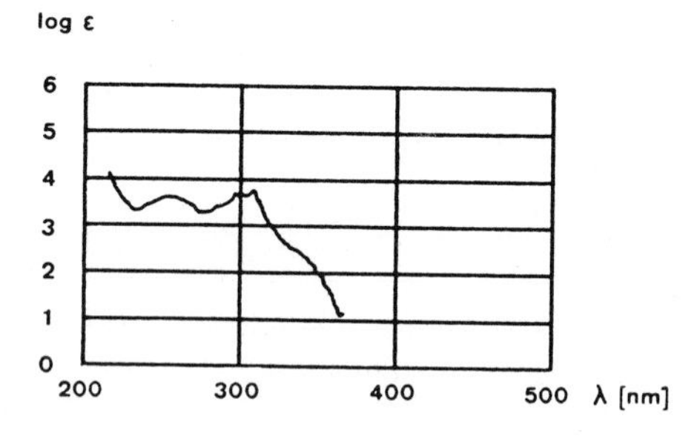

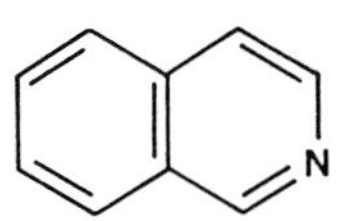

(in petroleum ether)

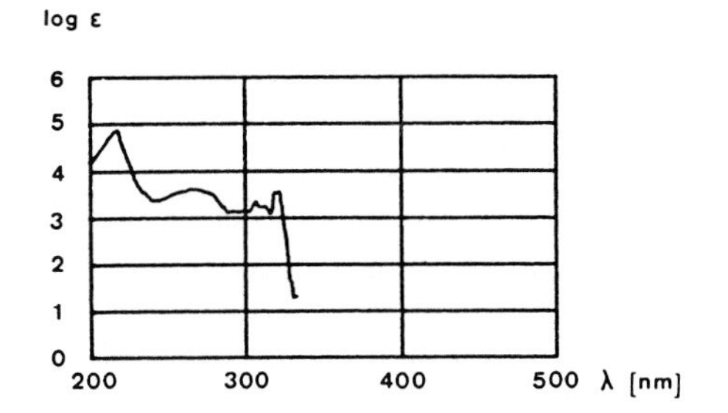

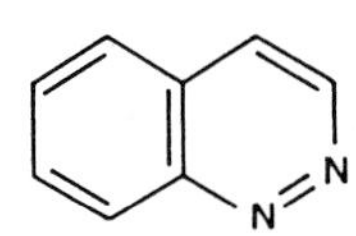

(in cyclohexane)

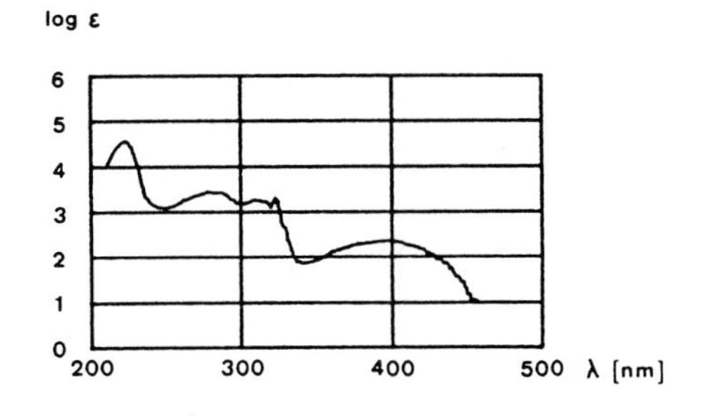

(in cyclohexane)

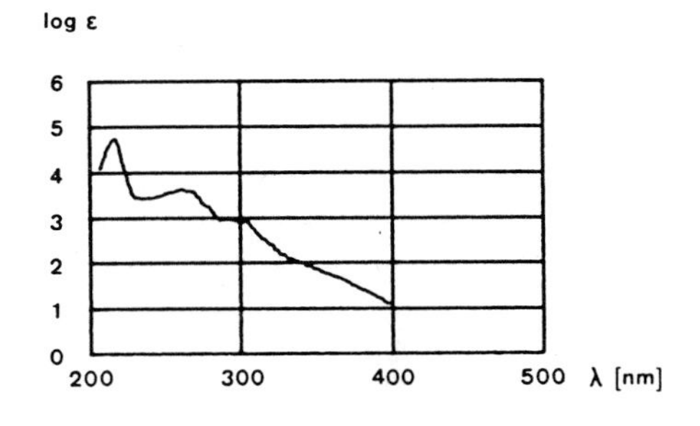

(in cyclohexane)

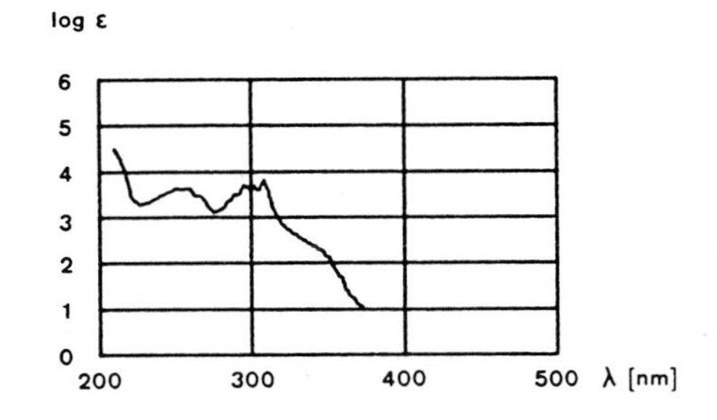

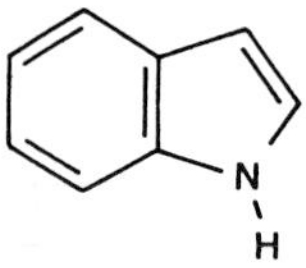

(in heptane)

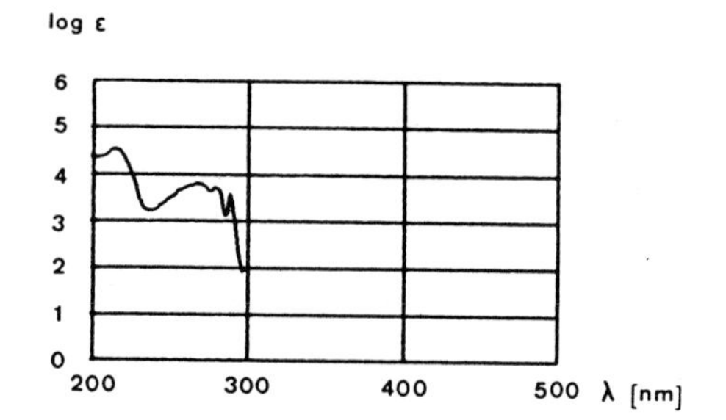

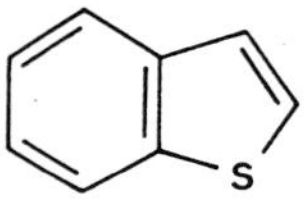

(in petroleum ether)

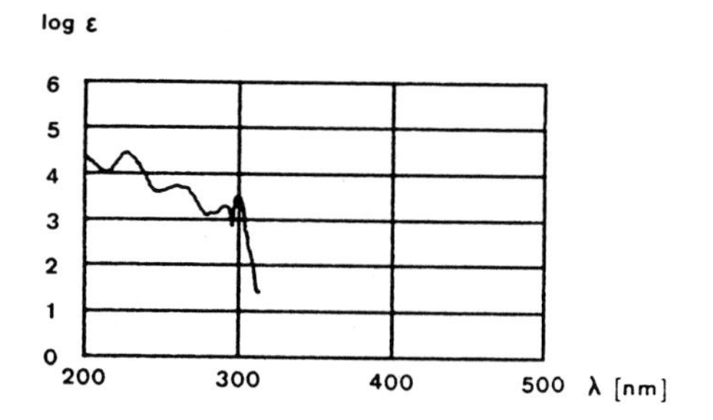

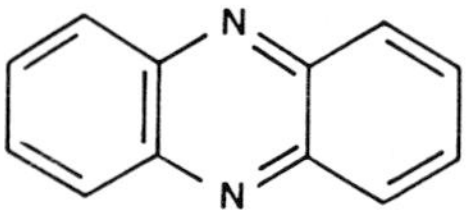

(in methanol)

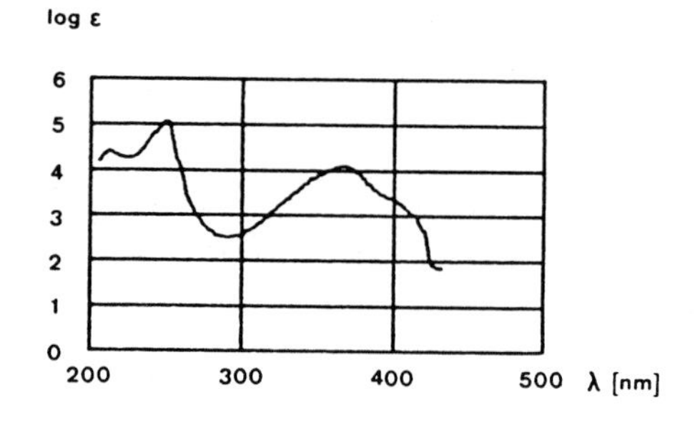

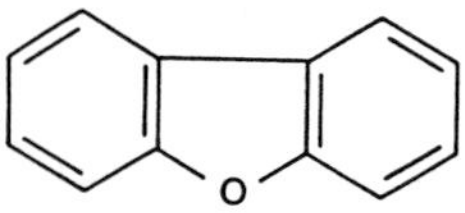

(in petroleum ether)

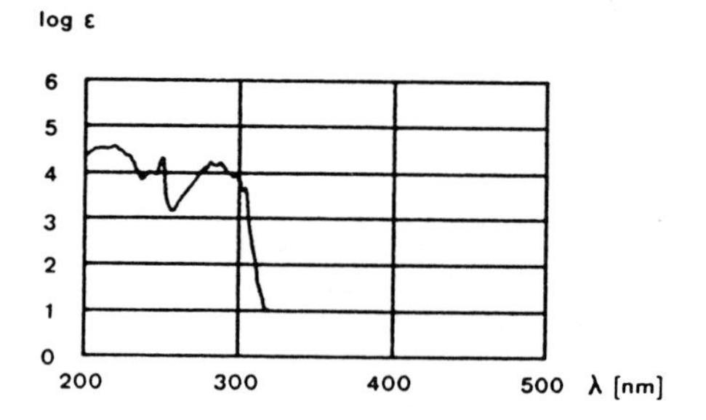

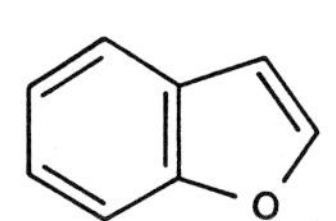

(in ethanol)

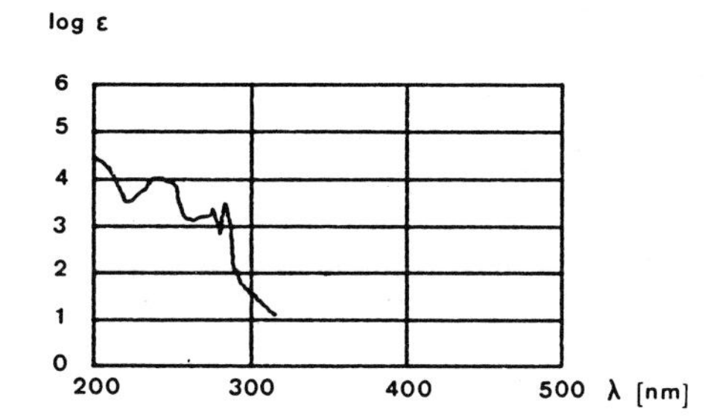

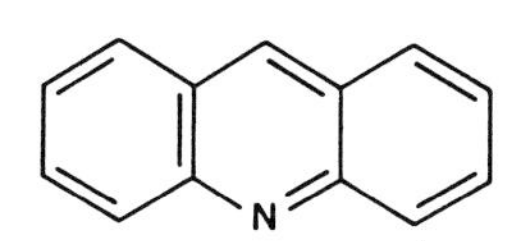

(in ethanol)

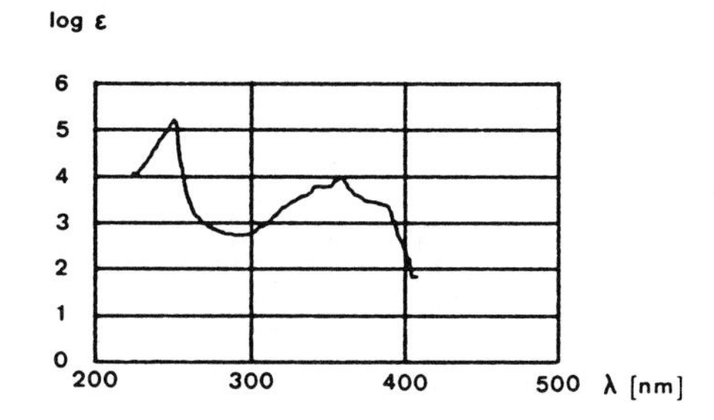

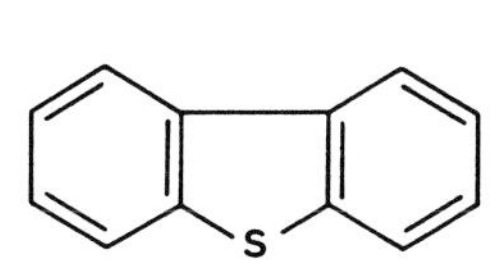

(in ethanol)

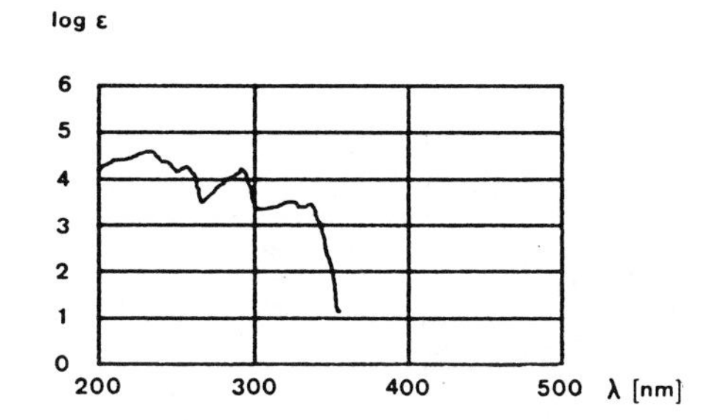

S

(in heptane)

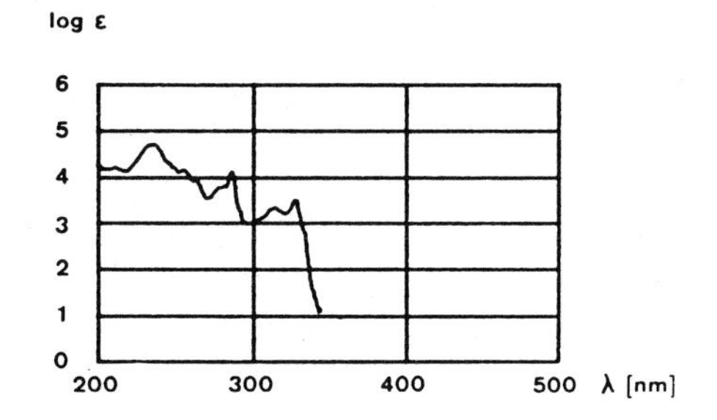

Calculation of the Number of Double Bond Equivalents (DBE) from the Molecular Formula

General Equation

$$\mathrm{DBE} = \frac{2 + \sum_i n_i (v_i - 2)}{2}$$

v_i: formal valency of element i
n_i: number of atoms of element i in molecular formula
i : identifier of the elements

Short Cut

For compounds containing only C, H, O, N, S and halogens, the following steps permit a quick and simple calculation of the double bond equivalent:

1. O and divalent S are deleted from the molecular formula
2. Halogens are replaced by hydrogen
3. Trivalent N is replaced by CH
4. The resulting hydrocarbon C_nH_x is compared with the saturated hydrocarbon C_nH_{2n+2}:

$$\mathrm{DBE} = \frac{(2n + 2) - x}{2}$$

Correction for Volume Susceptibility when Measuring Nuclear Magnetic Resonance Spectra Using a Cylindrical External Reference

When the reference material is not added directly (internal reference) to the sample solution but instead is enclosed in a capillary (external reference), the local fields differ for sample and reference because of the difference in volume susceptibility. This leads to a shift of the reference signal and, therefore, a displacement of the zero point of the x-axis scale. This displacement can be corrected for cylindrical sample tubes as follows:

For permanent and electromagnets where the magnetic field is perpendicular to the sample probe:

$$\delta_{int} = \delta^{\perp}_{observed} - \frac{2\pi}{3}(\chi_{ref} - \chi_{sample})$$

For superconducting magnets the magnetic field is oriented coaxially to the probe. In this case the following relationship holds:

$$\delta_{int} = \delta^{\parallel}_{observed} + \frac{4\pi}{3}(\chi_{ref} - \chi_{sample})$$

The volume susceptibility χ for some practically important solvents are listed in the table on p. V3.

The following holds for D_2O as the solvent and tetramethylsilane in a capillary as an external reference:

$$\delta_{int} = \delta^{\perp}_{observed} - 0.34 \text{ ppm (permanent and electro magnets)}$$

$$\delta_{int} = \delta^{\parallel}_{observed} + 0.68 \text{ ppm (superconducting magnets)}$$

The chemical shifts relative to an internal reference can be calculated without the knowledge of the volume susceptibility if the spectrum is measured with both a superconducting magnet and an electromagnet:

$$\delta_{int} = \frac{1}{3}(\delta^{\parallel}_{observed} + 2\delta^{\perp}_{observed})$$

Volume Susceptibility χ of Selected Solvents

Solvent	$\chi\cdot10^6$	Ref.
Acetone	-0.460	a
Acetonitrile	-0.534	a
Formic acid	-0.527	a
Benzene	-0.611	a
Bromoform	-0.948	a
Chlorobenzene	-0.688	a
Chloroform	-0.740	a
Cyclohexane	-0.627	a
Cyclopentane	-0.629	a
Diethylether	-0.534	a
1,1-Dichloroethylene	-0.635	a
o-Dichlorobenzene	-0.748	a
Dichlorodifluormethane	-0.642	a
Dimethylformamide	-0.503	b
Dimethylsulfoxide	-0.609	b
Dioxane	-0.606	a
Acetic acid	-0.551	a
Methyl acetate	-0.537	a
Ethanol	-0.575	a
Fluorotrichloromethane	-0.638	a
Formamide	-0.551	a
n-Hexane	-0.565	a
Methanol	-0.530	a
Nitromethane	-0.391	a
n-Octane	-0.595	a
Pyridine	-0.611	a
Carbon disulfide	-0.699	a
Sulfuric acid	-0.808	b
Tetrachloroethylene	-0.802	a
Carbon tetrachloride	-0.691	a
Tetramethylsilane	-0.543	b
Toluene	-0.618	a
Water, H_2O	-0.719	a
Water, D_2O	-0.705	b

[a] Handbook of Chemistry and Physics, 63. ed., published by R.C. Weast, CRC Press, Cleveland (Ohio) 1982, p. E-123.

[b] H. Suhr, Anwendungen der kernmagnetischen Resonanz in der organischen Chemie, Springer-Verlag Berlin, Heidelberg, New York 1965, p. 25.

Effect of Symmetry and Rapid Conformational Changes of Molecules on Nuclear Magnetic Resonance Spectra

1. Nomenclature for spin systems

A spin system consists of a closed set of nuclei which are coupled to each other. The definitions listed below should be used to characterize such systems and to predict the features of the spectrum expected. It should be noted that various definitions can be found in the literature (e.g., isochronicity), but the ones presented here are particularly useful for the routine interpretation of spectra:

Isochronicity

- Nuclei are isochronous, if they do not exhibit any measurable difference in their chemical shifts under the experimental conditions employed. Isochronous nuclei may become nonisochronous upon change of the experimental conditions.

Magnetic Equivalence

- Nuclei are magnetically equivalent under the given experimental conditions if they are isochronous and if their individual coupling constant with each one of the other nuclei, which are nonisochronous with them, is the same.

  Isochronous and accidentally isochronous nuclei can become magnetically equivalent by chance, if their individual coupling constants with each one of the other nuclei, which are nonisochronous with them, is the same.

Characterization of spin systems

In accordance with these definitions spin systems are characterized as follows:

- A set of magnetically equivalent nuclei is denoted by a capital letter and a subscript indicating the number of magnetically equivalent nuclei within the group. If a set of isochronous nuclei contains sub-sets of magnetically equivalent nuclei, these are denoted by primes. Nonisochronous nuclei are denoted by letters adjacent in the alphabet if they are strongly coupled such that higher order spectra result. If the chemical shift differences are large, i.e., the spins are weakly coupled (Δ [Hz] > ca. 10J [Hz]) the nuclei are indicated by letters far apart in the alphabet.

2. Relationship between identical atomic nuclei within a molecule in the broadest sense (long lifetime, at the NMR time scale, of the species under discussion)

- Nuclei are constitutionally equivalent if they have the same connectivity (bondedness).
- Nuclei are diastereotopic, if they are constitutionally equivalent but not equivalent by symmetry.
- Nuclei are enantiotopic (equivalent by reflection), if they can only be interchanged into each other by an improper rotation. Note that symmetry plane and inversion center are special cases of improper axes.
- Nuclei are homotopic (equivalent by rotation), if they can be interchanged into each other by a proper rotation.

The hierarchy of these properties may be illustrated by the following scheme:

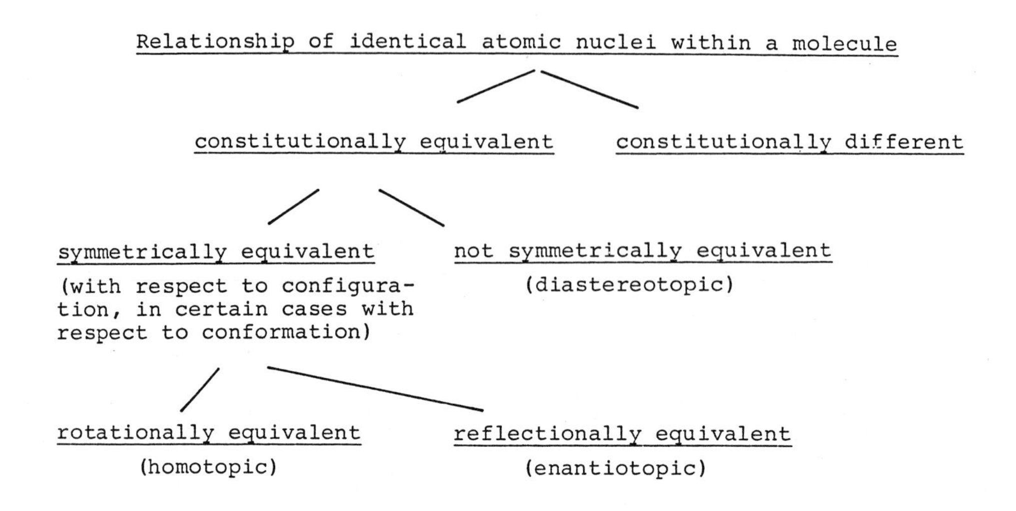

3. Relationship between the topical nature, isochronicity and the magnetic equivalence in rigid systems

Isochronicity

- Constitutionally different nuclei are at best accidentally isochronous.
- Homotopic nuclei are always isochronous.
- Enantiotopic nuclei are always isochronous in a non-chiral environment. In a chiral environment they are in principle nonisochronous (they may, however, be accidentally isochronous).
- Diastereotopic nuclei are in principle nonisochronous (they may, however, be accidentally isochronous).

Magnetic equivalence

- Isochronous nuclei are always magnetically equivalent in achiral media, if there exists a symmetry operation which permutes the isochronous nuclei but leaves unchanged all coupling partners (the "coupling pathways" are symmetrically equivalent). In chiral environments they are magnetically equivalent, if that symmetry operation (which permutes the nuclei under consideration but leaves unchanged all coupling partners) is a proper rotation (the "coupling pathways" are rotationally equivalent).

4. Relationship between symmetry properties, isochronicity and magnetic equivalence for fast conformational equilibria

If a group of atoms rotates fast relative to the rest of the molecule around an axis, which is a symmetry axis for this group, all nuclei permuted by this rotation become isochronous and, with respect to all couplings to nuclei of the rest of the molecule, equivalent.

If the rest of the molecule is in fast conformational equilibrium the rules described above for rigid systems may be applied for an arbitrarily chosen conformation of the highest possible symmetry, regardless of whether or not this conformation is energetically reasonable.

5. Examples

Molecule	Spectral type in achiral environment	Spectral type in chiral environment
H, R_3, ^{13}C, R_1, R_2	AX	AX
H, H, ^{13}C, R_1, R_2	A_2X	ABX
H, H, C, F, F	A_2X_2	AA'XX'

VARIA

EFFECTS OF SYMMETRY

Molecule	Spectral type in achiral environment	Spectral type in chiral environment
H H H H (cyclopropene)	A_2B_2	AA'BB'
R_1, H, H, H, H, R_2 (para-disubstituted benzene)	AA'BB'	AA'BB'
$F_2C=CH_2$	AA'XX'	AA'XX'
$CH_3-C(R_1)(R_2)(R_3)$	A_3	A_3
$R_1-CH_2-C(R_2)(R_3)(R_4)$	AB	AB
$R_1-CH_2-C(R_2)_2-C(R_3)(R_4)(R_5)$	AB	AB
$(CH_3)_2CH-C(R_1)(R_2)(R_3)$	A_3B_3C	A_3B_3C

Molecule		Spectral type in achiral environment	Spectral type in chiral environment
$R_1-CH_2-CH_2-R_2$		AA'BB'	ABCD
$(R_1)_2CH-CH_2-CH(R_2)_2$		A_2BC	ABCD
$(R_1)(R_2)CH-CH_2-CH(R_1)(R_2)$	(RS)	A_2BC	ABCD

Data for Various Spin Systems [1]

Spin system	Maximum number of lines	Parameter	Parameters directly apparent from the spectrum	Parameters not defined by the spectrum [2]
AX	4	δ_A, δ_X, J_{AX}	δ_A, δ_X, $\lvert J_{AX}\rvert$	
AB	4	δ_A, δ_B, J_{AB}	$\lvert J_{AB}\rvert$	
A_2	1	δ_A, J_{AA}	δ_A	J_{AA}
AMX	12	δ_A, δ_M, δ_X, J_{AM}, J_{AX}, J_{MX}	δ_A, δ_M, δ_X, $\lvert J_{AM}\rvert$, $\lvert J_{AX}\rvert$, $\lvert J_{MX}\rvert$	relative sign of J_{AM}, J_{AX}, J_{MX}
ABX	14	δ_A, δ_B, δ_X, J_{AB}, J_{AX}, J_{BX}	δ_X, $\lvert J_{AB}\rvert$, $\lvert J_{AX} + J_{BX}\rvert$	relative sign of J_{AB}
ABC	15	δ_A, δ_B, δ_C, J_{AB}, J_{AC}, J_{BC}		
A_2X	5	δ_A, δ_X, J_{AA}, J_{AX}	δ_A, δ_X, $\lvert J_{AX}\rvert$	J_{AA}, relative sign of J_{AX}
A_2B	9	δ_A, δ_B, J_{AA}, J_{AB}	δ_A, δ_B	J_{AA}, relative sign of J_{AB}
A_3	1	δ_A, J_{AA}	δ_A	J_{AA}
A_3X	6	δ_A, δ_X, J_{AA}, J_{AX}	δ_A, δ_X, $\lvert J_{AX}\rvert$	J_{AA}, relative sign of J_{AX}
A_3B	16	δ_A, δ_X, J_{AA}, J_{AB}	J_{AB}	J_{AA}, relative sign of J_{AB}
AA'XX'	20	δ_A, δ_X, J_{AX}, $J_{AX'}$, $J_{AA'}$, $J_{XX'}$	δ_A, δ_X, $\lvert J_{AX} + J_{AX'}\rvert$, $\lvert J_{AA'} + J_{XX'}\rvert$, $\lvert J_{AA'} - J_{XX'}\rvert$	
AA'BB'	24	δ_A, δ_B, J_{AB}, $J_{AB'}$, $J_{AA'}$, $J_{BB'}$	$(\delta_A + \delta_B)/2$	
ABCD	56	δ_A, δ_B, δ_C, δ_D, J_{AB}, J_{AC}, J_{AD}, J_{BC}, J_{BD}, J_{CD}		

[1] for nuclei with nuclear spin quantum I=1/2

[2] the absolute sign of a coupling constant is never defined by the spectrum.

Subject Index

Acenapththene C115, H250

Acenaphthylene C115, H250

Acetaldehyde C30, C186, C187, H5, H120

Acetals I90
- alkyl groups I5
- cyclic C172, C173, H70, I90
- methyl M35

Acetamides C183, C186, I145, M20, M120
- aliphatic C10, C183, C196, C197, H5, H155, I145, I151, U10
- aromatic C120, C196, H160, H260, H315, I145, I151

Acetanilide C120, C196, H150, H160, H260, I151, U85

Acetates I140, I141, I186, M20, M30, M115, U10
- alicyclic C70, C194, H195, I140
- aliphatic C30-C46, C183, C186, C194, H5, H140, I140, M160, M165, U10
- aromatic C120-C130, C183, C194, H255, H315, I140, I142
- olefinic C90, H225, I140, I141

Acetic acid C30, C40, C186, C191, H5, H135, I171, M155, U10

Acetone C30, C40, C183, C186, C188, C250, H125, M155, U10

Acetonitrile C200, C250, H10, H180, M155, U15

Acetophenone C120, C183, C186, C189, H5, H125, H255, I131, U50, U90

Acetyl acetate C183, C186, H170

Acetylacetone C97, C189, H125, I132

Acetyl chloride C183, C186, H170

Acetylene (see also Alkynes) C110, C230, H5, H225, U15

Acid chlorides C183, C186, C198, I180, M30, M120
- alicyclic C70, C186, I180
- aliphatic C10, C30-C46, C186, H170, I180
- alkynyl C186
- aromatic C120, C126, C186, H255, H285, H305, H310, I180
- olefinic C90, C186, H215, H225, I180

Acid halides C183, C198, H170, H255, I180

Acridine C165, H350, U155
- -9-one H350

Acrolein C187, H215, I37

Acrylonitrile C200, H215, I37

Acrylic acid C191, H215, I37

Adamantane C75, H185

Additivity rules
- alicycles C65
- aliphatics C10-C25, H15-H17
- alkenes C85-C95, H215
- benzene C120, H255
- carbonyl groups C184
- furan H285
- haloalkanes C168
- methylcyclohexanes C50-C65
- naphthalenes C126-C130
- perfluoroalkenes C168
- pyridines C140-C150, H315
- pyrroles H295
- thiophenes H305

Alcohols H50, H55, I85, M5, M35, M40, M220
- acyclic C70, C71, H50-H60, H195, H200, I85, M20, M115, M220
- aliphatic C30-C46, C170, C260, H5-H17, H50, H55, H370, I85, M115, M155, M220, U10
- aromatic C120-C130, C140-C150, H50, H55, H255,

H315, I85, M220, U50, U80, U85
combination table B135
primary M10, M20
substituted C171
tertiary M5

Aldehydes C186, C187, C230, H120, I120, M5, M10, M215
alicyclic C70, C71, C186, C187
aliphatic C10, C40, C46, C184, C187, C188, H5, H15, H120, I120, M30, M35, M40, M115, M215
-α-methyl- M25
allyl- M10
aromatic C120-C130, C140-C150, C184, C187, H120, H255, H285, H305, H315, I120, M215, U40, U50, U95
carbonyl group C186, C187
additivity rule C184
combination table B185
halogenated C186, C187
olefinic C90, C184, C187, H120, H220, H225, I120, U20, U65

Alicycles M30, M40, M205
additivity rule C50-C65, C71
combination table B75, B85
condensed C72, C105, H190, H240
perfluorinated M25
polycyclic C75, H185, H240, M15, M35, M115, M205
saturated C30, C47-C75, C255, C260, H185-H200, I5, I15, M20, M25, M30, M40, M115, M160, M205
unsaturated C100, H230, I20, I35, M115, M210, U30

Aliphatics C10-C46, H5-H25, I5-I15, M205
additivity rule C10, H15
aromatic substituted C10, C30, H30
conformational effects C15, C20
steric effects C15

Alkanals (see Alkdehydes, aliphatic)

Alkanes C5, C30-C46, H5-H45, I5, M25, M30, M40, M115, M160, M2o5, U10
additivity rules C10, H15
aromatic substituted H30
combination table B75
coupling constants H20
ethyl- M10
halogenated C40, C167, H45
- additivity rules C168
monosubstituted C40, C46, H5, M40
polysubstituted H15-H17
propyl- M20

Alkanols (see Alcohols, aliphatic)

Alkanones (see Ketones, aliphatic)

Alkenes C10, C80, H5, H205, I20-I30, I36, I37, M15, M30, M40, M115, M160, M205, U10, U30, U60, U85, U100
additivity rule C85, H215
combination table B85
1,2-disubstituted C96
substituted C80, C90-C96, H215, H225

Alkynes C10, C110, C230, H5, H225, I40, M10, M115, M210, U15
combination table B95

Allenes C105, H210, I70, U15

Amides C183, C185, C186, C196, H150, I145-I152
alicyclic C183, C196
aliphatic C10, C183, C185, C186, C196, H5, H16, H17, H155, I145, I151, M120, M240, U10
aromatic C120, C126, C140-C150, C183, C185, C186, C196, H255, H315, I145, I151, M240, U85
N-butyl- M25
combination table B235-B245
N-ethyl- M10
olefinic I151

Amidines I190

Amines C174, H75, I100, I105, M30, M125, M225
alicyclic C70, C71, C176, H75, H195, I100, I105, M20, M225
aliphatic C10, C30-C46, C174, H5-H17, H75, I100, I105, M115, M225, U10, U75
aromatic C120-C130, C140-C150, C176, H75, H225, H315, I100, M230, U80, U85, U100
cyclic C177, H85, I100, M225
N,N-dimethyl- I5, I10, M20
combination table B155
N-ethyl- M5, M20

hydrochlorides C174-C176
N-methyl- M10
primary M5
olefinic C90, C176, H215, I37, I100

Amino acids C205, H351, I171, I175

Ammonium compounds I100
alicyclic C70, I100
aliphatic C10, C30, C174, H16, H17, H75, I100
aromatic C120-C130, H260, I100, U85
olefinic C90, I100

Anhydrides C186, C198, H170, I185, M20, M230, M235

Anilines C120, C176, H75, H260, M40, M45, M120, M230, U50, U80, U85
N-diethyl- C120, U80
N-methyl- C120, C176, H75, M15

Anisole C172, H60, H245, H255, U80

Anthracene C115, H250, U110
benzo[a]- C115, H250, U115
dibenzo[a,h]- U120
dibenzo[a,j]- U120
9,10-dihydro- H240

Anthraquinone I132

Anthrone I132

Aromatics C115, C250, C255, H245, H370, I45, M10, M15, M35, M40, M45, M120, M160, M165, M210, U40, U50, U110-U125
alkyl- C10, C30-C46, C120, H5-H17, H30, H255, I45, M45, M210
butyl- C120, M20
diphenyl- C120, H255, M50
disubstituted M40
combination table B105
pentyl- M25
propyl- M10

N-Aromatics C135-C166, H265-H300, H315-H350, H370, M10, M160
chlorinated M45
2-methyl- M15

O-Aromatics C135, C160-C166, H265-H290, H325-H350, M10, M155

Azetidine C177, H85
N-methyl- C177

Azides C71, I80

Azines H180, I190

Aziridines C177, H85
alkyl groups I5
N-methyl- C177

Azobenzene C120, H255, U100

Azomethane H90, U15

Azomethines C201, H180, H265, I190

Azothio compounds I200

Azo compounds I200
aliphatic H90, I200
aromatic C120, H255, I200, U100

Azoxy compounds H90, I200

Azulene C115

Barbituric acid I160

Benzaldehyde C120, C186, C187, H120, H255, I120, U40, U50, U95

Benzamide H150, H255, I151

Benzenes C115, C120, C250, H245, I45, M160, U50, U110
alkyl- C120, H255, M40, M120
disubstituted α-keto- M45
halomethyl substituted C169
monosubstituted C120, C169, C186, H255, I55
perhalogenated M35
polysubstituted C130, I55
substitution pattern I55

Benzene sulfonates C120, H255, U90

Benzhydrole H55

Benzimidazole C125, C160, H325

Benzoates C71, C120, C186, C192, C193, H5, H140, H145, H255, I142, I170, U95

Benzo[a]anthracene U115

Benzodioxol C172, H70, I90

Benzoic acid C120, C186, C191, H135, H260, I172, U40, U50, U90
esters C120, C186, C193, H5, H140, H145, H255, I142, U40

2,3-Benzofuran C160, H325, U155

2,1,3-Benzofurodiazole C161, H330

Benzo[a]naphthacene U125

Benzonitrile C200, H255, I65, U50

Benzo[c]phenanthrene U115

Benzophenone C120, C189, H255, I131, U105

Benzoquinone C190, H130, I132

2,1,3-Benzothiadiazole C161, H330

Benzothiazole C161, H330

Benzo[b]thiophene C160, H325, U150

Benzotriazole H330

Benzoxazole C160, H330

Benzoyl derivatives C120, C186, H255, M45

γ-Benzpyrones M45

Benzyl alcohol C120, C171, H55, H255

Biphenyl C120, H255, U90
  2,2',6,6'-tetramethyl- U95

Boron compounds I260

Bromides (see also Halogen compounds) M35, M40, M100-M110, M245
  alkyl- M45

2-Bromo-2-methylpropane C40, H5, U70

Bromoform C167, C255, H45

1-Bromopropane C40, H5, H45, U10

Butadiene C90, C105, H210, U30

2-Butanone C30, C40, C186, C188, H10, H125, U65

Butyl acetate I141, M165

tert-Butyl alcohol C40, C170, H5, H55

Butyl carbonate C199, H170

tert-Butyl group C5, C40, H5, I10, I15

tert-Butyl nitrite U65

Butyraldehyde C40, C187, H5, H120

γ- Butyrolactam C197, H160, I150
  N-acetyl- H165
  N-methyl- H160

γ-Butyrolactone C195, H145, I140

ε-Caprolactam C197, I150

ε-Caprolactone C195

Carbamate C199, H170, I160, I161,
  phenyl- M45

Carbazole C165, H350, U155

Carboiimides C70, C202, I80, U15

Carbonates C199, H170, I155, I161
  imino- I190

Carbon dioxide C199, I265

Carbon disulfide C199, C265, I270, I275, M160

Carbonic acid derivatives C199, H170, I155-I162

Carbon tetrachloride C167, C265, I270, I275, M165

Carbonyl compounds B45-B60, C186, U10
  additivity rules C184
  aromatic U40
  α,β-unsaturated U20

Carboxylates C10, C30, C70, C120, C186, C192

Carboxylic acids C186, H135, I165-I172, M5, M20, M35, M40, M230
  additivity rule C184
  alicyclic C70, C186, C191, I165
  aliphatic C10, C30-C46, C186, C191, H5-H17, H135, I165, I171, M115, M155, M230, U10
  alkynyl- C110, C186, C191
  aromatic C120-C130, C186, C191, H255, H285, H290, H305-H315, I165, I172, M5, M230, U40, U90, U95
  α-C_2- M40
  combination table B205
  halogenated C169, C191, C265, I171
  α-methyl- M35
  salts C10, C30, C70, C120, C186, I170, U65, U95
  α,β-unsaturated C90, C186, C191, H215, I165, I171, U20, U60, U65

Chelates I85

Chlorides (see also Halogen compounds) M15, M35, M100-M110, M245 alkyl- M40, M120

Chlorobenzene C120, H255, U50, U70

Chloroethyl group C40, C167, H5, H45, M30

Chloromethane C30, C40, C167, H5, H45, U10

Chloromethyl group C40, C46, C167, C169, H5, H45, M25

Chloroform C167, C255, H45, I270, I275, M165

Chromen-4-one H345

Chromophore U10

Chrysene U110

Cinnoline C162, H340, U145

Combination tables
- alcohols B135
- aldehydes B185
- alicycles
  - saturated B75
  - unsaturated B85
- alkanes B75
- alkenes B85
- alkynes B95
- amides B235
- amines B155
- aromatics B105
- carboxylic acids B205
- carboxylic esters B215
- cycloalkanes B75
- cycloalkenes B85
- esters B215
- ethers B145
- halogen compounds B125
- heteroaromatics B115

Complementary colors U5

Conformational effect
- for alcohols H60
- for aliphatics C15
- for alkenes C80
- for allylic couplings H205
- for amides H150
- for methylcyclohexanes H50, C55
- on vicinal couplings H20, H25, H60, H80, H150

Constitutional equivalence V5, V6

Contaminants, common I265, M170

Coronene U125

Coupling constants
- ^{13}C-^{13}C- C169, C240
- ^{13}C-^{19}F- C169, C171, C245
- ^{13}C-^{1}H- C220-C235
- additivity rule C225
- ^{13}C-^{14}N- C200
- ^{13}C-^{31}P- C215
- ^{1}H-^{19}F- H356
- ^{1}H-^{14}N- H80, H85, H180
- ^{1}H-^{31}P- H360

^{1}H-^{1}H-Coupling constants
- aldehydes H120
- alicyclics H185, H230-H240
- alkanes H20
- alkenes H205
- alkynes H225
- alcohols H60
- amides H150, H155
- amines H80, H85
- amino acids H351-H355
- ammonium compounds H80
- aromatics H245
- ethers, cyclic H65, H70
- formates H140
- heteroaromatics H265-H280, H325-H350
- isonitriles H180
- ketones H125, H130
- lactones H145
- sulfides, cyclic H100, H105
- survey B25
- thioethers (see sulfides)
- thiols H95

Crotonaldehyde I165, U65

Crotonamide I151

Crotonate I141, U65

γ-Crotonolactone H145, I140

Crotonic acid I171, U60, U65
- esters I141

Coumarine H345

Cubane C75

Cyanates H180, I65

Cyanides I65, M15, M25

Cyanoalkyl groups C30-C46, C200, H5-H17, H180, I10, M15, M25

Cyanurates I190

Cycloalkanes (see Alicyclics, saturated)

Cycloalkanols (see Alcohols, alicyclic)

Cycloalkanones (see Ketones, cyclic)

Cycloalkenes (see Alicyclics, unsaturated)

Cycloalkylamines (see Amines, alicyclic)

Cyclobutane C47, H185
alkenyl- I35
-1,2-dione H130

Cyclobutanone C190, H130, I125

Cyclobutene C100, H230, I20, I35

Cycloheptane C47

Cycloheptanone C190, I125

Cycloheptatriene C100, H235

Cycloheptene C100, H235

Cyclohexadienes C100, H210, H235

Cyclohexane C47, C70, C255, H185, I5, I15, M35, M40, M160, M205
alkenyl- I35
-1,3-dione H130
methylsubstituted, additivity rules, C50-C65
monosubstituted C70, C71, C186, C245, H55, H60, H75, H195, H200

Cyclohexanol C70, C71, H55, H60, H195, H200

Cyclohexanone C190, H130, I125, I131

Cyclohexenes C100, H230, I20, I35, M10, M25, M30, M35, M210
methyl- I35, M20

Cyclohexenones C190, I131, M20, M30, U20
methyl- M25

1,5-Cyclooctadiene C100

Cyclooctatetraene C100

Cyclooctene C100, H235

Cyclopentadiene C100, H230

Cyclopentane C47, H185
alkenyl- I35

Cyclopentanone C190, H130, I125

Cyclopentenes C100, H230, I20, I35 M30

Cyclopentenone C190, M35, U20

Cyclopropane C47, H185, I5
alkenyl- I30, I35
ketones B195
lactams B235
lactones B215
nitro compounds B165
phenols B135
sulfides B175
thiols B175

Cyclopropanone H130

Cyclopropene C100, H230, I20, I35

Decalins C72

Deuterium oxide H370

Deuterobromoform C255

Deuterochloroform C255, H365, M165

Diacetamide I152

Diastreotopy V5, V6

Diethyl ether C40, H5, H60, M160

Diethyl sulfide C180, H5, H95, U10, U75

Diazo compounds I65, I75, M10, M250

Diazophenyl derivatives M45

Dibenzo[a,h]anthracene U120

Dibenzo[a,j]anthracene U120

Dibenzofuran C165, H350, U150

9H-Dibenzo[b,d]pyrrole C165, H350, U155

Dibenzo[b,d]thiophene U155

2,6-Di-t-butyl-p-cresol M170

Dibutyl phthalate I142, M170

Dicarbonyl compounds C97, C189, H125, I132
β-enol form C97, H50, I85, I132

Dicarboxylic acids C191, H135, I165, I171, I172

Dichloroethylene C96, C167, H215, I36

Dichloromethane C167, H45, M160

Dienes C90, C105, H210, H215, I30, I37, M35, M115, U30, U60

Diesters C194, I135, I142, M45

9,10-Dihydroanthracene H240

Dihydrofurans C173, H65, H210

9,10-Dihydrophenanthrene H240

Dihydropyran C173, H70, M35

Diisopropyl ether C40, H5, H60, M165

Diketones I130, I132
aliphatic C97, C189, H125, I130, I132
cyclic C97, C190, H130, I130, I132, U100, U105

1,2-Dimethoxyethane M165

Dimethyl ether C30, C40, C172, H5, H60, M155, U10

N,N-Dimethylamines I5, I10, I100, I105
alicyclic C70, C176, H75
aliphatic C30-C46, C174, H5-H17, H75, I100, I105
aromatic C120-C130, C176, H75, H255, I100, I105
methyl groups C30, C40, C174-C176, H75, I5, I10

Dimethyl carbonate C199, I161

Dimethyl disulfide C30, C40, C180, H5, H110, U70

N,N-Dimethylformamide C30, C196, H155, I151, M160

Dimethylnitrosamine C178, H90

Dimethyl sulfoxide C40, C181, C260, H110
as solvents for alcohols H60

Dioctyl phthalate M170

Diols C171, M40, M155

1,3-Dioxane C173, H70, I90

1,4-Dioxane C173, C260, H70, I90, M165

1,3-Dioxolane C173, H70, I90

Diphenyl ether C120, H255, U105

Diphenylamine C120, C176, H75, H255, U100

Diphenyl ketone C120, C189, H255, I131, U105

Diphenylmethane, derivatives M50

Diphenyl oxide C120, H255, U105

Diphenyl sulfide C120, H255, U105

Disulfides I215, M30
aliphatic C30, C40, C180, H5, H110, I125, U10, U70
cyclic H105, I215

Dithiocarboxylic acids I220

Dodecadeuterocyclohexane C260

Double bond (C=C) (see Alkenes; alicycles, unsaturated)

Double bond equivalents V1

Enantiotopy V5, V6

Enols C97, H50, I85, I132

Epoxides C10, C120, C173, H65, I5, I30, I35, I90, M5

Esters C186, I135-I142, M35, M40, M45, M235
alicyclic C70, C71, C194, H195, H200, I135
aliphatic C10, C30-C46, C186, C193, C194, H5-H17, H140, I135, M115, M235, U10
aromatic C120-C130, C194, H255, I135, I140, I142, U40
butyl I141, M40
carbonyl group C184, C186
combination table B215-B230
ethyl M10, M20
methyl C186, I141, I142, M10, M15, M25, M30, M35
α-methyl M40
of aromatic acids C120-C130, C140, H255, H285-H320, I142, M235
of inorganic acids H115, H140
olefinic C90, C194, H215, H225, I135, I141, U20, U30
phenol I140, I142, M40
propyl M20, M25, M30

Ethane C5, C30, H5, U10
halogen substituted H45

Ethanol C30, C40, C170, H5, H55, M155

Ethanethiol C40, C179, H5, H95, U70

Ethers I90, M35, M40, M220
alicyclic C70, C71
aliphatic C10, C30-C46, C172, H5, H60, I90, M115, M155, M160 M165, M220, U10
aromatic C120-C130, C140-C150, C172, H255, I90, M120, M225, U80, U105
aryl-alkyl M225
cyclic C172, C173, C260, H65, I5, I30, I35, I90, M155, M165, M220

combination table B145
ethyl M20, M160
methyl M10, M15, M155
- aromatic M20
olefinic C90, C96, H5, H215, I30, I36, I90
phenol M40
propyl M25, M30

Ethyl acetate C40, C194, H5, H145, M165

Ethylbenzene C30, C40, C120, H30, H255

Ethylene C90, H205, U10
halogenated C167, H215, I36

Ethylene carbonate C199, H170, I155, I161

Ethylene glycol C171, M155
dimethyl ether C172, M165

Ethylene urea I162
N,N-dimethyl- C199

Ethyl group C40, H5, M10

Exchange reactions
between amino protons H80
between hydroxyl protons H50, H60

External reference V2

FAB, matrices M175-M190

Fatty acids
derivatives M40
ethyl esters M40

Fluorene H240

Fluorides (see also Halogen compounds) M5, M245

Fluoromethyl group C40, C46, C167, C169, H5, H45, M15

Fluorolube I280

Formaldehyde C186, C187, H120

Formamides C196, H155, I151, M160

Formates C186, H140, I140, I141, I170

Formic acid C186, C191, I171

Fragmentation of compound classes M205-M225

Fumaric acid C191
diesters C194, I142

Furan C135, H265, I50, M10, M35, M155, M225, U135
2,3-benzo- C160, H325, U155
dibenzo- C165, H350, U150
dihydro- C175, H65, H210
methyl- C30, H30, H285, H290
monosubstituted H285, H290
tetrahydro- C173, H65, I90, M30, M155

Furazan H275

Glutarimid I152

Glycerols C171, M175

Glycols C171, M30, M35, M120, M155
isopropylidene M25

Glycol ethers M40, M120, M165

Guanidines I160, I162, U15

Haloethanes C167, C168, H45, I110, I115

Halogenated benzenes I110, I115
disubstituted I115
monosubstituted C120, C245, H255, U50, U70

Halogen compounds I110, I115, M100, M105, M245
alicyclic C70, C71, H195, H200, U30
aliphatic C10, C30-C46, C167-C169, C255, C265, H5-H17, H45, H365, I110, I115, M160, M165, M245
aromatic C120-C130, C140-C150, C169, H255, H285, H290, H305-H315, I110, I115, U50, U70
combination table B125
olefinic C90, C96, H215, H225, I110, M165, U30

Holomethanes C30, C40, C167, C168, C255, H45, H365, I110, I115, M15, M25, M160, M165, U10

Heptacosatrifluorotributylamine M170

Heteroaromatics C135-C166, C265, H265-H350, H370, I45, I50, M10, M155, M160, U130-U155
combination table B115
condensed C160-C166, H325-H350, U140-U155
sixmembered C140-C155, C265, H275, H315, H370, I45, I50, U130, U135

Heteroatoms
  indications of M125
  F-indicator M5
  N-indicator M10, M25, M35, M40
  O-indicator M5-M40
  P-indicator M20
  S-indicator M15, M20, M30, M35

Hexadeuteroacetone C250, H365

Hexadeuterobenzene C255, H370, M160

Hexadeuterodimethylsulfoxide C260, H365

Hexane C5, M160

1-Hexene C90, M160

Homotopy V5, V6

Hydantoins I162

Hydrazides I145, I150, M15

Hydrazines C120, H255

Hydrazones C202, H175

Hydrogen bonds
  aldehydes I120
  alcohols H50, I85
  boron compounds I260
  carboxylic acids I165
  conjugated esters I135
  β-dicarbonyl compounds H50, I130

Hydroperoxides I95

Hydroxybenzene derivatives M45
  chlorinated M45

Hydroxylamines M5

Imidazole C135, H270, M35
  benz- C161, H325
  methyl- C30, H35

Imides H165, I145, I150, I152, U10

Imines C201, H180, H255, I190

Imino carbonates I190

Indane C105, H240

Indanones H130

Indazole C161, H325

Indene H240

Indole C160, H325, M45, M230, U150
  methyl- C30, H40

Indolizine C161, H330

Interferences I265

Iodides (see also Halogen compounds) M45, M245

Iodomethane C30, C40, C167, H5, H45, U10

2-Iodopropane C40, H5, U75

Isoalkanes C5, M205

Isobutenes H225

Isobutyraldehyde C40, C186, C187, H5, H120

Isobutyrates C40, C186, C193, H5, H140

Isochronicity V4, V6

Isocyanates C120, C201, H180, I75

Isocyanurates I150

Isonitriles I65
  alicyclic C71
  aliphatic C10, C30, C40, C200, H180, I65
  aromatic C120, C200, I65
  olefinic C90, C200, I65

Isopropanol C40, C170, H5, H55

Isopropyl group C5, C40, H5, I10, I15

Isoquinoline C162, H340, U145
  -N-oxide H340

Isothiazole C135, H270

Isothiozyanates C30, C71, C120, C204, H115, I80, U15

Isotopes
  distribution patterns M55
  masses and abundances M60

Isotope combinations
  calculation of M90
  chlorine and bromine M100, M105

Isoxazole C135

Ketals I90
  ethylene M20, M30, M40
  thioethylene M45

Ketenes I70, U15

Ketenimines I70

Keto esters I135, I141

Ketones I125, M5
additivity rule C184
alicyclic C70, C186, C189, I131, M215
aliphatic C10, C30-C46, C186, C188, C250, H5-H17, H125, H365, I125, M30, M35, M40, M115, M155, M210, U10, U20, U65
alkynyl C186, C188, C131
aromatic C120-C130, C140-C150, C186, C189, H255, H285-H320, I125, I131, M120, M215, U40, U90, U105
butyl M20
combination table B195
cyclic C190, H125, H130, H190, I125, I131, M10, M20, M30, M40, M115, U20
furyl M30, M40
methyl M20, M25
pentyl M25
propyl M10
pyrryl M40
tolyl M45
α,β-unsaturated C90, C186, C188, H215, H225, I37, I125, I131, M20, U20, U60

Lactams C197, H160, I145, M10, M240
combination table B235-B245

Lactones C195, H145, I140, M5, M10, M20, M35, M115, M240
combination table B215-B230

Magnetic equivalence V4, V6

Maleic acid C191
anhydride C198, H170
diester C194, I142

Malonic acid C193, H135, I171
diester C194, I142

Mass correlation table M5-M50

Mercaptans (see Thiols)

"Metastable Peaks" M145

Methane C5, C30, C40, H5, U10

Methanol C30, C40, C260, H5, H55, M155, U10

Methyl acetate C30, C40, C186, C194, H5, H140, M160

Methyl ether C30, C40, C172, H5, H60, I90, M10, M15
alkyl groups I5

Methylcyclohexane C50-C71, H195, H200
additivity rules C50-C65

Methylene chloride C167, H45, M160

1,2-Methylenedioxybenzenes C172, H70, I90

Methylenedioxy group C172, H70, I5, I90

Methylene groups C90, H15-H17, H205, I10-I35

N-Methylethyl carbamate C199, H170, I161

Methylfurans C30, H30, H285, H290

Methyl groups C5-C46, H5, I5-I15, M5
in methylcyclohexanes (additivity rules) C50-C65
geminal I10, I15
in the solid phase I10

Methylimidazoles C30, H35

Methylindoles C30, H40

Methylphenyl ketone C30, C40, C120, C183, C186, C189, H5, H125, H255, I131, U50, U90

Methylphenyl sulfide C30, C40, C120, C180, H95, H255, U95

N-Methylpiperidine C171, H75

Methylpyrroles C30, H30, H295, H300

Methylthiophenes C30, H35, H305, H310

Molecular ions
determination of the relative mass M135

Monocyclanes M115

Monosaccharides C210

Morpholine C173, C177, H70, H85
N-methyl- H85

Naphthacene U110
benzo[a]- U125

Naphthalene C115, H245, M45, M50, U115
methyl- C30, H30

monosubstituted C126-C130
tetrahydro- C105, H240, M25, M30, M45

Naphthoquinones C190, I130, I132, M45, U100, U105

1,5-Naphthyridine U145

Nitramines C179, I210

Nitrates I210, U15

Nitriles C200, I65, M10, M245
alicyclic C70, C71, C200, I65
aliphatic C10, C30-C46, C200, C250, H15-H17, H180, H365, I10, I65, M155, M245, U15
alkynyl- I140
aromatic C120-C130, C140-C150, C200, H255, H285-H320, I65, M250, U50
olefinic C90, C200, H220, I30, I65

Nitriloxides I65

Nitrites C10, C46, I205, U15, U65

Nitrobenzene C120, C178, H255, U50, U75

Nitromethane C30, C40, C178, H5, H90, U60

Nitrosamines C178

C-Nitroso compounds I205, M10, U15
aromatic C125, H260, I205
dimers I200

N-Nitroso compounds C178, H90, I205, M10

Nitro compounds I210, M5, M10, M15, M20, M250
alicyclic C70, C178, H195, I210
aliphatic C10, C30-C46, C178, H5, H15, H90, I10, I210, M250, U15, U60
aromatic C120-C130, C140, H260, H285-H315, I210, M250, U50, U75
combination table B165
olefinic C90, C178, H215, I210

Nonyl aldehyde C186, I120

Norbornadiene C100

Norbornane C75

2-Norbornanone (norcamphor) H190

Norbornene (norbornylene) C100, I35

Nucleosides C211-C212

Nujol I280

Octanes C5, C30, C46

Olefins (see Alkenes)

Opaque regions I265-I275

Ovalene U125

Oxalic acid C191, I171
diester C194, I142

Oxane C173, H70, I90, M40

Oxazole C135, H270
benz- C160, H330
methyl H35

Oxetane C173, H65, I90

N-Oxides H275, H280, H335, H340, M5

Oximes C10, C201, H5, H175, H315, H320, I195, U15

Oxiranes B145, C10, C173, H65, I30, I35, I90, M5

Oxolan (see Tetrahydrofuran)

Oxolenes C173, H65, H210

Oxopyrans (see Pyrones)

Pelargonaldehyde C186, I120

Pentadeuteropyridine C265, H370

1,3-Pentadiene U60

Pentane C5, M160

3-Penten-2-one U60

Peracids I95

Perfluoroethyl derivatives M45

Perfluoroalkanes C168, M45

Peroxides I95
cyclic M15
diacyl I95

Perylene U120

Phenanthrene C115, H250, U115
benzo[c]- U115
9,10-dihydro- H240

Phenanthroquinone I132

Phenazine C165, U150

Phenetole C40, H60, H255

Phenol C120, H50, H255, I85, M10, M40, M120, M220, U50, U80
 derivatives M40, M220
 combination table B135

Phenolate C120, U50, U85

Phenylacetylene C110, C120, H225, H255, U50

Phenylethyl derivatives M45

Phosphates C120, C215, H215, I235, M130
 alkyl M45, M250
 ethyl M10, M250

Phosphines C120, C215, I135, M250

Phosphinoxides C215, I235, M250

Phosphinsulfides C215, I245

Phosphonates C120, H215, H255, I235

Phosphorous compounds C215, H215, H255, I235, M250

Phthalates I142, I265, M50, M170, M235

Phthalazine C162, H345, U145

Phthalimide I152
 N-methyl- H165, I152

Phthalic acid I172
 anhydride C198, I265
 diesters I142

Piperazine C177
 N-methyl- H85

Piperidine C177, H85
 N-alkyl- M40
 N-methyl- C177, H75, M40

Pivalaldehyde C40, C186, C187, H5, H120

Pivalic acid C40, C186, C191, H5, H135, I171

Pivalonitrile C200, H180

Polyenes U30

Polymethyl compounds M10

Polyols C171

Polypeptides I145, I150, I170

Potassium bromide I280

1,3-Propane sultone H110

β-Propiolactam I150
 N-methyl- H160

β-Propiolactone C195, H145, I140

Propionates C40, C186, C193, H5, H140, I140, I141

Propyl group C40, H5, M20

Propyl ketones C40, C188, H125, M10

Proteins I145

Purine C162, H335

Purine bases C212

Pyran C173, H70, M35
 dihydro- C173, H70, M35
 oxo- H70, H145, I140
 tetrahydro- C173, H70, I90, M40

Pyrazine C155, H280, U130

Pyrazole C135, H270, M35
 methyl- C30, H35

Pyrene C115, U120

Pyridazine C155, H280, U135
 -N-oxide H280

Pyridine C155, C265, H275, I45, M35, M45, M160, M230, U130
 alkyl- M40
 methylsubstituted C30, C140, C145, H40, H315, H320
 monosubstituted C140, C145, C245, H55, H315, H320
 -N-oxide H275
 polysubstituted C150
 tetrahydro- H85, M35

Pyridones, derivatives M40

Pyrimidine C155, H280, U135
 -N-oxide H280

Pyrimidine bases C211

Pyrones H70, H145, I140

Pyrrole C135, H265, H295, I50, M35, U135
 methyl- C30, H30, H295, H300
 monosubstituted H295, H300

Pyrrolidine C177, H85, M30
 N-acetyl- H160
 N-methyl- C177, M40

2-Pyrrolidone C197, H160, I150
 N-acetyl- H165
 N-methyl- H160

Pyruvic acid I171

Quinazoline C162, H340, U140

Quinoline C162, H335, M45, M50, U140
  -N-oxide H335

Quinone C190, H130, I130, I132, M10
  naphtho- C190, I130, I132, M45, U100, U105

Quinoxime I195

Quinoxaline C162, H340, U140

Salicyl aldehyde I120

Salicylic acid I172
  derivatives M45
  esters I142

Schiff's bases C201, H180, H255, I190

Scott rules U40

Selenophene H265

Silicones I265

Silicon compounds I250, I255, M35, M165, M255
  aliphatic C30, C183, I141, I250
  aromatic C120-C130, C140-C150, H255
  ethers C71, C120, I250, M35, M255
  olefinic C90

Solvent spectra C250-C265, H365, H370, I270, I275, M155-M165

D-Sorbitol C171

Spiro compounds C72

Steric effects
  in aliphatics C15
  in alkenes C85
  in cyclohexanes C55

trans-Stilbene H215, H255, U100

Structural type, indicators M115

Styrene C90, C120, H215, H255, I37, U50, U85

Succinic acid C191, H135, I171
  anhydride C198, H170
  diesters C194, I142

Succinimide H165, I152, U10
  N-bromo- H165, I152

Sulfates C182, H115, I230

Sulfides I215, M15, M35, M40, M120, M245
  aliphatic C10, C30-C46, H5-H17, H95, I215, M245, U10, U75, U95
  alkynyl C110, I215
  aromatic C125, C180, H95, H260, H285-H320, I215, U95, U105
  combination table B175
  cyclic C182, H100, I215
  ethyl C40, C180, H5-H17, H95, M30
  methyl C30-C46, H5-H17, H95, I10, I215, M20
  olefinic C90, H205, H215, I215

Sulfinates I225

Sulfines C10, C30-C46, C125, C181, H215, I225

Sulfinic acids C181, I225
  derivatives H110, I225
  esters I225

Sulfinyl chlorides I225

Sulfites C182, H115, I225

Sulfolane C182, H100

3-Sulfolene H100

Sulfonamides H110, I230

Sulfonates C30, H140, H145, I230, M20, M30, U90
  aromatic C125, H225, H260, I230
  ethyl M20

Sulfones I230, M5, M20, M30, M245
  aliphatic C10, C30, C40, C181, H110, I230
  aromatic C125, H305-H320, I230
  cyclic C182, H100, H105, I230
  ethyl C40, C181, M20
  methyl C30, C40, I10
  olefinic C90, H215, I230

Sulfonic acids H110, I230
  aromatic C125, H110, H320, I230
  derivatives H110
  esters C120, H110, H140, H145, H255, H260, I230, M20
  hydrated I230
  thioesters I230

Sulfonium salts C181

Sulfonyl chlorides I230
  aliphatic C10, C30, C40, C181, H110, I230
  aromatic C125, H260, H305, I230

Sulfonyl halogenides I230

Sulfoxides I225, M5, M20, M245

aliphatic C10, C30-C46, C181, C260, H110, H365, I225
cyclic C182
sulfur compounds, fragmentation M245

Summary tables
^{13}C-chemical shifts B5
^{1}H-chemical shifts B15
^{1}H-^{1}H-coupling constants B25
IR-absorption bands B35
- carbonyl groups B45-B60
UV/VIS-absorption bands B65

Suspension media I265, I280

Tetrachloroethylene C167, M165

Tetradeuteromethanol C260, H370

Tetrahydrofuran C173, H65, I90, M30, M155

Tetrahydropyran C173, H70, I90, M40

Tetrahydropyridines C177, H85, M35

Tetralin C105, H240, M25, M30, M45
derivatives M45

α-Tetralone H130

β-Tetralones M25, M30

2,2',6,6'-Tetramethylbiphenyl U95

Tetramethylsilane M165

1,2,4,5-Tetrazine C155

2,1,3-Thiadiazole H275
benzo[b]- H330

Thiairane C182, H100

Thiane C182, H105

Thiazole C135, H270
benzo- C161, H330
methyl- H35

Thietane C182, H100

Thioaldehydes I220

Thioamides C10, C203, H315, H320, I220

Thiocarboxylic acids
aliphatic C10, C203, H115
derivatives H115
esters C10, C30, C203, H115, I220
halides I220
olefinic H220

Thiocyanates I65
aliphatic C10, C204, H115, I65
aromatic C125, H285-H310
olefinic H215, I65

Thioethers (see Sulfides)

Thioketones C120, I220, U15

Thiolactams I220

Thiolane C182, H100

Thiol carboxylic acids C30, C183, C203, H115
esters C30, C183, C203, H115, H215

Thiolenes H100

Thiols I215, M35, M120, M245
alicyclic C70, C71, C179, H195, H200, I215
aliphatic C10, C30-C40, C179, H5, H16, H17, H95, I215, U10, U70
aromatic C125, C179, H260, H305, I215
combination table B175

Thiophene C135, H265, I50, U130
alkyl- H305, H310, M40
benzo[b]- C160, H325, U150
dibenzo[b,d]- C165, U155
methyl- C30, H35, H305, H310
monosubstituted H305, H310

Thiophenoyl derivatives M45

Thiopyrans H105

Thiourea
derivatives I160
tetramethyl C199

1,4-Thioxane C173, H70

Toluene C30, C120, H30, H255, M165, U50

p-Toluylsulfonates (tosylates) C40, H5, H140, H145

Triethylamine C174, U75

Triazines C155, H280, U130

Triazoles C135, H275

Trichloroethylene C167, I36, M165

Trideuteroacetonitrile C250, H365

Trienes I37

Trifluoroacetates H145, I141, I170

Trifluoroacetic acid C169, C191, C245, C265
esters H145, I141

Trifluoromethyl aromatics C120, C169, C245, H255, M25

Trifluoromethyl group C167-C169, C245, I110, M30

Trimethyl-phenyl lead C125

Trimethylsiloxyl compounds C71, C120, I250, M35, M255

Trimethylsilyl compounds C90, C120-C130, C140-C150, C183, H255, I141, I250, M35, M255

1,3,5-Trioxane C173, H70

Triphenylene U125

Twistane C75

Ureas C199, H170, I160, I162

Urethanes (see Carbamates)

∂-Valerolactam C197, H160, I150
N-acetyl- H165
N-methyl- H165

∂-Valerolactone C195, H145, I140

Vinyl compounds C90, C186, H205, H215, I25, I30, M10
alkyl substituted C20-C46, H5
aryl substituted C120, H255, H315, U85

Volume susceptibility V3
correction V2

Water H370, I265, M155

Woodward rules U20

Woodward-Fieser rules U30

Xanthen-9-one H350

Zinc compounds C10, C120, C140-C150

D. J. Gardiner, P. R. Graves (Eds.)

Practical Raman Spectroscopy

With contributions by H. J. Bowley, D. J. Gardiner, D. L. Gerrard, P. R. Graves, J. D. Louden, G. Turrell

1989. VIII, 157 pp. 87 figs. 11 tabs.
ISBN 3-540-50254-8

Contents: Introduction to Raman Scattering. – Raman Sampling. – Instrumentation for Raman Spectroscopy. – Calibration and Data Handling. – Non-Standard Physical and Chemical Environments. – Raman Microscopy. – Further Reading. – Subject Index.

The book provides a practical guide to important and frequently encountered techniques in Raman spectroscopy. It comprises a valuable working reference as well as a useful introduction to the technique; emphasis throughout the book is on advice from experienced workers in the subject. The theoretical content of the book has been kept to a minimum and chapters dealing with instrumentation, sample handling, data acquisition and analysis, calibration, and microscopy are treated to emphasize the practical aspects of the various branches of the subject. Tables of useful data are included.

Springer-Verlag Berlin
Heidelberg New York London
Paris Tokyo Hong Kong

H. J. Fischbeck, K. H. Fischbeck

Formulas, Facts and Constants

for Students and Professionals in Engineering, Chemistry and Physics

2nd, rev. and enl. ed. 1987. XV, 260 pp.
ISBN 3-540-17610-1

Contents: Basic mathematical facts and figures. – Units, conversion factors and constants. – Spectroscopy and atomic structure. – Basic wave mechanics. – Facts, figures and data useful in the laboratory. – Subject Index.

This book provides a handy and convenient source of formulas, conversion factors and constants for the student as well as the professional. An extensive collection of the fundamental tools of mathematics, constantly needed in every area of the physical sciences and engineering, has been compiled. The International System of Units of measurement (SI) is given in detail and convenient tables for conversion between different units have been included. As a manual in the laboratory and for problem-solving a collection of frequently needed data and constants is given. Formulas and tables have been amended by examples in all of those cases where their use might not be self-explanatory.
For the second edition, a chapter on "Error Analysis" as well as an Index for easy entry have been added.

Springer-Verlag Berlin
Heidelberg New York London
Paris Tokyo Hong Kong